ENGINEERING CONSTRUCTION SPECIFICATIONS

ENGINEERING CONSTRUCTION SPECIFICATIONS

The Road to Better Quality,
Lower Cost, Reduced Litigation

Joseph Goldbloom, FASCE, PE

VNR Van Nostrand Reinhold
New York

Softcover reprint of the hardcover 1st edition 1989
Library of Congress Catalog Card Number 89-5551
ISBN-13: 978-1-4684-1454-7 e-ISBN-13: 978-1-4684-1452-3
DOI: 10.1007/978-1-4684-1452-3

Van Nostrand Reinhold
115 Fifth Avenue
New York, NY 10003

Van Nostrand Reinhold International Company Limited
11 New Fetter Lane
London EC4P 4EE, England

Van Nostrand Reinhold
480 La Trobe Street
Melbourne, Victoria 3000, Australia

Nelson Canada
1120 Birchmount Road
Scarborough, Ontario M1K 5G4, Canada

16 15 14 13 12 11 10 9 8 7 6 5 4 3 2 1

Library of Congress Cataloging-in-Publication Data

Goldbloom Joseph
Engineering construction specifications.
Bibliography: p.
Includes index
1. Engineering—Contracts and specifications. I. Title.
TA180.G63 1989 620.1 89-5551
ISBN-13: 978-1-4684-1454-7

To
The Engineering Profession

Foreword

For the past 25 years, Joe Goldbloom and I have conducted a running debate over whether specifications writers engage in the unlawful practice of law. Joe's position is that lawyers have no business writing specifications, that being the designer's province. Having been given the honor to write this foreword, I have the opportunity for the last word, at least for now.

Joe Goldbloom and I first met in 1964, while serving together on the ASCE Committee on Contract Administration. Joe became my teacher, mentor, and friend. Underlying our good natured debate was the serious issue of the technical qualifications required of a specifications writer. As a matter of fact, specifications writing traditionally has fallen in a crack between the two professions. Specifications writing typically is neither taught in engineering school nor in law school. Engineers are taught how to design; lawyers are taught how to draft contracts. Specifications writing requires mastery of the technical elements of design as well as the skills of contract drafting. Specifications writing is neither glamorous nor sexy; it is often viewed as a necessary evil of the designer's job.

Having professional training in both engineering and law, and being engaged in legal practice specializing in construction, I nevertheless feel unqualified to write specifications because I lack a further necessary element of training: namely, practical field experience in construction. After all, specifications are intended to be the written communication between the designer and the constructor, expressing how the project is to be built. Specifications must be written for the person in the field who is charged with the responsibility of building the project. Ironically, all too often specifications are a stepchild, written by a designer with little or no practical field experience, and, after a dispute has arisen, are ultimately interpreted by a judge, possessing no engineering training or construction experience. It is then a matter of the blind leading the blind.

As the reader will soon see, Joe Goldbloom has abundant skill in the art of communication. Moreover, he has a lifetime of experience in construction, starting as a field engineer and superintendent for a contractor, then as an inspector and resident engineer, and finally as the chief specifications writer for the internationally preeminent design firm of Parsons Brinckerhoff. The book is not merely a primer in how technically to specify the bricks and mortar of construction. It extends to the philosophical underpinnings of the construction process, involving issues of fundamental fairness and risk allocation as between owner and constructor.

The book is scholarly and at the same time intensely practical. It describes current specifications practices, defines good practices, and admonishes against specifically defined poor practices. The book lays out potential pitfalls in specifications and how to avoid them. It defines when, and under what circumstances, to seek legal counsel. The book is filled with illustrative case histories from the author's extensive experience in construction.

In the same manner that the book instructs on how to prepare well organized, well written specifications, the book, itself, is exceptionally well organized and well written. Explanations are clear and concise; technical jargon is avoided. The book is a pleasure to read.

The book is nearly encyclopedic. It contains a wealth of information, including check lists, sources of information for specifications writers, a bibliography, and specimen provisions. It is extensively cross-referenced. It is both a valuable primer for students and beginning specifications writers and an invaluable reference for the seasoned practitioner. Joe Goldbloom is to be congratulated on producing an absolutely superb, if not monumental, book that fills a void in current construction literature.

Robert A. Rubin, Esq., P.E.
Partner, Postner & Rubin, NYC
Fellow, American Society of Civil Engineers

Preface

There are two major areas of designed construction:

(1) Building (architectural) construction, whose design is performed and supervised by architects. This type of construction is sometimes referred to as vertical construction.

(2) Engineered construction, which this book is concerned with, includes the construction of bridges, highways, tunnels, dams, pipelines, airfields, rapid transit facilities, and other types of construction that utilize the designs of engineers. This type of construction is generally referred to as heavy construction.

When a construction contract is signed, the Specifications become the rule book that governs performance of the Work and controls the official relations between the Contractor, Owner, and Engineer. This book has been prepared for both the practicing engineer and the student of engineering. It is presented in two parts:

Part I, Explaining Engineering Construction Specifications, is directed to the student. It describes and explains the elements that make up the set of engineering construction specifications. Part I presents the "whys" and "wherefores" of the various requirements and instructions encountered in the Specifications. It also explains construction contracts and the relationship of the Plans and Specifications.

Part II, Preparing and Presenting Engineering Construction Specifications, is devoted to the specification writer. It presents guidelines and recommendations, "do's" and "don'ts," and pitfalls to be avoided in preparing Specifications. Among the items discussed are: the difference in responsibilities represented by "quality control" and "quality assurance"; identifying and controlling the risks in construction; full disclosure of known information; and the Engineer's responsibility when assigned authority to supervise the Contractor's work, or to suspend his operation.

The reader will benefit by reviewing related material that is presented in both Part I and Part II. This will particularly hold true for Chapters 4 and 11, dealing with the General Conditions, and Chapters 5 and 12, dealing with the Bidding Documents.

Throughout the book, actual case histories are presented to illustrate problems on the job because of inadequate or poorly prepared Specifications, and how the

Specifications enabled the Engineer to control and handle other situations such as unanticipated subsurface conditions; an uncooperative contractor; and questionable work which had to be uncovered for reexamination. In addition to benefitting the student and the specification writer, material presented in this book will be found useful by the project engineer, the Designer, the Owner's site representative, the construction Contractor, and the construction claims lawyer.

It has been said that over 50 percent of the construction claims that occur, are caused by Drawings and Specifications that are unclear, ambiguous, or contradictory. When these claims wind up in court and there are questions concerning the intent of the Contract, the court will most likely turn to the Specifications rather than to the Drawings. It is much easier for judges and juries to interpret Specifications which are the written word, than it is to comprehend a technical drawing.

There appears to be no recently prepared text on engineered construction specifications. The authors of current texts on Specifications portray an architectural background. It was this author's goal to produce a text that could be used in a classroom by the engineering student, as well as in a design office by the engineer who writes Specifications.

This book reflects the experience and general knowledge acquired by the author in a professional career spanning more than 45 years. Of the first 25 years, which were spent in the field on construction projects, the author served 21 years as a representative of the Owner. One of his many responsibilities required the interpretation, enforcement, and defense, of the Contract Specifications. In the second half of his career, the author transferred to the design office where his responsibilities progressed from the preparation of Specifications, to the supervision of their preparation, and to the final review of completed Contract Specifications.

In preparing this book, permission has been granted for the use of quotations and reproductions from previous publications. It is the author's sincere hope that the material presented herein will be of help to both the practicing engineer and the student, in developing a sound knowledge of the subject of engineering construction specifications.

The author wishes to express appreciation to his employer, Parsons Brinckerhoff, Inc., New York, New York, and in particular to Henry L. Michel, Chief Executive Officer, for making the office facilities available throughout the years of research and preparation of this book. My sincere gratitude goes to Mrs. Susan La Regina of the Specifications Department for her tremendous assistance in transforming my pencilled drafts into the final copy. And last but not least, I owe a tender debt of gratitude to my wife Doris for her patience, understanding, and encouragement, during the years 1981–1988 of my so-called retirement, in the preparation of this book.

Contents

CHAPTER 3, THE TECHNICAL SECTIONS

CHAPTER 7, NATIONAL REFERENCE STANDARDS

Part II—Preparing and Presenting Engineering Construction Specifications

CHAPTER 8, PROCEDURES AND PRACTICES IN SPECIFICATION WRITING

APPENDIX A—SAMPLES OF CITED DOCUMENTS AND FORMS

APPENDIX B—SOURCES OF INFORMATION FOR THE SPECIFICATION WRITER

ENGINEERING CONSTRUCTION SPECIFICATIONS

Part I

Explaining Engineering Construction Specifications

Chapter 1

Construction Contracts

1.1 Introduction

Engineered construction is accomplished by written contract in which the Contractor and the Owner each agree to perform in accordance with the terms spelled out in the Contract Documents. The Contractor promises to construct the project in accordance with the Plans and Specifications, and complete it ready for use within the time specified; all for an agreed sum of money. The Owner promises to furnish the rights-of-way, preliminary information and data, and make periodic payments to the Contractor during the progress of the Work.

The purpose of a contract is to define the rights and responsibilities of both parties to the transaction. Legal, financial, and engineering considerations are involved in construction contracts. The designations "Party of the first Part" and "Party of the second Part" formerly used in construction contracts are now rarely encountered. Instead, reference is made to the Owner, the Contractor, and the Engineer. Other commonly used references to the Owner are the Government, the State, the City, the Authority, or the Company, as the case may be.

The Owner has three primary goals: most economical cost; specified quality; and completion on schedule. All of these goals may not be attainable in any one contract. The most economical cost may conflict with specified quality and completion time. Quality usually means higher cost and slower completion. Most rapid possible completion will work against low cost and high quality. The Owner may therefore have to compromise among his three goals to achieve that which is most important to him. The strongest factor influencing a contractor's performance is the profit motive. This motivation for the Contractor should present no problem to the Owner if the Contract Documents have been carefully prepared and administered, and no unforseeable events develop during the Work.

1.2 Proposal Solicitation

Proposals for construction contracts are received from interested contractors by competitive bidding or through the process of negotiation. The procedure for awarding a public works construction contract is generally outlined by law. In those instances, sealed bids are publicly invited by advertising in local, technical, and construction publications (see Article 5.2A, Notice to Contractors). The Contract is awarded to the lowest qualified bidder. Projects funded by private capital are not bound by any restrictions in the solicitation of proposals. The private owner need not accept a bid from anyone who wishes to submit one; nor need he accept the lowest bid. Proposals may be obtained by:

A. Open competitive bidding, as in publicly funded projects.
B. Competitive bidding by invitation to a group of selected contractors.
C. Direct negotiation with a selected contractor. The fundamental concept of negotiating a contract is mutual faith and confidence between Owner, Engineer, and Contractor. The Contractor is selected on the basis of his past performance, reputation, financial standing, and the personal character of the principals of the firm. In the negotiation process, factors other than price are considered.

1.3 Types of Contracts

Two basic types of construction contracts are in general use; the fixed price contract, and the cost reimbursable contract.

A. Fixed Price Contract.

Competitively bid construction contracts are of the fixed price type of contract. The Contract contains an agreement for work to be performed at a fixed price, regardless of the cost to the Contractor. This type of contract provides the greater degree of risk for the Contractor. On the other hand, it provides the ultimate incentive for the Contractor to control costs and increase productivity. The fixed price contract is common to major public works, heavy engineering construction, and commercial and residential construction. Before a fixed price type of contract can be awarded, the Owner must allow for the time required to produce a completely designed project including final Plans and Specifications. There are two variations to the fixed price type of contract; lump sum contract and unit price contract.

1. A lump sum contract provides for one overall price to pay for the entire Work of the Contract. This type of contract can be used where it is possible to accurately define on the Plans, all the limits of the proposed Work. This Plan data enables a bidder to determine the quantities of the various items of work. With this information it enables him to prepare and submit a more realistic bid price. Types of construction that would qualify for a lump sum contract include buildings, bridge superstructures, and most any kind of aboveground construction.

One advantage for the Engineer working with a lump sum contract is that it eliminates the need to take field measurements of completed work for determining payment quantities, were it a unit price contract. On the other hand, a lump sum contract has no built-in flexibility to accomodate the inevitable minor changes that have to be made to adjust to field conditions. For example, a change like moving the location of a drainage catch basin or manhole a few feet from that shown on the Plans can involve a written authorization and cost negotiations.

2. A unit price contract is applicable when the construction limits of items of work to be performed cannot be defined on the Plans. Types of construction falling into this category would include highways, foundations, pipelines, tunnels, and most any type of project involving work below the ground surface. The unit price contract provides for bidders to insert their prices in a Bid Schedule listing the various items of work and their estimated quantities. The Bid Schedule may also include individual lump sum payment items, when limits of the particular item of work can be completely defined on the Plans.

In the unit price contract, the Work is broken down into elements for payment purposes according to the type of work and trades involved. Payment is made on the basis of the number of units of each item of work completed and accepted. The unit price contract has built-in flexibility which allows the Engineer to order minor changes in the Work, without having to negotiate cost differences with the Contractor. A nominal change in the quantity of a unit price item is automatically taken care of, since the Contractor gets paid for the actual number of units he constructs.

B. Cost Reimbursable Contract.

A cost reimbursable contract is one in which the Owner and a selected contractor enter into direct negotiations to establish the terms of agreement between them and the amount of fee to be paid. This type of arrangement can be used to handle special construction projects where risks involved cannot be determined beforehand. It is also used in situations where the Owner wishes to begin construction before Plans and Specifications are 100 percent complete. The Contractor is reimbursed for his actual costs in constructing the project, plus an agreed upon fee. Various types of arrangements may be made for the fee. Some of the more commonly negotiated arrangements are listed below.

1. Cost Plus a Fixed Fee, in which the Contractor is reimbursed for his actual costs and is paid a fixed fee of an agreed upon amount. This type of contract is suited for projects involving research and development. Under this arrangement there is no real incentive for the Contractor to control his expenses since his fee remains fixed, regardless of the final cost.

2. Cost Plus a Percentage Fee, in which the Contractor is reimbursed for his actual costs plus an agreed upon percentage of those costs as his profit. The Owner may make any desired changes as the work progresses, without having to negotiate change orders. The higher the costs, the greater the Contractor's

profit. Here again there is no incentive for the Contractor to control expenses. This type of arrangement may be successfully used with a contractor of high integrity.

3. Cost With a Ceiling Price Plus a Fixed Fee, in which the Contractor is reimbursed for all costs up to the maximum figure specified in the Agreement. He also receives a fixed fee which will not vary, regardless of the ultimate cost of the work. Under this arrangement the Owner is assured that the cost of his project will not exceed the guaranteed maximum price. The Contractor on the other hand must closely control his operations if he is to keep his costs within the specified limit. Should he exceed the ceiling price, he receives no additional reimbursement for his excess costs.

4. Incentive Type, of which there are various incentive type contracts. The basic feature is that the Owner will pay a bonus for economic construction and for earlier completion. On the other hand, the Contractor may be penalized for his inefficiency and for late completion.

1.4 Contractual Relationships

There are four broad categories of contractual relationship in construction:

A. Construction by a Single Prime Contractor.

This is the normal relationship of construction contracting and is the best known. The Owner prepares detailed Plans and Specifications for the Work, using either his own staff or engaging the services of a design professional. Bids are solicited from contractors as outlined earlier in Article 1.2, Proposal Solicitation, and then an award is made. The Contractor is responsible for the successful completion of the project within the specified Contract time. He subcontracts work of a specialized nature that he is not equipped to perform. The Contractor has the duty to manage and coordinate all construction operations, including the work of his subcontractors. Under this arrangement, the Owner has to deal with only one Contractor.

The major disadvantage of this type of contractual relationship is the time element. Construction does not begin until the design has been completed, bids received, and the Contract awarded. This may not be the best arrangement for an Owner desiring early completion of his project. On the other hand, this arrangement can result in a lower cost to the Owner. Contractors are better able to submit a bid with fewer contingencies for protection against unknown risks, when they are provided with a completely detailed design.

B. Design-Build Contract.

In the design-build form of contractual relationship, the Owner engages a single organization with the responsibility for both design and construction. The Owner provides only basic performance criteria such as heating and cooling requirements, electrical capacities, and space and layout requirements. The Contractor may either be a firm having the capability to both design and construct; a joint venture combining a construction contractor and a design professional; or a construction contractor who subcontracts the detailed design (plans and specifications) to a design professional. Contracts may be competitively bid or negotiated. In a design-build contract, the Contractor takes on the added responsibility and risk of preparing an adequate design.

This form of relationship has been most successful in projects funded with private capital, where minimizing the construction cost is secondary to advancing the date of completion and the start of production. Publicly funded projects require public competitive bidding on a definite design furnished by the Owner. The design-build contract has been commonly used in the construction of chemical process plants, mills, thermal and nuclear power plants, cement plants, warehouses, and other similar types of projects.

A design-build contract is more difficult to price because the Contractor has to include in his bid a figure for a design that he has not yet prepared. Consequently, he incorporates a contingency in his bid to allow for these unknowns. As a result, construction of a project by design-build contract will generally cost more than if it were constructed under the normal procedure of having a contractor submit a bid on a completed set of Plans and Specifications. One feature of the design-build form of Contract is that the Owner gains the advantage of time. The Contractor can begin construction before final design is complete. This overlapping of construction and design allows for completion of the project at an earlier date.

C. Multiple Prime Contracts.

Many governmental agencies are required by statute or regulations to award multiple prime contracts in cases where the construction of a building will exceed a specified cost. Separate contracts are established for principal portions of the Work, namely General Construction (sitework, structural, and architectural); Electrical Work; Plumbing Work; and Heating, Ventilating, and Air Conditioning Work.

It has been maintained that under this arrangement, the Owner saves the cost of coordination of subcontractors that a general contractor would include in his bid price, if all the work were to be done under one contract. On the other hand, coordination of the work of these separate prime contracts now becomes the responsibility of the Owner.

D. Construction Manager.

Construction management may involve the development of a construction project during any phase, from its inception until it is a complete usable facility. It may include feasibility studies, preliminary design, scheduling, cost control, value engineering, phased construction, contract award, and technical inspection. There appears to be no singular common definition of construction management. The Construction Manager is a professional; he represents the Owner; and he works for an agreed-upon fee.

An owner who does not have the staff or the capability to develop his project may contract with a construction manager for either the complete development of his project or just some part of it. A contract for construction management may be given to a design professional; a construction contractor; a management firm; or to a combination of these firms. This type of arrangement can be more efficient than contracting separately for the different steps in the development of the project. A construction manager can manage and coordinate all activities to control time, cost, and the quality of the work.

Many times, an owner will engage a construction manager to provide (during the design phase) advice on design improvements, construction procedures, construction economies, and schedules. The Construction Manager can continue on into the construction phase, awarding contracts for the Owner and coordinating the work of the various prime Contractors. At other times, a construction manager may be retained for the construction phase only. Under this arrangement, he becomes the Owner's representative on the site, coordinating and inspecting the work of the several prime construction Contractors.

Chapter 2

Function, Composition, and Arrangement of the Specifications

2.1 Introduction

The two principal documents of the construction Contract are the Plans and Specifications. They provide the information necessary to guide the Contractor in constructing the project and fulfilling his contractual obligations. Information that can be effectively portrayed graphically is shown on the Plans. Information and instructions which can be more readily described in words are presented in the Specifications.

The Plans present a graphic representation of the Contract. Their primary purpose is to illustrate the construction. They show existing site conditions; the limits of Work; and location, shape, dimensions, size, and geometric relationship between the various construction components. The Specifications, presented in written form, transmit information that cannot be shown on the Plans. They present instructions that will enable the Contractor to perform the Work and fulfill the intent of the design. The Plans and Specifications complement each other; each conveys its own part of the story. Generally what is shown in one document is not repeated in the other.

Webster's Third New International Dictionary defines Specifications as "A written or printed description of constructional work to be done forming part of the contract, describing qualities of material and mode of construction, and giving dimensions and other information not shown in drawings."

The term "Specifications" as used in this book applies to the written Contract document, which includes bidding procedures, legal requirements, insurance requirements, material and workmanship requirements, inspection and testing procedures, and procedures for measurement and payment of the Work.

2.2 Function of the Specifications

In addition to defining the scope of Work, the Specifications establish obligations of the contracting parties. With respect to the Owner and his Engineer, it is the obligation to clearly define what is required; to establish a plan for its enforcement to the extent required during the period of execution; and to indicate how the

Work will be measured and paid for. With respect to the Contractor, it is the obligation of complying with the Contract requirements during the construction period.

The Specifications provide detailed requirements for the Work that is to be performed. The courts will usually require that the Specifications must be fulfilled by the Contractor, if performance is possible. If the Specifications are loosely worded and inconclusive as to what is required, the courts generally will make a determination based on: A) A fair interpretation of the Specifications; and B) What is required in the Specifications must be reasonable.

The role of Specifications is not only to comprehensively delineate the product that the Owner wishes to buy, and to set forth the terms of the Contractor's compensation, but also to establish a contractual relationship between the Contractor and the Owner with respect to the consequences of risks which may materialize. Risks have to be recognized and acknowledged, and each of the contracting parties must assume under the Contract, and without ambiguity, his share of the risks.

One of the basic principles in the preparation of a proper set of Specifications is to equitably distribute the risks inherent in the construction process among the Owner, Contractor, and Engineer. Depending on the type of construction involved, these risks can include labor strikes, unanticipated subsurface conditions, unexpected weather conditions, problems with underground utilities, and the hazards presented by existing structures situated next to deep excavations. The equitable distribution of risk is discussed in Article 8.13, Identifying and Controlling Risks.

Once the Contract has been signed, the Specifications become the rule book that governs the performance of the Work. It is the instrument that governs the official relations between Contractor, Owner, and Engineer.

2.3 Composition of the Specifications

The book of Specifications for a construction contract contains information covering the various phases of the Contract. It includes three major categories: the Bidding Documents, the General Conditions, and the Technical Sections. The Bidding Documents and the General Conditions are sometimes referred to as the "boilerplate", or "up front" documents, because they are located in the front part of the book.

A. The Bidding Documents.

The Bidding Documents (see Chapter 5) are of prime interest to the prospective bidder. They contain a general description of the proposed Work, detailed instructions and requirements under which a bidder is to prepare and submit his bid, and the conditions under which the Contract will be awarded and executed. These documents provide information relating to:

1. The Advertisement for Bids
2. Instructions to Bidders
3. Preparation of Proposal
4. Submittal of Proposal
5. Award of Contract
6. Execution of Contract

B. The General Conditions.

The General Conditions (see Chapter 4) set forth the rights and responsibilities of the parties to the Contract, the requirements governing their business and legal relationships, and the authority of the Engineer. They encompass several subsections as follows:

1. Definitions and Terms
2. Scope of Work
3. Control of Work
4. Control of Material
5. Legal Relations and Responsibility to the Public
6. Prosecution and Progress
7. Measurement and Payment

C. The Technical Sections.

The Technical Sections (see Chapter 3), sometimes called the Technical Provisions or Technical Requirements, define the minimum requirements of quality of materials and of workmanship. They also outline how the Work is to be measured and paid for. Requirements for items of construction like Excavation and Embankment, Bearing Piles, Concrete, and Structural Steel, are presented in separate, individual Specification Sections. Unlike the Plans, the Technical Sections concentrate the requirements for specific items of the Work, thus eliminating the possibility of overlooking related requirements placed in other Sections of the Specifications. This arrangement also facilitates specifying measurement and payment procedures for the items of work.

2.4 Arrangement of the Specifications

A. Users of the Specifications.

The initial user is the prospective bidder. He reads the Advertisement for Bids which gives him a general description of the proposed Work, including estimated quantities of the major items of construction. If the type and extent of the proposed

Work falls within the capabilities of a prospective bidder's organization, he will most likely take the next step and obtain a set of the Plans and Specifications.

The next portion of the Specifications that the prospective bidder will want to review, would be the Bidding Documents. These contain instructions on how the bid is to be prepared and submitted. In the course of preparing his bid, the bidder will solicit price quotations from material suppliers and subcontractors, both of whom also become users of the Specifications. The material supplier must review the Technical Sections in order to determine requirements for the material on which he is being requested to quote a price. The subcontractor must review the General Conditions and those Technical Sections that present requirements for the item or items of construction on which he will be submitting his price.

When the Contract is awarded, the successful low bidder becomes the Contractor, and he is now responsible for complying with all requirements of the Specifications. The Engineer, who is the Owner's representative on the site, is responsible for assuring that the Work will conform to the Contract requirements. By virtue of their complete involvement, both Contractor and Engineer become the major users of the Specifications, since they have to become familiar with all of its contents.

B. Optimum Arrangement for Users.

The most suitable arrangement in presenting Specifications is that which will best satisfy the needs of a user. For the benefit of prospective bidders, the Advertisement for Bids is the first item in the Book of Specifications, following the Table of Contents. The next major category in the Book of Specifications will be the Bidding Documents, to assist those prospective bidders who have decided to bid for the Work. The material presented in the Bidding Documents will guide the bidder through the different stages in the preparation and submittal of his bid. In the course of preparing his bid, the bidder will find it necessary to consult the rest of the Specifications, namely the General Conditions and the Technical Sections. The General Conditions will be presented before the Technical Sections, since items that concern the Contract as a whole are generally first considered, before delving into specific construction items.

It should be noted that the seven subsections of the General Conditions listed earlier in this Chapter follow a progressive pattern in the course of the Contract. Beginning with Definitions and Terms, the reader is guided through Prosecution and Progress to Measurement and Payment for completed work.

Similarly, the Technical Sections are arranged in the order of the prosecution of the Work. Thus the user would expect to find the requirements for clearing the site ahead of those for roadway pavement, or find the requirements for concrete foundations ahead of those for structural steel. An illustration of this arrangement is presented in Article 13.5, Table of Contents.

Chapter 3

The Technical Sections

3.1 Introduction

In preparing Contract Specifications, the Technical Sections are generally prepared first, with the General Conditions completed next, and then the Bidding Documents.

Whereas the Bidding Documents introduce the project to the bidder, and the General Conditions outline the general requirements and legal responsibilities of the parties to the Contract, the Technical Sections represent the "meat" of the Specifications. They provide the Contractor with the information that will guide him in constructing the facility. The primary objective of the Technical Sections is to present the desired quality of materials, acceptable level of workmanship, and how work is to be measured and paid for.

This information is not only for the benefit of the Contractor, but also for his purchasing agent, his material and equipment suppliers, and his subcontractors; and last but not least, for the benefit of the Owner's site representative, who will have the responsibility for administering the construction contract.

3.2 The Technical Section (Five Part Format)

The technical portion of the Specifications is arranged in Sections. Each Section deals with the work of a specific item of construction, such as Excavation or Concrete. All items of work that are to be permanently incorporated in the project are accounted for in the Technical Sections.

The Technical Section provides instructions to guide the Contractor in performing the work required for a specified item or items of construction. This information is generally arranged in a pattern established by the American Association of State Highway and Transportation Officials (AASHTO). The pattern is frequently referred to as the AASHTO format or five part format (see Article 3.4, Sample Technical Section). It is called a five part format because the Section is divided into five parts or subsections. Each subsection is given a numerical number and title, as follows:

1. DESCRIPTION
2. MATERIALS

3. CONSTRUCTION REQUIREMENTS
4. METHOD OF MEASUREMENT
5. BASIS OF PAYMENT

An explanation of each of these subsections follows below.

1. DESCRIPTION.

This subsection serves as an introduction to the Section, presenting a description of the type and extent of work required. On a highway project for example, a Section on Concrete would open with the following statement: The work specified in this Section includes furnishing, placing, finishing, and curing cast-in-place concrete for bridges, culverts, and other miscellaneous structures, as shown on the Plans and specified herein.

Following the opening statement, additional information bearing on the work of the Section is presented. This would include instructions covering the following requirements, when applicable to the Work of the Section:

A. Shop Drawings.

These drawings are to be prepared by the Contractor and submitted to the Engineer for approval. They present details of the Contractor's proposed fabrication and erection procedures for items of permanent construction. Additional information on this subject is presented in Article 4.4D, Plans and Contractor's Drawings.

B. Working Drawings.

These drawings are also prepared by the Contractor and submitted to the Engineer. They are concerned however, with temporary construction such as concrete formwork, falsework, temporary cofferdams, and support of deep excavations (see Article 4.4D, Plans and Contractor's Drawings).

C. Design Criteria.

The Contractor is normally required to design his temporary installations, such as the support of deep excavations, roadway decking, and groundwater control systems. The designer will usually spell out the design criteria that is to be followed by the Contractor in preparing his design.

D. Warranty.

When a warranty is required for the Work of the entire Contract, it is specified in the General Conditions (see Article 4.3K, Warranty of Construction). The normal duration of this warranty against defective materials and faulty workmanship is one year from the date of final acceptance. If a warranty is specified in a Technical Section, it is for a greater duration than the overall warranty. When applied to a piece of equipment, it may be for a period of five years. A specific warranty may also be required in the form of a roofing bond, covering a period of ten years.

E. Permits and Notices.

Before a contractor may be permitted to begin certain phases of work, such as dredging or building demolition, he is usually required to obtain certain permits and serve notice to affected utility owners (see Article 4.6C, Permits, Licenses, and Taxes).

F. Work by Others.

There may be situations where certain phases of work related to or affecting the work of the Contractor is to be performed by others. Examples of work that may be performed by others, are:

(1) Work Related to Existing Site Utilities. Utility owners will make the connections of new work to their existing systems and repair any damage to their systems caused by the Contractor's operations.
(2) Furnishing of Equipment. On multicontract projects like rapid transit systems, the Owner will, in the interest of standardizing maintenance procedures, purchase items of equipment like lighting fixtures and portable fire extinguishers, from the same manufacturer for the entire project.
(3) In times of scarcity of certain materials like structural steel, the owner of a multicontract highway project may, in the interest of preventing delays, purchase rolled structural sections for the bridges, under an earlier procurement contract. This steel would then be furnished to each bridge contractor.

G. Salvage.

On projects involving reconstruction, the Owner may require that certain removed items be salvaged. For example, in roadway reconstruction, a Section on Excavation would require that frames and castings removed from existing manholes and catch basins, and not reused in the Work, be delivered to the storage yard of the Municipality.

2. MATERIALS.

Now that the Contractor has been presented with the general information and instructions concerning the work of a particular Section, the specifics are considered next.

The work specified in Sections dealing with such items as Embankment, Concrete, Roadway Base Course, and Storm Sewers cannot really get underway until the Contractor first obtains approval of his proposed materials. Material requirements are therefore presented next, to enable the Contractor to establish his source of supply and then proceed to obtain the necessary approvals. Material requirements can cover many areas. In addition to the physical and chemical characteristics, there are requirements governing samples, tolerances, mix proportions, tests, certification, fabrication, shop painting, shipping, and storage of material on the site.

A. Physical and Chemical Requirements.

The most commonly used construction materials are those which originate in the soil and in the underlying rock. To illustrate physical and chemical requirements, consider the fine and coarse aggregate used in producing concrete. Requirements for aggregate are almost universally referenced to an industry Standard prepared and published by the American Society for Testing and Materials (see Chapter 7). This particular Standard is ASTM C33, Standard Specification for Concrete Aggregates. Requirements governing gradation, deleterious substances, soundness, and the methods of sampling and testing for both fine and coarse aggregate will all be found in ASTM C33.

B. Samples.

Samples are generally required from the Contractor before his proposed materials will be considered for approval. Until the furnished samples have been tested and accepted, the Contractor does not have the approval to use his proposed materials. The general procedure for submitting samples is presented in the General Conditions (see Article 4.5D, Samples, Tests, Cited Specifications).

C. Tolerances.

Variations exist in all natural and manufactured materials. The Designer establishes the limits of variation within which the desired results can be produced. These limits are then specified, to guide the Contractor in his material procurement. An example of variation and allowable tolerances is illustrated in the following Table of gradation limits for aggregate base course material:

Sieve Size (square openings)	Percent Passing (by Weight)
2 inch	100
1 inch	75–95
3/8 inch	40–75
No. 4	30–60
No. 10	20–45
No. 40	15–30
No. 200	5–20

D. Mix Proportions.

Mix proportions of concrete materials have to be designed to ensure that the proportions will be able to produce concrete meeting the specified requirements. Many contracts will require the Contractor to design the mix proportions for certain items of construction such as portland cement concrete, and bituminous concrete. In line with this requirement, it is necessary for the Designer to provide guidelines for the Contractor to follow.

Concerning portland cement concrete, it is now no longer necessary to specify in detail the procedure for determining mix proportions. This can now be referred to an industry Standard established by the American Concrete Institute (ACI). The Standard is ACI 211.1, Recommended Practice for Selecting Proportions for Normal, Heavyweight, and Mass Concrete. This subject is discussed in more detail in Article 10.12, Cast-in-Place Concrete.

E. Tests (Contractor).

In the process of getting approval for his proposed materials, the Contractor is expected to have performed certain preliminary qualifying tests prior to submitting his samples. This testing is considered as being part of the Contractor's quality control responsibility, to prevent any undue delays which can result from the rejection of material that does not conform to the requirements. Accordingly, the Contractor may be required to submit along with his samples, results of specified tests indicating conformance to requirements.

F. Fabrication.

Many of the items to be incorporated in the Work require fabrication in a shop or yard, prior to delivery to the site. This would normally include such items as structural steel, metal stairs and ladders, precast concrete items, and highway signs. When fabrication is specified, requirements can concern adequacy of the fabricator's facilities and his experience; conformance to approved shop drawings; conformance to welding, galvanizing, and other industry standards; fabrication of test units (precast concrete); shop assembly and workmanship; and surface finish.

G. Certification.

Some manufactured and fabricated items may not warrant fulltime inspection in the shop by a representative of the Owner, during the manufacturing or fabrication process. To cover this situation, the Specifications may require the Contractor to submit a notarized Certificate of Compliance. This would apply to individual items or to relatively minor components produced in limited quantity. A Certificate of Compliance would be required for: (1) Filter fabric specified for use in protecting the roadway subsurface drainage course of granular material, against the intrusion of silt or soil particles; (2) Units of precast concrete lagging which are to be placed inside flanges of steel H-piles, installed for retaining wall construction; or (3) Precast concrete bumpers specified for use in vehicle parking areas.

More information on this subject is available in Article 4.5E, Certification of Compliance.

H. Shop Painting.

Painting requirements are generally specified or referred to under the two headings of Shop Painting and Field Painting. Work that is to be performed off-site is specified in the Materials subsection, and this would include shop painting. Field painting, which is to be performed on the site, is specified in the Construction Requirements subsection. Shop painting is usually specified for items of uncoated ferrous metal such as structural steel, miscellaneous metal fabrications, and bridge guard railing.

I. Shipping.

Many fabricated items are subject to damage from handling during shipment to the site, if proper precautions are not taken. Structural steel members and precast concrete girders particularly qualify for this precaution. Requirements pertaining to the loading and handling of members during shipment to the site, would be specified.

J. Storage on Site.

Because of the limitations of some construction procedures, and the timing of certain operations, it is often necessary to store materials and fabricated items on the site until they can be incorporated into the Work. During the period of storage, the contamination of materials and damage to fabricated items have to be prevented.

Portland cement in bags would have to be stored above the ground, within suitable weatherproof enclosures. Structural steel members and reinforcing steel would have to be stored above the ground on platforms or on other supports, to keep them free of mud and other foreign matter.

3. CONSTRUCTION REQUIREMENTS.

This subsection deals with requirements governing the Contractor's construction operations on the site. It covers such areas as preliminary preparations, disposition of removed materials, procedures, allowable tolerances, finishes, end results, inspections, and tests. Each item of construction has its own particular requirements.

Construction requirements are generally presented in the same sequence in which the related work will be performed. For example, when specifying requirements for concrete construction, the requirements for curing concrete would follow the requirements governing the placement of concrete, which in turn would follow the requirements for formwork.

A. Preliminary Preparations.

Some items of construction require that certain preliminary preparations be completed, before the main operation may begin. To illustrate: In the demolition of a building, the Contractor would be required to first have all utility services disconnected before demolition could get underway. In addition, if the building was a multistoried structure situated in a populated area, the Contractor would also be required to construct a covered pedestrian walkway before starting the demolition. This subject is covered in more detail in Article 10.5, Demolition.

B. Disposition of Removed Material.

In the course of constructing a project, various materials and equipment items may be removed as a result of clearing the site, excavating, demolishing structures, or other operations. Disposition can include the reuse in the Work of suitable materials; disposal off the site of unsuitable and surplus suitable material, or the delivery of designated removed items to the Owner's storage yard. Instructions on the disposition of removed material or equipment have to be given to the Contractor. Additional information on this subject can be found in the following Articles:

(1) Article 3.2.1G, Salvage.
(2) Article 4.3L, Disposal of Material Outside the Work Site.
(3) Article 10.2A, Disposition of Removed Structures and Materials.
(4) Article 10.7C, Excavation; Paragraph 2, Disposition of Materials.

C. Procedures.

A procedure is defined as a particular or established way of doing things, or a particular course of action.

Generally, the Engineer refrains from telling the Contractor how to perform his work; otherwise he may be relieving the Contractor of his responsibility for producing the required results. However, the design profession has over the years acquired time-tested knowledge on correct and incorrect procedures concerning specific operations. In these instances, the Designer will specify the procedure because he knows from past experience that the end result can be achieved.

The procedure for many construction operations is obvious and familiar to those involved. For example, in concrete construction it is obvious that concrete cannot be deposited until forms are in place. It is also recognized and accepted that reinforcing steel and other embedded items must be in position before concrete is placed. However, in the interest of safety or to ensure that a desired result will be obtained, certain time-tested procedures and precautions will be specified for certain operations.

To illustrate, the demolition of a multistoried building would prompt the following precautionary requirements: Demolition shall begin at the top of the structure. It shall be completed above each floor before any supporting members on the lower levels are disturbed. Structural framing members shall be removed and lowered to the ground by means of hoists, derricks, or other suitable equipment. Demolition equipment and stored removed materials shall be located so as not to impose excessive loads on supporting walls, floors, or framing.

D. Allowable Tolerances.

Variations in materials were previously discussed in Article 3.2.2C, Tolerances. Allowable tolerances are required for many construction operations. It is physically impossible to produce a finished product to the exact dimensions and slopes, and at the precise locations, shown on the Plans. Recognizing this fact, the Designer will where necessary, establish the allowable tolerances or deviations from the indicated dimensions, slopes, and locations. Additional information on this subject can be found in Article 4.4E, Conformity with Plans and Specifications.

A good illustration concerns the installation of driven bearing piles for bridge foundations. The Plans indicate the required position of each pile in the foundation. If batter piles are required, the slope of these piles will also be shown. Many factors can affect the final position and slope of a driven pile. Unknown obstructions encountered in the soil being penetrated can cause the pile to shift from its intended location. In the driving of piles in water, it may be difficult to maintain the pile driver barge in position at all times during the driving of a pile. Allowable deviations would read as follows: Piles shall be driven with a maximum deviation from the vertical or the batter shown on the Plans, of 1/4 inch per foot of pile length. The final location of the top of each pile at cut-off elevation, shall be not more than three inches in any direction from the location shown on the Plans.

E. Finishes.

The quality of a finished surface for items of construction like concrete or structural steel can be an integral phase of the overall operation. Concrete surfaces can be given various finishes, depending on their locations and on their intended usage. For example, a requirement for finishing the surface of a concrete sidewalk would read: After screeding, and as soon as the water sheen has disappeared, the surface shall be given a wood float finish. All outside edges and joints, shall be finished with an edging tool having a 1/4 inch radius.

Additional information on concrete finishes can be found in Article 10.12H, Concrete Finishes.

F. End Results.

A Technical Section will have achieved its intended purpose if the Contractor is successful in producing the desired results. Except for those situations described earlier herein under Article 3.2.3C. Procedures, the Contractor is generally permitted to use his ingenuity and "know how" to better enable him to produce the results intended by the design.

In the construction of a roadway embankment for example, the Contractor would be required to produce (as an end result) the specified minimum density for the embankment in place. This minimum density would be expressed as a percentage of the maximum dry density for the particular material. Maximum dry density is predetermined in the laboratory, from a sample of the same material that is to be placed in the embankment. Many factors can directly affect attainment of the required in-place density. Some of them are listed below.

(1) Thickness of the uncompacted layer of the material: The thicker the layer, the greater the number of passes of compaction equipment are required to produce the same result. It may also require equipment of a larger capacity.

(2) Moisture content: The maximum dry density of a soil's material is directly related to the optimum moisture content for that material. Any appreciable variation from this optimum moisture content will make it more difficult to obtain the desired density.

(3) Method of depositing the material: Material deposited from bottom dump vehicles is easier to spread in a uniform thickness than material deposited from end dump vehicles.

(4) Type and capacity of compacting equipment: Certain types of equipment are more effective in compacting finer grained soils than granular types of soils, and vice versa. Also, compaction equipment of a greater capacity will require fewer passes to obtain the needed density.

(5) Number of passes of compaction equipment: Generally the more passes, the greater the compaction.

Except for specifying limitations on the loose thickness of each layer of material deposited, and moisture content of the material, the Contractor is free to develop his own procedures for hauling, dumping, spreading, controlling moisture, and selecting the type and size of his compaction equipment. To verify the end result, the Engineer will take a sample from the compacted embankment and determine the density of the material in place. Additional discussion of this subject is presented in Article 8.16, Methods and Results.

G. Inspections.

Items of construction which involve step-by-step-procedures may require interim inspections and approvals, as the work progresses. This would generally apply to work that is to be covered over in subsequent operations, and be hidden from view when that item of work is completed. To ensure that this work has been performed in accordance with the Contract requirements, it must be inspected before the Contractor will be permitted to cover it over.

To illustrate; in storm drain, sanitary sewer, and similar pipeline construction, a trench is first excavated, the supporting subgrade is prepared, the pipe is laid and joined, and then the next operation consists of backfilling the trench. If the pipeline work in the trench was not inspected before backfilling begins, the Engineer can only assume that the work, now hidden from view, was properly executed. Improper work in buried pipeline construction may not manifest itself until years after the project has been completed. By then it is difficult, if not impossible, to get the Contractor to remedy the fault, and the Engineer is often held responsible. To prevent this from occurring, the Specifications will state that no backfill shall be placed until the laid pipe and other underground construction has been inspected by the Engineer and approved. Additional discussion of this subject is presented in Article 4.4M, Inspection of the Work.

H. Tests (Engineer).

Inspection can be considered as a method of assuring compliance of the Contractor's work, by visual observations. For some items of work, compliance cannot be verified by visual observation alone. Items like concrete, compacted embankment, and pavement courses, also require tests to assure compliance. Concrete requires the breaking of test cylinders to verify required compressive strengths. Compacted embankment requires tests on samples taken from the compacted layers, to verify the specified in-place density. For pavement courses, cores are taken to verify thickness of the course. Cores from asphaltic concrete pavement are tested to verify required density of pavement.

In addition to outlining details of the tests, the Specifications will indicate frequency of the test, as this also will affect the Contractor's operation. Additional information on this subject is presented in Article 8.12, Quality Control and Quality Assurance.

4. METHOD OF MEASUREMENT.

This subsection of the Technical Section outlines the method of measurement to be used to determine the payment quantity for each completed item of work specified in the Section. This subsection also spells out the conditions under which work will be considered for measurement. Since these quantities will determine the amount of money to be paid the Contractor, methods of measurement are spelled out in simple, easily understandable language.

Engineered construction contracts are generally of the unit price type of contract, for the reasons outlined earlier in Article 1.3, Types of Contracts. In order to be able to determine the amount of money due the Contractor for work performed, the number of completed units of each payment item has to be determined. Methods of measurement and computation used in arriving at the payment quantities are those generally recognized in good engineering practice. A listing and explanation of these methods is presented at the beginning of Article 4.8, Measurement and Payment.

A unit price type of contract is equitable for both Contractor and Owner. The Contractor knows that he will receive payment for all work he performs in accordance with the Contract. The Owner, on the other hand, knows that he will be paying only for work and services performed in accordance with Contract requirements, no more or no less. The quantities of work for which payment will be made are determined through the use of various units of measurement. Some of the more commonly used units are:

A. Linear Foot (linear)
B. Square Yard (area)
C. Cubic Yard (volume)
D. Pound or Ton (weight)
E. Each (complete units)
F. Lump Sum

Examples of Methods of Measurement illustrating these units, are:

A. Linear Foot.

A method for measurement of storm drain pipelines, would read: The quantity of Reinforced Concrete Storm Drain Pipe to be paid for will be the number of linear feet of each size of pipe furnished and acceptably placed within the limits shown on the Plans or ordered by the Engineer. Measurement will be made along the centerline of pipe from the inside face of drainage structure or face of headwall.

B. Square Yard.

A method for measuring concrete sidewalk would read: The quantity of Concrete Sidewalk, 4-Inches Thick, to be paid for will be the number of square yards of sidewalk acceptably constructed within the limits shown on the Plans or ordered by the Engineer. No measurement for payment will be made for areas greater than one square yard occupied by other construction.

This method of measurement takes into consideration those situations where subsurface structures like vaults or subway vents, may extend through the sidewalk.

C. Cubic Yard.

A method for measuring roadway excavation would read: The quantity of Roadway Excavation, Unclassified to be paid for will be the number of cubic yards of material acceptably removed from within the limits shown on the Plans or ordered by the Engineer. The payment volume will be computed by the average end area method from original cross-sections taken before the surface is disturbed, and final cross-sections taken after excavation is completed. Cross-sections will normally be taken at fifty foot stations except at locations where it may be necessary to take intermediate sections to obtain a more realistic result.

The designation of unclassified excavation is used where no distinction will be made between the removal of earth and of rock, as far as payment for the work is concerned. Unclassified excavation includes the removal of all materials of whatever nature encountered in the Work. This designation is generally used where it is anticipated that little or no rock will be encountered. Where the borings indicate that an appreciable quantity of rock will have to be removed, the Designer will establish two separate payment items; Roadway Excavation, Common, and Roadway Excavation, Rock.

D. Pound.

A method for measuring reinforcing steel would read: The quantity of Reinforcing Steel to be paid for will be the number of pounds of reinforcing steel furnished and placed in accordance with the Contract requirements. The quantity will be determined from computations based on the nominal weights as listed in ASTM A615, Standard Specification for Deformed and Plain Billet-Steel Bars for Concrete Reinforcement. Laps of bars for all approved splices will be measured for payment. When bars are spliced by welding, the weight for payment will be computed as for lapped splices.

E. Each.

This method of measurement is established where the Contract calls for identical units or minor structures, of the same construction and size. In the case of identical drainage manhole structures, it is far more convenient to count the number of structures than it is to determine separately for each manhole unit the volume of concrete, weight of reinforcing steel, and weight of manhole frame and casting.

A method of measurement for completed units of a standard type of drainage structure, would read: The quantity of Manholes Type A to be paid for, will be the number of individual manholes complete in place and accepted, measured on a per each basis.

F. Lump Sum.

When the limits of work of an individual item of construction are clearly defined on the Plans and no changes are anticipated, payment for that item of work is sometimes established on the basis of a lump sum price. For such cases, it is not necessary to take measurements to determine payment quantities. After the Contractor satisfactorily completes that particular item of work, he is paid the lump sum price for that item. If the work of that item should extend over more than one progress payment period, the Contractor will be paid a part of the lump sum price, in proportion to the work completed. A method of measurement for the demolition of a specific building that has been designated as a lump sum item, would read: Demolition of Building No. 28 acceptably completed, will be measured for payment on a lump sum basis.

5. BASIS OF PAYMENT.

This last subsection of the Technical Section outlines, in detail, those costs that are included in the Contract price for a specific item of work. By the same token, it will also outline those costs of related work that are not included in the Contract price for that item. These costs are paid for under other items of work. By so specifying, points of disagreement are minimized between the Contractor and the Engineer as to whether certain costs for related work have been accounted for.

An explanation of the conditions governing these payments to the Contractor, and their significance, are presented in Article 4.8, Measurement and Payment.

The following examples of the Basis of Payment correspond to the same items of work presented earlier under 3.2.4, METHOD OF MEASUREMENT.

A. Linear Foot Basis.

A method of specifying payment on a linear foot basis would read: Reinforced Concrete Storm Drain Pipe will be paid for at the Contract Unit Price per Linear Foot, for each size of pipe listed below:

12-INCH REINFORCED CONCRETE STORM DRAIN PIPE
18-INCH REINFORCED CONCRETE STORM DRAIN PIPE

The Contract price shall include the costs of excavation, disposal of excavated materials, sheeting and bracing, dewatering, and backfilling.

B. Area Basis.

A method of specifying payment on the basis of area would read: Concrete Sidewalk, 4-Inches Thick, will be paid for at the Contract Unit Price Per Square Yard, as listed in the Unit Price Schedule. The price shall include the costs of excavation, disposal of surplus and unsuitable materials, bed course, expansion joints, and backfilling.

C. Volume Basis.

A method of specifying payment on the basis of volume would read: Roadway Excavation, Unclassified will be paid for at the Contract Unit Price Per Cubic Yard, as listed in the Unit Price Schedule. The price shall include the costs of disposing of surplus and unsuitable materials, and cost of disposal sites.

The intent of a design is to utilize suitable excavated material in the permanent construction, wherever possible. This would include use of the material for roadway embankment, and as backfill for trenches and other excavated spaces. The Engineer determines the suitability of this material. Should there be a surplus of suitable excavated material over and above that needed for the Work, then the Contractor would be required to dispose of the surplus material off the site.

D. Weight Basis.

A method of specifying payment on the basis of weight would read: Reinforcing Steel will be paid for at the Contract Unit Price Per Pound, as listed in the Unit Price Schedule.

E. Per Each Basis.

A method of specifying payment on a per each basis would read: Manhole Type A will be paid for at the Contract Unit Price Per Each, as listed in the Unit Price Schedule. The price shall include the costs of excavation and disposal of unsuitable material, sheeting and bracing, control of water, bedding course, backfilling, brickwork for adjustment to grade, reinforcing steel, and frame and cover.

F. Lump Sum Basis

A method of specifying payment on a lump sum basis would read: Demolition of Building No. 28 will be paid for at the Contract Lump Sum Price, as listed in the Unit Price Schedule. The price shall include the cost of permits; rodent control; protection of the public including construction of sidewalk canopies; capping and plugging of disconnected utilities; removal and disposal of all debris, materials, equipment, and other contents; removal of foundations; and backfilling to grade.

3.3 Section Arrangement

Technical Sections are generally arranged in the order of execution of the Work. The reason for this is obvious. The reader looking for Clearing of Site requirements would normally expect to find this information at the beginning of the Technical Sections, since this work is performed in the initial stage of construction. Placing this Section after Sections that specify work to be performed in later stages, such as concrete construction, could cause the reader to lose precious time searching for this information. The following examples illustrate an arrangement of Technical Sections for specific types of projects:

A. Highway Project.

This type of project would involve earthwork; roadway pavement courses; storm drainage facilities; grade separation structures; and incidental construction such as guard railing, fencing, curbs, and sidewalks. Sections would be arranged in the following order:

Clearing Site
Demolition of Buildings
Roadway Excavation and Embankment
Structure Excavation
Storm Drains
Manholes, Inlets, and Catch Basins
Aggregate Base Course
Portland Cement Concrete Pavement
Shoulder Construction

Bearing Piles
Structural Concrete
Reinforcing Steel
Structural Steel
Roadway Deck
Bridge Railing
Guard Railing
Fencing
Concrete Curb
Concrete Sidewalk
Pavement Traffic Markings
Traffic Control Signs
Topsoiling and Seeding

In highway work, construction of the roadway and of the grade separation structures would proceed concurrently.

B. Large Bridge Project.

A multispan structure over water will normally be constructed under two prime contracts; one for the substructure and another for the superstructure. The substructure contract could include Technical Sections arranged in the following order:

Clearing and Grubbing
Removal and Disposal of Existing Buildings
Detour Road
Structure Excavation
Cofferdams
Bearing Piles
Storm Drainage
Water Mains
Portland Cement Concrete
Reinforcing Steel
Timber Fender System
Structural Steel
Dampproofing and Membrane Waterproofing
Protective Coating on Concrete

C. Concrete Dam.

A dam contract for water storage could include Technical Sections arranged in the following order:

Diversion and Care of Water During Construction
Clearing and Grubbing
Excavation
Concrete Masonry
Steel Reinforcement
Plastic Waterstops
Backfilling of Dam (this operation is generally timed with the completion of each concrete lift)
Rock Fill
Dumped Riprap
Stone Block Paving (used in paved channels and gutters)
Crushed Stone Filter
Foundation Drilling and Grouting
Corrugated Metal Pipe
Foundation Drains
Piping Systems
Metalwork
Miscellaneous Aluminum Work
Timber Stop Planks
Discharge Valves
Cone Valves
Sluice Gates
Broome Gate
Bascule Gate
Traveling Crane
Water Level Indicator and Recorder
Emergency Generating Plant
Gate House Superstructure

3.4 Sample Technical Section

The following sample illustrates a Technical Section prepared in the AASHTO or five part format. Additional information on this format is presented in Article 13.2A, Five Part (AASHTO) Format.

SECTION 32, REINFORCING STEEL

1. DESCRIPTION

A. The work specified in this Section includes furnishing and placing reinforcing steel for concrete construction, in accordance with these requirements and as shown on the Plans.

B. The top layer of reinforcement in bridge roadway decks shall be epoxy coated. All other reinforcement shall be uncoated.

2. MATERIALS

A. Uncoated bars shall conform to the requirements of ASTM A615, Grade 60.

B. Epoxy coated bars shall conform to the requirements of ASTM A775, Grade 60. At the time of shipment, the Contractor shall furnish written certification from the supplier that the coated reinforcing bars meet the requirements of this Specification.

C. Before fabricating any material, the Contractor shall furnish bar lists, placement drawings, and bending diagrams to the Engineer for approval. Detailing of reinforcing bars shall be in accordance with the requirements of ACI 315, Manual of Standard Practice for Detailing Reinforced Concrete Structures. Each bar list shall show the calculated individual weight of each bar, total weight of each bar size, and total weight of all bars on the list. Calculated weights shall be based upon the theoretical unit weights shown in the Table in ASTM A615. No materials shall be fabricated until such approval is given.

D. Fabrication and Shipment—Cutting and bending shall be performed before shipment to the work site. All bars shall be bent cold. Bars shall be shipped in standard bundles, tagged and marked in accordance with the Code of Standard Practice of the Concrete Reinforcing Steel Institute.

E. Spacers and chairs for use in concrete surfaces to be exposed to view shall have plastic tipped legs or other non-staining type. Chairs, metal supports, and tie wire, for use with epoxy coated bars shall be plastic coated. Tie wire shall be black, annealed steel, 16 gauge or heavier.

F. Materials for patching damaged surfaces of coated bars shall be compatible with the coating, and inert in concrete. Patching materials shall be supplied by the epoxy resin manufacturer.

3. CONSTRUCTION REQUIREMENTS

A. Storage and Protection—Reinforcing steel shall be stored above ground on blocking, racks, or platforms. Bundles shall be plainly marked to facilitate inspection and checking. Bars shall be protected against damage and detrimental coatings of loose rust or foreign matter.

B. Handling Coated Bars—In handling coated reinforcing bars, the Contractor shall use padded or nonmetallic slings and padded straps, to avoid damaging the coating. Bundled bars shall not be dropped or dragged. Coatings damaged in handling shall be repaired in accordance with the instructions of the supplier.

C. Bar Supports and Spacers—Distance from forms shall be maintained by means of stays, blocks, ties, hangers, or other approved supports, so that the distance does not vary from the position shown on the Plans by more than 1/4 inch. Blocks for holding reinforcing from contact with the forms shall be precast mortar blocks of approved shape and dimensions, and shall have tie wires embedded in them. Supports for reinforcing steel in slabs shall be spaced not more than four feet apart transversely or longitudinally. Metal chairs resting on slab forms shall not be used unless otherwise approved by the Engineer. The use of pebbles, broken stone or brick, metal pipe, or wood blocks as spacers, will not be permitted.

D. Placing and Fastening—Bars shall be free from kinks and from bends that cannot be readily and fully straightened. All uncoated bars shall be free of loose mill scale and rust, and of oil, dirt, or other substances that may reduce the bond with concrete. Oiling of forms shall be completed before the placing of reinforcement.

Reinforcement shall be accurately placed in the positions shown on the Plans and firmly held in position during concrete placement. Bars shall be firmly tied with wire or secured with approved metal clips at each intersection, except that when spacing is less than 12 inches in each direction, alternate intersections need not be tied.

The placing of unsupported bars or fabricated mats on layers of fresh concrete as the work progresses, and adjusting bars during the placing of concrete, will not be permitted.

E. Splicing—Bars shall be spliced only at the locations shown on the Plans unless otherwise approved by the Engineer. Bars in lapped splices shall be placed in full contact and securely wired together. The lengths of laps shall be as shown on the Plans. Splices shall be staggered where possible.

F. Protection of Dowels—Exposed portions of dowels expected to be exposed to the weather for more than 60 days shall be protected against corrosion with a brush coat of neat cement grout.

4. METHOD OF MEASUREMENT

A. The quantity of reinforcing steel to be paid for will be the number of pounds of reinforcing steel furnished and placed in accordance with the Contract requirements. The quantity will be determined from computations based on the nominal weights as listed in ASTM A615. Laps of bars for all approved splices will be measured for payment.

B. Reinforcing steel included for payment in other items will not be measured for payment under this Section.

5. BASIS OF PAYMENT

A. Reinforcing steel will be paid for at the Contract Unit Price Per Pound for the following items listed in the Unit Price Schedule:

1. REINFORCING STEEL, UNCOATED
2. REINFORCING STEEL, EPOXY COATED

B. The Contract Unit Prices shall include the costs of bar supports, spacers, chairs, ties, and other devices for holding the reinforcing steel in place, coating of exposed dowels in uncoated reinforcing steel, and all else necessary to complete the work as specified.

C. Reinforcing steel used in concrete bearing piles, storm drainage structures, and precast concrete items, will be included for payment under their respective items in the Unit Price Schedule.

Chapter 4

The General Conditions of the Contract

4.1 Introduction

General Conditions cover the nontechnical requirements of the construction Contract and apply to the Work as a whole. Some owners title their documents the General Provisions. In this book we will use the designation of General Conditions, as adopted by the Engineers' Joint Contract Documents Committee.[1]

General Conditions of the Contract have been formulated and developed over the years by special committees representing various industry and professional groups, and public agencies. They include The American Association of State Highway and Transportation Officials (AASHTO), The Engineers' Joint Contract Documents Committee (EJCDC), The American Institute of Architects (AIA), The Associated General Contractors of America (AGC), The General Services Administration (GSA), and The U.S. Corps of Engineers (C. of E.).

In addition to outlining the rights and responsibilities of the parties to the Contract, the General Conditions present requirements governing their business and legal relationships, and include guidelines to be used in administering the Contract. The General Conditions also define duties and responsibilities of other parties affected by the Contract, such as subcontractors, the Engineer, and the Construction Manager. Other nontechnical matters presented in the General Conditions include scope of work, control of work, legal relations and responsibility to the public, prosecution and progress, and measurement and payment for the Work.

While Technical Sections of the Specifications present requirements for the Work to be constructed, the General Conditions provide an orderly procedure for handling situations and problems that may arise in the day-to-day dealings with the Contractor. The Articles which comprise the General Conditions are sometimes miscalled "the legals" and the "boilerplate."

---

[1]The Committee is composed of representatives from the National Society of Professional Engineers, American Consulting Engineers Council, American Society of Civil Engineers, and the Construction Specifications Institute.

Additional rights and responsibilities of the contracting parties are outlined in the Technical Sections. They, however, relate principally to obligations on the part of the Contractor, specifying minimum requirements for materials and workmanship in the performance of the Work.

The arrangement and titles of Subsections and most Articles presented in this Chapter follow the pattern established many years ago by AASHTO in their published Guide Specifications for Highway Construction. This document was adopted by the Federal Highway Administration and the State Departments of Transportation, and modified as necessary to serve the particular requirements of each owner. The AASHTO Guide Specifications are updated periodically. The Articles not directly related to the AASHTO Guide Specifications merit consideration because of their frequent occurrence in construction contract documents.

SECTION 100—GENERAL PROVISIONS of the Guide Specifications for Highway Construction, dated 1985, is reproduced with the permission of AASHTO. Sections 101 and 104 through 109, deal with the General Conditions. They are inserted at the end of this Chapter for direct reference. These Sections are referred to and discussed herein.

Further discussion, and guidelines for the specification writer when presenting Articles of the General Conditions, can be found in Chapter 11, Presenting the General Conditions.

4.2 Definitions and Terms

A. General.

A misinterpretation or misunderstanding of abbreviations, definitions, or terms used in the Contract Documents, can lead to unnecessary problems. To minimize these possibilities, abbreviations and definitions of the terms that may be used in a set of Contract Documents, are presented early in the General Conditions.

B. Abbreviations.

Refer to abbreviations presented in Article 101.01, same title, in the AASHTO Sample at the end of this Chapter. Commonly used abbreviations which are not listed in the AASHTO Sample, are here presented:

ACI	American Concrete Institute
ACIL	American Council of Independent Laboratories
AED	Associated Equipment Distributors
AISC	American Institute of Steel Construction
AWG	American Wire Gauge

CBR	California Bearing Ratio
CPM	Critical Path Method
CRSI	Concrete Reinforcing Steel Institute
C.Y. or Cu. Yd.	Cubic Yard
Dia. or Diam.	Diameter
DOT	Department of Transportation
EPA	Environmental Protection Agency
FAA	Federal Aviation Administration
FRA	Federal Railway Administration
Ft.	Foot or Feet
GSA	General Services Administration
In.	Inch or Inches
Lb.	Pound
L.F. or Lin. Ft.	Linear Foot (Feet)
L.S.	Lump Sum
MBM	Thousand Feet Board Measure
MUTCD	Manual on Uniform Traffic Control Devices for Streets and Highways
NBFU	National Board of Fire Underwriters
NBS	National Bureau of Standards
NEC	National Electric Code
NEMA	National Electrical Manufacturers Association
NRMCA	National Ready Mixed Concrete Association
OSHA	Occupational Safety and Health Administration
PCA	Portland Cement Association
PCI	Prestressed Concrete Institute
PVC	Polyvinyl Chloride
SSPC	Steel Structures Painting Council
S.Y. or Sq. Yd.	Square Yard(s)
UMTA	Urban Mass Transportation Administration
USSG	United States Standard Gauge (uncoated sheets and thin plates)

C. Definitions.

Refer to definitions presented in Articles 101.02 through 101.63 in the AASHTO Sample at the end of this Chapter. Commonly used terms that have not been listed in the AASHTO Sample are presented here.

Addendum—Written interpretations or revisions to the Bidding Documents, issued by the Owner within a specified time before the opening of bids (see Article 5.2E, Examination of Plans, Specifications, Special Provisions, and Site of Work; Paragraph 1, Addenda).

Agreement—See Article 101.08, Contract, in the AASHTO Sample.

Bid—See Article 101.34, Proposal, in the AASHTO Sample.

Bid Bond—See Article 101.36, Proposal Guarantee, in the AASHTO Sample.

Bid Form—See Article 101.35, Proposal Form, in the AASHTO Sample.

Bid Schedule (Unit Price Schedule)—A prepared Schedule included with the Proposal Forms, containing estimated quantities of unit price items for which bid prices are invited.

Construction Manager—An authorized representative of the Owner acting directly or through properly authorized agents, within the scope of the particular duties delegated to him. (see Article 4.4C, Construction Manager)

Contract Documents—These documents include Advertisement or Invitation for Bids; Specifications; Plans; Addenda; Proposal; Contract forms and bonds; Agreement; Notice of Award; Notice to Proceed; and all subsequent agreed-to modifications that are required to complete construction of the Work in an acceptable manner.

Contract Drawings—See Article 101.31, Plans, in the AASHTO Sample.

Day—As used in the Contract Documents, "day" shall be understood to mean calendar day, unless otherwise designated. (This definition is offered to prevent "day" being confused with "Working Day" as defined in Article 101.61 in the AASHTO Sample.)

Federal Specifications—Specifications and standards, supplements, amendments, and indices thereto, as prepared and issued by the General Services Administration of the Federal Government.

Fixed (Plan) Quantity—The quantity of an item of construction listed in the Bid Schedule, which is computed from Plan dimensions. Payment for the item will be made on the basis of the Fixed Quantity. (see Article 4.8C, Fixed (Plan) Quantities)

Interim Work—Work provided by the Contractor for use during construction and which is to be left in place and in service, upon completion of the Contract. An example would be an interim pumping system to dispose of water inside a tunnel during its construction, and which is left in place until the installation of a permanent system under a subsequent contract.

Mobilization—Preparatory work and operations of the Contractor necessary for the initial movement of personnel, equipment, supplies, and incidentals to the project site. This would include establishment of the Contractor's field offices, shops, plant, storage areas, sanitary and other facilities required by the Contract, as well as by Local or State law and regulation, and all other work and operations which must be performed or costs incurred prior to beginning work on compensable items of work at the site. (see Article 10.4, Mobilization)

Notice of Award—A written notice from the Owner to the successful Bidder informing him that his bid has been accepted and that he is required to execute the Contract and furnish satisfactory Contract bonds. (see Article 5.3B, Award of Contract)

Payment Bond—A Contract bond in an approved form, executed by the Contractor and his Surety as a guarantee by the Contractor that he will pay in full, all bills and accounts for materials and labor used in the construction of the Work. (see Article 5.3E, Requirements of Contract Bonds)

Performance Bond—A Contract bond in an approved form, executed by the Contractor and his Surety as a guarantee by the Contractor that he will complete the Work in accordance with the Contract. (see Article 5.3E)

Prequalification—A procedure used to assure the Owner of a Bidder's ability to perform the Work, has experience in similar work, and that his financial condition is acceptable. (see Article 5.2B, Prequalification of Bidders)

Provide—When referring to work to be performed by the Contractor, the word "provide" as used in these Specifications, means to furnish and install complete in place.

Reference Drawings—Supplemental drawings which contain information necessary for conduct of the Work, but which are not included in the Contract Drawings. Reference drawings can include boring logs, drawings of existing structures, and utility location plans.

Shop Drawings—Drawings prepared by the Contractor and which include fabrication, erection, and setting drawings; schedule drawings; manufacturers' scale drawings; wiring and control diagrams; cuts or entire catalogs; pamphlets; descriptive literature; and performance and test data; all relating to the permanent construction and which the Contractor is required to submit to the Engineer for his approval. (see Article 4.4D, Plans and Contractor's Drawings)

Specialty Item—A designated item of work that requires highly specialized knowledge, craftsmanship or equipment, not ordinarily available in contracting organizations qualified to bid on the Contract as a whole, and in general is usually limited to minor components of the overall Contract. (see Article 4.7B, Subletting of Contract)

Substantial Completion—The date as determined by the Engineer, when the Work or a specified portion thereof is sufficiently complete in accordance with the Contract Documents, so that the Owner may occupy the Work or designated portion, for the use for which it is intended.

Supplemental Agreement—A written agreement between the Contractor and the Owner, covering added work which is beyond the scope of the Contract. When approved and properly executed, the supplemental agreement becomes a part of the Contract.

Temporary Work—Work provided by the Contractor for use during construction and which is to be removed before final acceptance of the Contract.

Working Drawings—Drawings prepared by the Contractor for temporary structures or construction, such as concrete formwork, falsework, support of excavation, cofferdams, temporary bridges, support of utilities, groundwater control systems, or such other work as may be required for construction but which do not become an integral part of the completed project. Working drawings have to be submitted to the Engineer for his review. (see Article 4.4D, Plans and Contractor's Drawings)

4.3 Scope of Work

A. Intent of Contract.

Refer to Article 104.01, same title, in the AASHTO Sample at the end of this Chapter. The basic premise of this Article is to remind the Contractor that he has the responsibility to provide for the construction and completion of the Work in every detail in accordance with the Contract Documents.

B. Changes.

Refer to Article 104.02, Alteration of Plans or Character of Work in the AASHTO Sample at the end of this Chapter. It is a rarity for a contract to be completed in every detail as originally designed and contracted for, without changes having to be made during the course of construction. No matter how well planned, it is virtually impossible during the design phase to anticipate in advance, every detail necessary to produce a complete job. This is particularly true when trying to ascertain the conditions existing below the surface of the ground. Changes to the Contract can result from:

1. Unanticipated subsurface conditions encountered at the construction site. For example, the Contractor in his excavating operations may encounter rock, where none was indicated.
2. Changes made at the request of the Owner. This may involve changes in Owner furnished facilities such as the availability of right-of-way and work areas, and services such as furnishing electrical power and water. Changes may also involve elimination of a portion of the Work or of an entire pay item. (see Article 4.8G, Eliminated Items.)
3. Design changes due to omissions or error. The size of a proposed water main may have been omitted from the Plans, or the wrong pipe material may have been specified in the Specifications.
4. Introducing additional work; work that was not contemplated when the Contract was awarded.

Changes made in the Contract can be reflected in the Specifications or on the Plans; in the manner of performance of the Work; or in accelerating performance of the Work. Most changes will fall within the general scope of the Work of the Contract. Sometimes the Contractor is requested to perform work which is not provided for in the original Contract. This would then be considered as Extra Work. It is covered more fully in Article 4.3E, Extra Work. Changes usually result in an increase or decrease in the Contractor's costs, and may also impact on Contract time.

Changes to the Contract are initiated by issuance to the Contractor of a written Change Order signed by the Owner or the Engineer. The basis for computing adjustment in cost or contract time or both, will usually be agreed to before the Change Order is issued. The Change Order will include a statement describing the scope of the work involved in the change and the agreed-upon method for determining adjustment in price and contract time. Price adjustment is generally handled by one of the following methods:

a. Payment under applicable Contract unit prices.
b. Mutual agreement on a lump sum price.
c. Mutual agreement on unit prices which are not provided in the Contract.
d. If agreement cannot be reached on any of the above arrangements, then payment is made on a force account basis. In this instance, the Contractor is reimbursed for the actual cost to him of labor, materials, equipment, insurance and taxes, plus an allowance for profit and overhead. See Article 4.8F, Payment for Extra and Force Account Work.

An added note is introduced here to expand on the fourth paragraph in referenced AASHTO Article 104.02. This paragraph prohibits the Contractor from making a claim for any loss of anticipated profit because of variation between the "approximate" and actual quantities of work. In this paragraph, the term "approximate quantities" is used in referring to the quantities shown on the Proposal Form or Bid Schedule. The term "estimated quantities" is preferred, as illustrated in Article 5.2F, Interpretation of Quantities in Bid Schedule.

The last paragraph in referenced AASHTO Article 104.02 deals with differing subsurface conditions. Because this subject is frequently involved in damage claims, the next Article, 4.3C, is being devoted to it.

C. Differing Subsurface Conditions.

Refer to the last paragraph in Article 104.02, Alterations of Plans or Character of Work, in the AASHTO Sample at the end of this Chapter. One distinguishing characteristic of construction performed below the surface of the ground is that the actual subsurface conditions can never be completely realized until they are uncovered during construction. Because of the limited time and funds usually available to the Designer, it is virtually impossible for him to establish a complete, accurate record of existing subsurface conditions at the site. Borings may be taken and test pits dug, but they can only reveal the subsurface conditions existing at the precise locations of the boreholes or test pits. Subsurface conditions between these locations may be assumed to vary uniformly, but there is no guarantee that this assumption will prove to be correct. Available existing records of subsurface utilities are not always correct, nor are they always maintained up-to-date. Then again, the limited time given bidders to prepare their bid leaves them little or no time to conduct their own subsurface investigation. They therefore have to rely on the information made available to them by the Owner. Consequently, during the course of construction, a subsurface condition is encountered that was neither anticipated nor planned for by either Owner or Contractor, and therefore represents a changed condition.

A Differing Subsurface Conditions Article, sometimes referred to as a Changed Conditions Clause, provides that, if during the progress of the Work, the Contractor: 1) Encounters subsurface conditions at the site differing materially from those indicated in the Contract Documents, or 2) Encounters such subsurface conditions which could not reasonably have been anticipated by the Contractor, and 3) If these conditions will materially affect the cost of doing the Work, then he (the Contractor) is to immediately notify the Engineer before these conditions are disturbed.

If, after investigating these conditions, the Engineer agrees that they will affect the Contractor's costs, an equitable adjustment is negotiated and the Contract is modified accordingly in writing. The following case history illustrates an unanticipated subsurface condition and how it was resolved:

During the years of 1957–1961, the author was Resident Engineer on the construction of a water reservoir project consisting of 14 prime contracts. The major features included:

1. A mass concrete gravity dam 700 feet long with the top of spillway 85 feet above rock foundation, to form a storage reservoir of three billion gallons capacity.
2. Relocation of three miles of single track railroad.
3. Relocation of two miles of secondary road including a new three-span bridge crossing of the river.
4. A concrete gravity wall 2400 feet long, to protect the main tracks of a railroad adjacent to the proposed reservoir.
5. A screen chamber building.
6. An aerator basin.
7. A chemical building rising 85 feet above the ground to receive, store, and feed chemicals for treating the water supply.

Excavation for the dam foundation was paid for at a unit price per cubic yard, under two items; Earth Excavation for Dam, and Rock Excavation for Dam. Rock excavation was defined in the Specifications as excavation in solid ledge rock which requires drilling and blasting for its removal. All other excavation for the dam was classified as earth excavation.

In the process of excavating the earth overburden at one end of the dam, the Contractor encountered material that was extremely difficult to remove in its natural state with the 2-1/2 C.Y. shovel that he was using. He complained to the author that this condition was affecting his production considerably. The Contractor claimed this was a changed condition because the material did not meet the normal conception of earth excavation; nor did it qualify by the definition for rock excavation.

Accompanying the Superintendent to the scene, it was noted that the material consisted of a glacial till which had cemented together into a hard conglomerate. The shovel operator demonstrated the problem by trying to load his bucket, but the teeth of the bucket just bounced off this hard material. He then hooked his bucket under a ledge of the material and attempted to load, but the rear of the machine just raised off the ground. The Contractor stated that he would have to blast this material in order to remove it.

The soils boring samples had not indicated the existence of this conglomerate in the earth overburden. The Contractor therefore had anticipated no production problems in removing the overburden for which he had submitted a unit price of $0.65 per cubic yard. The Specifications included neither a disclaimer clause on the boring information, nor a changed conditions clause for unanticipated subsurface conditions. Yet this clearly was the case of a contractor being damaged by a subsurface condition unanticipated by both Owner and Contractor.

The Contractor requested that, in addition to being paid the Contract unit price for Earth Excavation, he be reimbursed for his costs of drilling and blasting. This seemed like a reasonable request and the Contractor's formal Notice of Claim for damages was forwarded with the author's favorable recommendation. The cemented conglomerate material was drilled and blasted on two consecutive Saturdays. The Contractor was reimbursed for this additional work on a force account basis, to the satisfaction of all concerned.

D. Variations in Estimated Quantities.

When the Work is completed, the final pay quantities of most unit price items in a Contract will, for various reasons differ from the estimated quantities originally listed in the Unit Price Schedule. This difference can be a result of changes made in the Work or of changes caused by unanticipated site conditions.

In developing his unit prices, a bidder will distribute his indirect fixed costs such as home office overhead, supervision and profit, over the number of units indicated in the Unit Price Schedule (see Article 5.2F, Interpretation of Quantities in Bid Schedule). If the number of units in an item should be reduced, he will not fully recover his fixed costs. On the other hand, if there is an increase in the number of units, the Contractor may recover more than his fixed costs. In either situation there is improper compensation.

Recognizing this fact, many Owners will provide in the Specifications a mechanism for renegotiating the Contract unit price of a major item when its final quantity differs from the estimated quantity by more than a specified percentage. A general definition of a major item can be found in Article 101.27, Major and Minor Contract Items, of the AASHTO Sample at the end of this Chapter. The limits defining a major item may vary slightly between different sets of General Conditions. A provision for handling variations in estimated quantities can be illustrated by the following Article quoted from the General Provisions for Construction Contracts, 1976, State of Maryland Department of Transportation, Mass Transit Administration, Baltimore Region Rapid Transit System.

"GP-4.03 Variations in Estimated Quantities

Where the quantity of a major pay item as designated in the Proposal in this Contract is an estimated quantity and where the actual quantity of such pay item varies more than twenty-five percent (25%) above or below the estimated quantity stated in this Contract, an equitable adjustment in the Contract price shall be made upon demand of either party. The equitable adjustment shall apply only to that quantity above one hundred twenty-five percent (125%) of the estimated quantity or that quantity below seventy-five percent (75%) of the estimated quantity. If the quantity variation is such as to cause an increase in the time necessary for completion, the Engineer shall, upon receipt of a written request for an extension of time, ascertain the facts and make such adjustment for extending the completion date as in his judgment the findings justify."

It should be noted that, in addition to providing for renegotiation of the Contract unit price, quoted Article GP-4.03 provides for adjustment of Contract time, if found justified.

E. Extra Work.

Refer to Article 104.03, same title, in the AASHTO Sample at the end of this Chapter. As presented earlier in Article 4.3B, Changes; after construction gets underway, changes to the Contract are almost inevitable. Sometimes an ordered change will involve work that was neither contemplated nor provided for when the Contract was awarded. This work which cannot qualify for payment under any of the prices in the Unit Price Schedule, is considered to be Extra Work. The Contractor will be ordered in writing to do such work in accordance with the Specifications and accompanying written instructions.

The method of reimbursement is generally mutually agreed to before implementation of the written order. It can take one of the following forms: 1) A lump sum price, or 2) A unit price or prices, or 3) If agreement cannot be reached by Method 1) or 2), or a combination of them, then payment may be made on a force account basis as provided in Article 4.8F, Payment for Extra and Force Account Work. Force account work is sometimes referred to as Cost-Plus Work or Time and Material Work.

F. Maintenance and Protection of Traffic.

Refer to Articles 104.04, Maintenance of Traffic, and 107.10, Barricades and Warning Signs, of the AASHTO Sample at the end of this Chapter. Most projects during their construction phase, will in some way or another affect the normal functioning of traffic in the area. Depending on the type of project and its location, the traffic affected may be vehicular, pedestrian, railroad, air borne, or water borne. Each type of traffic will have its particular requirements for maintenance and protection.

This Article will concern itself with requirements governing the maintenance and protection of traffic as it affects motor vehicles and pedestrians. The maintenance and protection of railroad traffic, and the protection of traffic in navigable waters, will each be discussed later in this Chapter under Articles 4.6I, Railway-Highway Provisions, and 4.6J, Construction Over, In, or Adjacent to Navigable Waters.

The requirements for maintenance and protection of vehicular and pedestrian traffic can in themselves vary greatly, depending on the type of project and its location. The referenced Articles in the AASHTO Sample present general requirements for the maintenance and protection of vehicular and pedestrian traffic as they are affected by road construction. The requirements for maintaining and protecting traffic in a populated area may be more demanding. This subject is discussed in more detail in Article 11.3F, same title.

G. Rights In and Use of Materials Found on the Work.

Refer to Article 104.05, same title, in the AASHTO Sample at the end of this Chapter. Some projects may require select material for use as backfill or fill in special areas or for special purposes. The term "select material" generally indicates that original material has to undergo some processing in order to satisfy the requirements. Processing may involve screening or blending with other material, in order to meet gradation requirements. Other physical requirements for select material may concern durability and hardness.

In preparing his proposal, a bidder would normally assume that select material would have to be obtained from a source off the site. It sometimes happens that excavated material on the site is suitable for select material. It would be quite arbitrary on the part of the Engineer to insist that this excavated material be placed in roadway embankment as originally planned in the design. No useful purpose would be served by this rigid interpretation other than to impose an unnecessary expense on the Contractor. AASHTO Article 104.05 prevents this situation from developing on the job. As a further step in eliminating any possible disagreement in the field, this Article defines how payment will be made and outlines the conditions that go along with it.

H. Cleaning Up.

Refer to Article 104.06, Final Cleaning Up, of the AASHTO Sample at the end of this Chapter. It will be noted that the AASHTO Article concerns cleanup before final acceptance. The author favors requirements that regulate cleanup throughout the life of the Contract. A sample requirement would read:

> Cleaning-Up. The Contractor shall throughout the life of the Contract keep his work and storage areas free of accumulations of waste materials, rubbish, trash, and debris. Before final inspection of the Work, he shall remove all waste and surplus material, trash, rubbish, debris, and equipment from the Project site, material sites, and other areas occupied by him in connection with the Work. All areas shall be left in a neat and presentable condition.

I. Value Engineering Proposals by Contractor.

Refer to Article 104.07, same title, of the AASHTO Sample at the end of this Chapter. Simply explained, value engineering is an endeavor to discover a less expensive way of getting the same results.

Construction contractors, particularly those of long establishment, have acquired over the years an expertise in their particular field of work. Endeavoring to fully utilize this experience, some owners will provide the Contractor an opportunity to submit a proposal for lowering the cost of the Contract. The inducement to the Contractor is a sharing with the Owner, usually on a 50–50 basis, in the net savings resulting from the accepted value engineering proposal.

J. Temporary Utility Services.

In order to perform his work, the construction Contractor will have a need for certain utilities long before they can be permanently provided under the Contract. Services most commonly needed are electrical power and water. Unless these services are already available at the site, it is customary to require the Contractor to provide temporary electrical power and water needed for construction purposes. Other temporary facilities such as sanitary provisions, first aid, and construction safety, are discussed in Subsection 4.6, Legal Relations and Responsibility to the Public.

A requirement for temporary services would read:

Temporary Utilities.

The Contractor shall provide, maintain and remove when no longer required, temporary utilities for the proper and expeditious prosecution of the Work. This shall include:

1. Temporary Light and Power.

The Contractor shall:

a. Provide and maintain sufficient lighting for the safety of his construction forces and to ensure the proper construction, inspection, and prosecution of the Work, in addition to any lighting necessary to protect the Work and the public;

b. Make application to the local private Utility Company for the necessary temporary electric service. He shall assume the cost of all fees and permits, all private Utility Company charges, and cost of all energy consumed;

c. Provide all temporary wiring, extension cords, sockets, and all lamps, both initial and replacement, used for the temporary power and lighting systems.

d. When the permanent electrical power and lighting systems are in operating condition, said systems or portions thereof may be used in lieu of the temporary service for construction purposes, provided that the Contractor 1) assumes full responsibility for the entire power and lighting systems, and 2) pays all costs for operation and restoration of the systems.

2. Temporary Water.

The Contractor shall:

a. Make all arrangements for obtaining temporary water connections and pay all costs incurred thereby. He shall furnish, install, and pay for all piping and equipment required to provide water for the execution of the Work;

b. Pay all costs of water until final acceptance of the Work;

c. Provide drinking water with suitable cups for all personnel on the job.

K. Warranty of Construction.

At one time, a one-year warranty for general construction usually applied to building construction contracts. However, it is now frequently included in contracts for civil works construction. The following illustration of a warranty requirement is quoted from the Supplementary General Provisions for Construction Contracts, 1976, State of Maryland Department of Transportation, Mass Transit Administration, Baltimore Region Rapid Transit System:

SGP-7.05 Warranty of Construction

A. The Contractor warrants all work done under the Contract to conform to the Contract requirements and to be free from faulty materials and workmanship for a period of one year from date of acceptance thereof, which one year period shall be covered by the Performance Bond and Payment Bond therefor as specified in General Provisions Articles GP-3.05 and GP-3.06, except that in the case of defects or failure in a part of the Work which the Administration takes possession prior to final acceptance, such period shall commence on the date the Administration takes possession.

B. The Administration will give the Contractor prompt written notice of any defects or failures following their discovery. The Contractor shall commence corrective work within ten days following notification by the Administration of the defect or failure and shall diligently prosecute such work to completion, provided that, the Administration shall have the right to use unsatisfactory materials and equipment until they can be taken out of service without injury to the Administration. If the Administration exercises the right to use unsatisfactory materials and equipment the Contractor will be permitted to commence corrective work within six months from time of discovery of the defect of failure requiring correction.

C. Replacement parts and repairs shall be subject to approval of the Administration. The Contractor shall bear all costs of corrective work, which shall include necessary disassembly, transportation, reassembly, and retesting, as well as repair or replacement of the defective materials or equipment, and any necessary disassembly and reassembly of adjacent work; provided that the Administration will disassemble and reassemble at its expense adjacent materials or equipment not furnished by the Contractor, where necessary to give access to the defective materials or equipment.

D. If the Contractor fails to perform corrective work in the manner and within the time stated, the Administration may proceed to have such work performed at the Contractor's expense and he will honor and pay the costs thereof upon demand, and his sureties will be liable therefor. The Administration will be entitled to all costs and expenses, including reasonable attorneys' fees, necessarily incurred upon the Contractor's refusal to honor and pay such costs.

E. The Contractor's performance bond and payment bond shall continue in full force and effect during the period of this guarantee, but during such period shall apply only to corrective work required to be performed hereunder.

F. The rights and remedies of the Administration under this Article are not intended to be exclusive, and do not preclude the exercise of any other rights or remedies provided by this Contract or by law with respect to unsatisfactory work performed by the Contractor.

Some Specifications have used the words "warranty" and "guarantee" interchangeably. These words do have different meanings, as pointed out in Article 11.3K, Warranty of Construction.

L. Disposal of Material Outside the Work Site.

Most jobs will produce waste or surplus material that cannot be utilized in the Work. In most cases, the Contractor is required to dispose of this material off the site, and he has to make his own arrangements with owners of private disposal sites. In making these arrangements, the Contractor has to conform to certain procedures. These procedures are illustrated in the following example quoted from the General Provisions for the Construction of the Stage 1 Light Rail Transit System, Port Authority of Allegheny County, Pittsburgh, Pennsylvania, July 1981:

GP4.18 *Disposal of Material Outside the Work Site.* Unless otherwise specified, the Contractor shall make his own arrangements for disposing of waste and excess materials outside the Worksite and shall pay all costs incurred therefor.

GP4.18.1 When any material is to be disposed of outside the Worksite, the Contractor shall notify the appropriate Regional Office of the Department of Environmental Resources and request approval of the proposed disposal site. The Contractor shall also obtain a written permit from the property owner on whose property the disposal is to be made and he shall file with the Engineer said permit or a certified copy thereof, together with a written release from the property owner absolving the Authority from responsibility in connection with the disposal of material on said property. No disposal operations may begin until the above approvals, permits, and releases have been obtained by the Contractor and evidence of same furnished to Engineer.

GP4.18.2 Contractor in disposing of waste and excess materials shall comply with all Federal, State and local governmental rules, regulations, laws and ordinances concerning such disposal of waste and excess materials.

4.4 Control of Work

A. Introduction.

An Owner may spend millions of dollars on the construction of his project. Consequently, he wants to assure himself that the Contract he awards will be faithfully performed. With this in mind, he will assemble on the site a team of field personnel to act as his authorized representatives to inspect the work of the Contractor for conformance to the requirements of the Contract. This responsibility of on-site representatives of the Owner to assure quality of work performed by the Contractor, is referred to as Quality Assurance. It is not to be confused with Quality Control, which concerns procedures and controls established by the Contractor within his organization, to produce the quality of work called for. Quality Assurance is a program conducted for or by the Owner, whereas Quality Control is the program established and conducted by the Contractor.

The intent of the following Articles in this Subsection, Control of Work is to present requirements in a manner that will protect the Owner's rights and interests, yet at the same time allow the Contractor the freedom to plan and conduct his work.

B. Authority of the Engineer.

Refer to Article 105.01, Authority of the Engineer, and Article 105.09, Authority and Duties of the Engineer, of the AASHTO Sample at the end of this Chapter. Generally, after the Contract has been signed, there are few direct contacts between the Contractor and the Owner. All details of administering the construction Contract are handled by the Engineer, who represents the Owner on the site.

The authority given the Engineer, as described in the first paragraph of referenced AASHTO Article 105.01, has been perceived by many as having him serve in a dual capacity. In addition to inspecting the Work and determining its acceptability, the Engineer also has authority to render decisions in matters concerning the adequacy of his own performance; namely the Plans and Specifications which he prepared. This situation, in which the authorized representative of one party to a contract may also act as the unbiased referee of disputes between the two parties, may be acceptable only because of the professional integrity of the Engineer. This dual role is discussed in more detail in Article 11.4B, Authority of the Engineer.

The second paragraph of referenced AASHTO Article 105.01, and the last sentence of referenced AASHTO Article 105.09, give the Engineer authority to suspend the Contractor's work wholly or in part. This right to stop the work came into being initially to prevent the continuance of installation of unsatisfactory materials or of poor workmanship. However, over the years this right to stop the work began to take on additional meaning. The courts started to interpret this authority also as an obligation on the part of the Engineer to stop an operation when unsafe conditions or circumstances existed that might endanger life or property. It was becoming evident that along with this authority went the responsibility to use it at the appropriate time, and not to abuse it. The authority to suspend the Contractor's work is now shifting from the Engineer to the Owner. This subject is further discussed in aforementioned Article 11.4B.

C. Construction Manager.

Refer to Article 1.4D, Construction Manager. Up until about 15 years ago, the field administration of construction contracts was handled by employees of the Designer or of the Owner. At about that time, another type of site representative for the Owner was being introduced. He was called the Construction Manager (CM) and he could be either a construction contractor or another engineer consultant. As pointed out in Article 1.4D, the area of construction management can cover many phases of a project, depending on the desire of the Owner. Employment of a CM has been more common in the field of commercial and industrial building construction than in civil works construction.

On some projects, the Engineer and CM may both be on the site as representatives of the Owner. In this instance, the responsibilities of each would be defined in the Specifications. For example, the CM could be responsible for inspecting and accepting the Work, and the Engineer could be responsible for reviewing and approving the Contractor's drawings and interpreting the Contract Documents.

D. Plans and Contractor's Drawings.

Refer to Article 105.02, Plans and Working Drawings, of the AASHTO Sample at the end of this Chapter. The Contract Drawings (Plans) cannot provide all the details for the proposed construction. Some procedures and methods are left unspecified, to allow contractors to best utilize their available plant and equipment. Other details like erection procedures, are also left to the Contractor, in order to benefit from his expertise and ingenuity. Also, specific information on equipment to be furnished by the Contractor can naturally be provided only by him. Drawings that have to be submitted by the Contractor involve both permanent and temporary construction, and include those submitted by his subcontractors.

The referenced AASHTO Article uses the designation "Working Drawings" to apply to all drawings submitted by the Contractor. Some project Specifications differentiate between drawings for permanent construction and those for temporary construction, by using the designation "Shop Drawings" for permanent construction and "Working Drawings" for temporary construction. (see their definitions in Article 4.2C.) The author favors the use of these two terms and they will be adhered to in this book.

Permanent construction requiring the Contractor to furnish shop drawings and related data, include:

1. Concrete Construction—Design mix proportions; sequence of concrete placement for bridge decks, and for dams; method of curing.
2. Reinforcing Steel—Bar schedules; bending details.
3. Structural Steel Work—Fabrication details; connection details; erection procedures.
4. Emergency Power System—Single line diagrams showing equipment ratings; schematic wiring diagrams; instruction manuals.

Design of temporary construction is generally left to the Contractor. He is responsible for its adequacy and its successful completion. Temporary construction which requires the Contractor to furnish working drawings, include:

1. Building Demolition—Details of sidewalk canopies for protection of the public; sequence of demolition.

2. Open Cut Excavations Including Pipe Trenches—Details for supporting excavation side walls; details for supporting street decking; details for control and disposal of subsurface water.
3. Concrete Construction—Details of formwork; details of falsework including design computations.
4. Bridges Over Water—Details and design computations of temporary cofferdams for constructing pier foundations.

Shop drawings for permanent construction receive a thorough check by the Engineer, and require his approval. Working drawings for temporary construction are usually reviewed for their general compliance with Contract requirements, since the Contractor is responsible for any design involved. In either case, the Contractor is advised that any work done or material ordered, before the related drawings have been approved or reviewed by the Engineer, will be at the Contractor's risk.

Drawings that have been approved or reviewed by the Engineer and returned to the Contractor, will generally contain one of the following notations:

1. APPROVED (REVIEWED).
2. APPROVED (REVIEWED) AS NOTED.
3. REVISE AND RESUBMIT.
4. DISAPPROVED.

E. Conformity with Plans and Specifications.

Refer to Article 105.03, same title, of the AASHTO Sample at the end of this Chapter. The intent of this Article is to remind the Contractor of his responsibility to conform the Work to the requirements of the Contract Plans and Specifications. It recognizes the presence of variability in materials and workmanship, by referring to allowable tolerances. The Article also outlines the terms and conditions under which minor deviations from the specified requirements may be acceptable, as determined by the Engineer. Materials or workmanship determined to be not acceptable are to be removed and replaced, or otherwise corrected. Additional information on this subject is presented in Article 3.2.3D, Allowable Tolerances and Article 11.4E, Conformity with Plans and Specifications.

F. Coordination of Plans and Specifications.

Refer to Article 105.04, Coordination of Plans, Specifications, Supplemental Specifications, and Special Provisions, of the AASHTO Sample at the end of this Chapter. In the preparation of Contract Plans and Specifications, the various parts that make up the Contract Documents (see definition in Article 4.2C) may be prepared by different individuals at different times. In addition to the involvement of the specification writer, input to the Specifications may be provided by the Geotechnical, Civil, Structural, Water Resources, Mechanical, Electrical, and Architectural designers, and legal and insurance consultants. No matter how much care may be exercised, discrepancies and ambiguities will manifest themselves in or between the various documents. The AASHTO Article shows how discrepancies and ambiguities arising between the various components of the Contract Documents may be resolved. This is further clarified in Article 11.4F, Coordination of Plans and Specifications.

G. Field Record Drawings.

Field record drawings are drawings upon which is recorded field information that is found to differ from that shown on the Contract Plans. For reasons outlined earlier in Article 4.3B, Changes, it is rare when the Work is constructed as originally shown on the Contract Plans. Another area for difference is in the location of existing subsurface utilities. When subsurface utilities are exposed, their location will in many cases differ from that indicated. This field information is superimposed on a set of the Contract Plans, for record purposes. It is particularly valuable for recording the location of concealed work.

The task of recording these changes is frequently handled by the Engineer. When it is made a responsibility of the Contractor, he is given an extra set of the Contract Plans on which to record this information. Upon completion of the Work, the set of field record drawings is delivered to the Engineer. The delivery of field record drawings to the Engineer is usually made one of the conditions for final acceptance of the Contract.

H. Cooperation by Contractor.

Refer to Article 105.05, same title, of the AASHTO Sample at the end of this Chapter. This Article sets forth the Contractor's responsibility for the items listed below.

1. Maintaining on the site, a complete set of the Contract Documents. This requirement is understandable. The absence of any of the Contract Documents can cause unnecessary problems and delays in the performance of the Work.

2. Having a competent Superintendent at the site, with full authority at all times during the progress of the Work. This requirement is needed to ensure that there will be someone in authority on the site, to whom directions and instructions can be issued with the same effect as if they were delivered to the Contractor in person. Some Specifications also specify that the Superintendent be capable of communicating in English.

3. Cooperating with the Engineer. A Superintendent who does not cooperate with the Engineer and keep him informed, fosters an adversary relationship that

will benefit no one. This requirement brings to mind a case history illustrating the climax to a series of noncooperative actions by a Superintendent, and experienced by the author on the water reservoir project described earlier in Article 4.3C, Differing Subsurface Conditions. This Superintendent was continually practicing probing actions to see how far he would be permitted to go in bypassing the Specifications. The latest incident involved backfilling operations around the ends of the dam. The Specifications called for impervious backfill material to be spread in 12-inch thick loose layers, and each layer compacted to 98 percent of maximum density. The largest dimension of stone permitted in the fill was six inches. Oversize stones and small boulders were numerous enough to keep a crew of five men busy removing them from the deposited fill. On this particular day, the Chief Inspector informed the author that the crew of five men had been reduced to three, with the result that all oversized stones were not being removed. The Contractor's foreman had informed the Chief Inspector that he had been instructed by the Superintendent not to interrupt the backfilling operation, even if it meant that some oversized stones would be left in the backfill. Some of the oversized stones not being removed were as large as 18 inches in dimension. As a result, the sheepsfoot roller was "walking up" on these boulders, thus giving no compaction to adjacent fill material. This latest incident exhausted the author's patience. A letter was immediately dispatched to the Superintendent advising him that if he was not more cooperative in having the work done as required, it would be necessary to request that he be replaced with a more cooperative Superintendent. A copy of this letter was also sent to the Contractor's home office. The letter accomplished its purpose, because the backfilling operation was soon corrected and it proceeded with no further problems.

I. Cooperation with Utilities.

Refer to Article 105.06, same title, of the AASHTO Sample at the end of this Chapter. Practically every contract has some involvement with utilities, whether it be by virtue of work being done by the Utility Owners; work by the Contractor on relocation, reconstruction, or new construction of utilities; or merely the work of supporting and protecting existing utilities encountered in the Work.

The progress of a project benefits when an owner can get the Utility Owners to complete their portion of the work before the Contractor comes on the site. The referenced AASHTO Article calls the Contractor's attention to the possibility of work being done on the site by the various Utility Owners. It also reminds the Contractor that he should make allowance for possible disruptions to his Work that may result from these existing utilities, and that he will receive no additional compensation for damages resulting from the presence of utilities shown on the Plans. This would not rule out consideration for damage resulting from existing utilities that were not shown on the Plans.

A contractor will normally be required to perform some work on utilities such as water distribution systems; installation of underground duct and construction of manholes for electrical and telephone systems; storm drain and sanitary sewer systems; and removal of abandoned utilities. This work would be specified in the applicable Technical Sections.

J. Cooperation Between Contractors.

Refer to Article 105.07, same title, of the AASHTO Sample at the end of this Chapter. Few projects consist of only one prime contract. Construction of highways, large bridges, rapid transit systems, airports, sanitary sewer systems, and water supply and treatment facilities, are usually multicontract projects in which prime contractors are working within the same area or in adjacent areas. A set of rules and guidelines are established to protect the rights of each prime contractor and enable the work to progress in an orderly manner.

The AASHTO Article:

1. Presents the right of the Owner to have other contractors performing work on or near the Work of the Contract. This alerts the Contractor that he may not be the sole occupant of the site.

2. Cautions the Contractor not to unduly interfere with the work being performed by other contractors. Interference is kept to a minimum by development of a joint schedule of operations by the contractors and approved by the Engineer. This is illustrated by the following third paragraph quoted from Section 105.08, Cooperation Between Contractors, of the Virginia Department of Transportation, Road and Bridge Specifications, January 1987:

> When Contracts are awarded to separate Contractors for concurrent construction within a common area, the Contractors, in conference with the Engineer, shall establish a written joint schedule of operations based on the limitations of the individual contracts and the joining of the work of one Contract with the others. The schedule shall set forth the approximate dates and sequences for the several items of work to be performed and will insure completion within the contract time. The schedule shall be submitted to the Engineer for review and approval no later than 30 days after the award date and prior to the first estimate. The schedule shall be mutually agreeable, signed by, and binding upon each Contractor. The Engineer may allow modifications of the schedule when mutual benefit to the Contractors and the State will result. Any modification of the schedule shall be in writing, mutually agreeable, signed by the Contractors, and shall be binding upon the Contractors in the same manner as the original agreement. Should the Contractors fail to agree upon a joint schedule of operations, they shall submit their individual schedules to the Engineer who will prepare a schedule which will be binding upon each Contractor.

3. Advises each contractor that the Owner is not to be held liable for any damages or claims that may arise as a result of the presence or operations of the other contractors. When more than one contractor is working within the same

limits of a project, it is almost inevitable that the operations of one contractor will at some time adversely affect the progress of the other contractor(s). The AASHTO Article advises the Contractor that if he suffers any damage, he must seek satisfaction from someone other than the Owner. Some Owners will grant the Contractor an extension of contract time for a delay caused by his failure to have access to the Work as scheduled, when it is due to no fault of his own.

4. Directs the Contractor to arrange his Work in a manner not to interfere with the operations of other contractors. This would be controlled by the joint schedule of operations. Some Specifications will state that, in the event of a dispute between the Contractor and other contractors involving cooperation, the Engineer will act as arbitrator.

K. Construction Stakes, Lines, and Grades.

Refer to Article 105.08, same title, of the AASHTO Sample at the end of this Chapter. Layout and control of the physical dimensions of the Work plays an important role in the successful construction of a project. There are two different approaches to this procedure. In one, the Engineer establishes the construction stakes for the Contractor to follow in performing the Work. This approach is illustrated in the referenced AASHTO Article. In the other approach, the Engineer furnishes only primary horizontal and vertical control points, from which the Contractor proceeds to perform his own detailed layout for constructing the different parts of the Work. The location of the control points established by the Engineer, are shown on the Plans.

There is a basic difference between these two procedures. When the Engineer establishes the construction stakes, the Contractor conforms his work to the location and limits defined by these stakes. If any of these stakes have been incorrectly set by the Engineer, the error may not be discovered until the work is well along, or nearing completion. When the Engineer furnishes control points only, he is afforded an opportunity to check the Contractor's layout as the work progresses. Thus an error in the Contractor's layout can be discovered by the Engineer's check. Many Specifications now make it the responsibility of the Contractor to perform the detailed construction stakeout, subject to check by the Engineer. The value of this independent check can be illustrated by the case history presented in Article 11.4K, Construction Stakes, Lines, and Grades.

L. Engineer's Field Office.

The Engineer requires a well-rounded field organization to properly perform the various tasks involved in administering a construction contract for the Owner. The three basic personnel classifications making up this organization are surveyors, inspectors, and office engineers. Depending on the complexity and size of a project, the Engineer's field organization can range in number from 10 to

50 people. Suitable field office facilities will enable this organization to better carry out its responsibilities.

In addition to providing individual office space for the Resident Engineer and for his secretary, space would be required for the office engineer and his assistants; survey and inspection personnel; furnace room; and toilet rooms. Light, heat, air conditioning, telephones, security, and operating and maintenance services are also needed, plus parking space for a specified minimum number of vehicles. Required furnishings would include office furniture, typewriter, copier, electric calculators, drafting tables, clothing lockers, water cooler, fire extinguishers, and first aid kit.

It is customary to have the Contractor provide and maintain these facilities for duration of the Contract. Depending on project requirements, the Contractor may either have to remove these facilities upon completion of the Contract or leave them in place for use by the Engineer on subsequent contracts. A mobile office trailer or trailers generally satisfy the requirements.

M. Inspection of the Work.

Refer to Articles 105.10, Duties of the Inspector, and 105.11, Inspection of the Work, of the AASHTO Sample at the end of this Chapter. The primary function of the Owner's resident site representatives is to ensure that the project will be constructed in accordance with the Plans and Specifications.

When the Owner such as a State DOT, has the inspection capability, the inspection team is composed of employees of the Owner. In the past, when the Owner did not have this capability, the inspection team consisted of employees of the Designer. However, with the involvement of the construction manager in the administration of construction contracts, this function is increasingly being removed from the control of the designer.

The construction inspector is the workhorse of the quality assurance program. He is at the scene of each operation of construction, from the depths of deep foundations to the heights of bridge towers. He represents the eyes of the Owner, for when work is covered over it may never be viewed again, unless there is a failure or the Contractor is specifically ordered to uncover it. The inspector's authority is limited, as indicated in referenced AASHTO Article 105.10. His basic responsibility is to see that materials and workmanship conform to Contract requirements.

It should be noted that, although work may have been inspected when it was performed, it is not officially accepted until the entire Contract has been completed. Thus, questionable work can be re-examined at any time before final acceptance and still be rejected if found not to comply with Contract requirements. An exception to this would apply to any portion of the Work that the Owner may desire to take possession of before completion of the Contract.

Construction contracting is motivated by the profit incentive. It is therefore not unusual for some contractors to be tempted to bypass certain required procedures when the Owner's representative is not present, particularly if the work is soon to be covered over. There are times when an inspector may not be present to observe an item of work before it is covered over. If the Engineer should have reason to question this piece of work and desire to have it uncovered, the second paragraph of referenced AASHTO Article 105.11 gives him the authority to order the Contractor to uncover it. It also provides a procedure for compensating the Contractor for costs involved, should it be later determined that the uncovered work did conform to Contract requirements.

The need for this authority can be illustrated by the following case history of damage caused to a concrete column and footing, resulting from an improper backfilling operation. The story relates to the construction of a chemical building, part of the water reservoir project described earlier in Article 4.3C, Differing Subsurface Conditions. The structure was to be supported by concrete columns resting on footings which bear on rock. The footings were approximately 15 feet below ground surface.

The Contractor had been reminded earlier that care would have to be exercised in placing the backfill around each column. The backfill was to be placed in layers uniformly around each column and then compacted. The backfilling operation began satisfactorily with the material being deposited with a clam bucket. The inspector normally assigned to this operation was out sick on the day that the Contractor was backfilling a column at a corner of the structure. One of the other inspectors had been instructed to look in on this operation from time to time. Later in the day, he reported back that the column looked out of plumb and appeared to be leaning inward toward the building.

When the author arrived on the scene, there was no backfill being placed and the column appeared to be plumb. The surrounding conditions however, looked very suspicious. Backfill at the outside face of the column had been removed to a depth of about ten feet. There were scrape marks and bruises on this face of the column, which could easily have been caused by an excavation bucket. A truck crane with a clam bucket attachment was parked nearby. In addition, timber struts had been placed between the inside face of the column and an adjacent concrete wall. This arrangement suggested that the column had been forced back to a vertical position. The Contractor however, strongly denied that anything was wrong. Nevertheless, he was ordered not to place any additional backfill around the column until further notice.

Placing the backfill material in layers and then compacting each layer as required, was a slow and costly operation. It was the author's belief that in the absence of the regular inspector, the Contractor decided to speed up his backfilling operation by pushing the material into the hole with a bulldozer. After filling the hole, the bulldozer must have ventured too close to the outside face of the column, forcing it to lean inward. Realizing what had happened, the Contractor must have hurriedly removed backfill at the outside face of the column with a clam bucket, installed the timber struts, and forced the column back into a vertical position.

On the following day the author informed the Contractor that the circumstantial evidence was too strong to ignore. He ordered the Contractor to remove the backfill from around the column, down to the footing. The Contractor objected and threatened to file a claim for damages if he had to proceed with this order, for it would also require sheeting and bracing, and dewatering. He was subsequently given a written order to proceed with this work and advised that if the work was found to be undamaged, he would be reimbursed for his costs.

The Contractor complied with the order and exposed the column down to the footing. Visually, the column appeared to have suffered no damage. However, by taking hold of the reinforcing dowels at the top of this square concrete column, one could disturb the column and feel the lateral movement. This movement was verified by survey instrument. Movement indicated that the footing had been disturbed and possibly separated from its rock foundation. To ensure that there would be full bearing under the footing, the Contractor was directed to drill three 1-1/2 inch diameter holes through the concrete footing into the supporting rock and inject a portland cement grout at a pressure of 25 psi. The Contractor did not file a claim for this corrective work, as he realized he had no case.

The last paragraph of referenced AASHTO Article 105.11 alerts the Contractor that representatives of other participating agencies are permitted to inspect work that is related to their jurisdiction. An example of this would be an underground duct or conduit installed by the Contractor, in which the local Utility Company would at a later date, install its electric cable. These representatives do not communicate directly to the Contractor. Any comments by them concerning the work are made to the Engineer.

N. Removal of Unacceptable and Unauthorized Work.

Refer to Article 105.12, same title, of the AASHTO Sample at the end of this Chapter. Work performed on a construction contract does not, for one reason or another, always conform to Contract requirements. Nor are all contractors cooperative and agreeable to removal and replacement of work considered by the Engineer to be not acceptable.

The requirements specified in the referenced AASHTO Article are set forth to discourage those contractors who might be tempted to procrastinate or those who do not comply with the Engineer's order for removal. It gives the Engineer the authority to enforce his order should the Contractor fail to comply. Few contractors will ignore the Engineer's order, for should the Engineer find it necessary to have this work done by others, it can become very costly to the Contractor.

In the third paragraph of the referenced AASHTO Article, the first sentence reads "No work shall be done without lines and grades having been given by the Engineer." This no longer holds true for many projects, as explained earlier in Article 4.4K, Construction Stakes, Lines, and Grades.

O. Load Restrictions.

Refer to Article 105.13, same title, of the AASHTO Sample at the end of this Chapter. Damage caused by excessive loads or heavy equipment is an ever present problem on construction projects. The referenced AASHTO Article reminds the Contractor that it is his responsibility to protect public roads and structures against damage when hauling materials or operating equipment over them. This responsibility also applies to the protection of existing construction. Damage can sometimes result from the premature operation of heavy paving or grading equipment too close to retaining walls, or over pipelines having insufficient cover.

P. Maintenance of the Work During Construction.

Refer to Article 105.14, Maintenance During Construction, of the AASHTO Sample at the end of this Chapter. In the day-to-day effort to expedite construction of the various pay items of work, a contractor may overlook or ignore work that is not being paid for directly, or fail to maintain work for which he has already been paid. Work that is not paid for directly would include maintaining the temporary and permanent drainage facilities to prevent damage to excavations, embankments, structures, and adjacent property, and maintaining detour and haul roads.

Referenced AASHTO Article 105.14 spells out the Contractor's responsibility to maintain and protect the different parts of the Work in their various stages of completion and even after completion, until final acceptance of the Contract.

Q. Failure to Maintain Project.

Refer to Article 105.15, Failure to Maintain Roadway or Structure, of the AASHTO Sample at the end of this Chapter. In the previous Article, the Contractor is reminded of his responsibilities in maintaining the Work. Suppose he becomes negligent and continues to ignore the Engineer's instructions. How can the Engineer enforce his instructions? He can issue a written order and if this too is ignored, the referenced AASHTO Article gives the Engineer the authority to have the maintenance work in question performed by others, and charge the costs to the Contractor. As stated previously, few contractors will allow a situation to deteriorate to this point, because paying for work done by others can be costly.

R. Acceptance of the Work.

Refer to Article 105.16, Acceptance, of the AASHTO Sample at the end of this Chapter. The goal of a Contractor is to obtain acceptance of the Work. Acceptance, whether it be partial or final, not only relieves the Contractor of the responsibility and expense of maintaining the work, it also brings him a step closer to the time that he will be eligible to receive his final payment.

Prior to the acceptance of the Contractor's Work, a final inspection by the Engineer is first made. Participating in this inspection are representatives of the Owner, Contractor, and Engineer. Work considered to be unfinished or unsatisfactory is noted. After the inspection is completed, the Engineer prepares a list of the items to be corrected and completed and then delivers this list, generally referred to as a Punch List, to the Contractor for execution.

Partial acceptance by an Owner will generally be considered when it benefits the Owner to accept part of the project. One example would be the acceptance of a portion of roadway for the purpose of opening it to relieve traffic congestion, as illustrated in Article 4.6P, Possession and Use Prior to Completion.

Final acceptance of the Work is a prerequisite step for final payment, as described in Article 4.8L, Acceptance and Final Payment.

S. Claims for Adjustment.

Refer to Article 105.17, same title, of the AASHTO Sample at the end of this Chapter. Claims seem to be a fact of life on construction contracts. A claim is a request by the Contractor for an equitable adjustment to the Contract, and is his response to an unanticipated situation. This is well explained in the following excerpt quoted from ASCE Paper 8781, Responsibilities of the Engineer and the Contractor Under Fixed-Price Construction Contracts, prepared on behalf of the Construction Committee of the United States Committee on Large Dams, March 1972:

> No matter how well a contract is written, how fair and equitable the conditions are, how competent the design is, and how explicit the specifications are, there will be disagreements between the parties. Disputes and claims can arise from a wide variety of real and imagined reasons. Some of the more common ones are disagreement between the Contractor and Engineer with respect to interpretation of the specifications, work required that is outside the scope of the original specifications, work required that is outside the scope of the original contract, delays caused by actions of the Owner or by the Owner's inaction, and inequitable evaluation of the cost and time effects of design changes.

Referenced AASHTO Article 105.17 establishes a procedure for the Contractor to follow when filing this type of claim. Note the requirement for the Contractor to notify the Engineer in writing of his intention to make a claim, before beginning the work on which he will base the claim. The amount of the Contractor's requested additional reimbursement is based on his costs for the labor, materials, and equipment involved. It is therefore essential that the Engineer be advised before the work in question is begun so that he can verify the Contractor's field records. The importance of this prior notification is illustrated by refusal to consider a claim, should the Contractor fail to comply with this requirement.

The mechanics for determining reimbursible costs are explained in Article 4.8F, Payment for Extra and Force Account Work.

T. Automatically Controlled Equipment.

Refer to Article 105.18, same title, of the AASHTO Sample at the end of this Chapter. Automatically controlled equipment is desired because more uniformity can be obtained in the resulting work, than with the use of manually operated equipment. Placing a time limit on the emergency use of manually operated equipment is therefore understandable.

When automatically controlled equipment is to be involved in a critical operation, the Specifications will often require that duplicate equipment be provided on the site to be available in case of a breakdown. For example, if concrete for an underwater tremie seal is to be supplied from a floating concrete plant, it is often required that another floating plant of similar capacity be on the site, as a standby.

4.5 Control of Materials

A. Introduction.

In civil works construction, natural materials constitute a sizeable portion of the finished work. Soils and aggregates play a major role in concrete and road construction. The processing of material for fabricated items is generally accomplished in the shop, where the control of quality presents no problem. The processing of soils and aggregates to have them conform to Contract requirements is done at the source of supply, where the control of quality is a function of human actions and decisions. It is therefore subject to human error.

B. Source of Supply and Quality Requirements.

Refer to Article 106.01, same title, of the AASHTO Sample at the end of this Chapter. The Contractor normally has the responsibility for selecting the source of materials to be used in the Work. There can be situations where the Owner will desire to specify a designated source known to contain material which meets

Contract requirements. This would be particularly true in earthwork projects, where it is desirable to have borrow pits close to the site of the Work.

Most materials require the prior approval of the Engineer (see Article 4.5D, Samples, Tests, Cited Specifications) before delivery to the site. In order to minimize delay in obtaining these approvals, the Contractor is required to notify the Engineer of his proposed sources of material as soon as possible after award of the Contract. This requirement of course would not apply in the case of a source of supply designated by the Owner.

Referenced AASHTO Article 106.01 indicates that the Engineer has the option of approving material either at the source of supply or upon its delivery to the site. This would depend on the location of the source, the quantity of material involved, and the extent of processing necessary for blending materials to meet gradation requirements. AASHTO Article 106.01 also reminds the Contractor that initial approval of a source of supply does not guarantee that all material from that source will be automatically accepted. This would be particularly true of material being taken out of a gravel pit, where a variation in gradation can be encountered at any time.

C. Local Material Sources.

Refer to Article 106.02, same title, of the AASHTO Sample at the end of this Chapter. The referenced AASHTO Article discusses material sources that may have been designated in the Contract Documents. It alerts the Contractor that, in designating a source of supply, the Owner does not guarantee that all material from that source will be acceptable in its natural state. The Contractor may have to process some of the material to meet gradation requirements.

If the Contractor decides to provide his own material source, he is reminded that he must obtain the necessary rights to remove such materials, and absorb all additional costs involved. The referenced AASHTO Article goes on to further specify the condition that borrow pits and quarries are to be left in, after all of the required material has been removed. The final condition concerns public safety, the environment, and final treatment of the site.

D. Samples, Tests, Cited Specifications.

Refer to Article 106.03, same title, of the AASHTO Sample at the end of this Chapter. Material approvals by the Engineer are generally based on the results of tests made on representative samples of materials proposed for use in the Work. The referenced AASHTO Article makes it clear at the very outset that any unapproved material incorporated in the Work is there at the Contractor's risk and will not be paid for. This statement is made to discourage any notion a contractor may have of bypassing the approval requirements.

Samples of manufactured items are supplied by the Contractor. Samples of natural materials are taken by the Engineer or by the Contractor in the presence of the Engineer. Thus, the Engineer is able to ensure that samples are truly representative of the material used. This makes it possible for the Engineer to maintain a complete history of the sample, should any question arise at a later date. This is particularly helpful in evaluating test results of concrete test cylinders, which are broken at various ages.

Tests are conducted by both the Contractor and the Engineer. Control testing performed by the Contractor serves to control his operation and thus ensure acceptability of the material or end product. Acceptance testing performed by the Engineer serves to verify acceptability of material furnished, or of work in place. When large quantities of material such as portland cement or structural steel are involved, provision is made for mill inspection and testing of samples taken at the mill. This work may be performed by the Engineer or by a private inspection and testing company retained by the Owner for this purpose. Wherever possible, test methods are referred to national or industry Standards, such as those established by ASTM, AASHTO, ANSI, or ACI (see Chapter 7, National Reference Standards). Test failures result in rejection of the proposed material, and can in some instances involve considerable expense for the Contractor. The last sentence of referenced AASHTO Article 106.03 rightfully states that at the Contractor's request he will be furnished a copy of the test results.

The next to last sentence of the AASHTO Article states that all materials are subject to tests at any time prior to incorporation in the Work. It is a straightforward statement, but a very important one, as illustrated in the following case history from the water reservoir project described earlier in Article 4.3C, Differing Subsurface Conditions.

> The interior of the concrete dam required a lean concrete mix containing coarse aggregate up to a maximum size of six inches. This aggregate was being delivered by truck from a quarry located eight miles from the site of the dam. A screening plant at the quarry produced the required graded aggregate size classifications.
>
> Problems were being experienced in consolidating the lean mix concrete after it was deposited in place. Some batches of concrete were harsh and unworkable, requiring additional effort and time for proper consolidation. A sample of the three-inch to six-inch size aggregate classification was taken from the hopper above the mixers at the concrete plant, and a sieve analysis performed. It was found that this size classification contained an excess of "fines"; almost twice the allowable. Investigation revealed that, from the time the coarse aggregate left the quarry to the time that it was placed in the steel hopper above the mixers, the aggregate had been rehandled a total of seven times. At each rehandling, small pieces would break off from the larger sized aggregate, thus adding to the amount of "fines." The Contractor solved the problem by adjusting his screening operation at the quarry to compensate for the additional "fines" being produced by the numerous rehandlings it was subjected to after leaving the quarry. Though the aggregate had tested satisfactorily at the quarry, it was still subject to retesting before going into the mixer. This particular case history is presented in more detail in Article 10.12J, Guidelines; Paragraph 7, Batching Plant and Mixing.

E. Certification of Compliance.

Refer to Article 106.04, same title, of the AASHTO Sample at the end of this Chapter. Initial sampling and testing of some materials or manufactured items may be impractical because of the small quantity involved, distance of the source from the project, or for other valid reasons. To prevent unnecessary delays in the Work, the Engineer will usually accept a written certification from the manufacturer or supplier that the item or material complies in all respects with the requirements of the Contract. This permits the Contractor to deliver and incorporate the item or material in the Work without delay. It does not, however, give the Contractor a "carte blanche" approval. The item or material is still subject to test by the Engineer at any time during the life of the Contract, and may still be rejected if it fails to meet the requirements.

F. Plant Inspection.

Refer to Article 106.05, same title, of the AASHTO Sample at the end of this Chapter. Plant inspection may involve inspection at: 1) A manufacturing facility removed from the site, or 2) A fabricating yard away from the site. This could be a facility for fabricating structural steel members or precasting concrete members, or 3) A concrete batching or mixing plant located either on the site or away from the site.

In most instances, the Contractor is required to provide a plant laboratory for the sole use of the inspector. The laboratory would require the following minimum furnishings: a desk, chair, work table, telephone, and file cabinet; basic testing equipment such as a set of vibrating screens, electric oven, weighing scale, and water supply.

G. Field Laboratory.

Refer to Article 106.06, same title, of the AASHTO Sample at the end of this Chapter. Previous Article 4.5F, Plant Inspection described the plant laboratory to be used in conjunction with plant inspection. This plant laboratory is situated in the manufacturing facility or batch plant. A field laboratory as its name implies, is located in the field, close to the related construction operation. Highway and dam construction projects in which large quantities of earthwork are involved would require use of a field laboratory. The handling of soils in the construction of embankments requires continuous monitoring for material suitability and compaction control. Results of sieve analyses and in-place density tests must be obtained and made known promptly, particularly if it should become necessary to modify the operation and correct deficiencies. To minimize the loss of time, the laboratory is stationed close to the scene of operations.

Since a field laboratory may have to be situated in an isolated location, its requirements are more extensive than those for a plant laboratory. Basic accommodations and services such as heat, lighting, air conditioning and sanitary facilities would be patterned after the requirements for an Engineer's field office. A field laboratory will generally be housed in a mobile office trailer since it may have to be moved several times during the life of the Contract.

Some public works contracts will not require the Contractor to provide a field laboratory because the Government Agency may maintain its own mobile field laboratory.

H. Foreign Materials.

Refer to Article 106.07, same title, of the AASHTO Sample at the end of this Chapter. Two reasons for the purchase of material or products outside the country are economics and availability. Structural steel and portland cement have been two of the more common foreign materials. Many contracts however, will specify that certain materials or equipment must be of United States origin or manufacture.

I. Storage of Materials.

Refer to Article 106.08, same title, of the AASHTO Sample at the end of this Chapter. The conditions under which materials are stored can play a major role in the preservation or impairment of their quality. For example, if portland cement is not protected against moisture, premature hydration sets in and the cement must be rejected. When reinforcing steel is stored directly on the ground it can become splattered with mud from the rains. This adversely affects the bond necessary between it and the concrete. Some materials require interior storage in weathertight spaces, while other materials, stored outside, have to be on platforms or blocking. Brick and concrete block have to be covered to protect them against frost and excess moisture.

In referenced AASHTO Article 106.08, the Contractor is again reminded that, even though his material may have been initially approved prior to its delivery and storage at the site, it is still subject to rejection prior to its incorporation in the Work. Throughout the Specifications, the Contractor is not permitted to forget that all materials and products are subject to additional inspection and testing at any time during the Contract, until the Work is finally accepted. AASHTO Article 106.08 also repeats the required conditions that must be met if the Contractor is to occupy private property. These requirements are included to protect the Owner against a damage claim by the property owner because of the Contractor's negligence.

J. Handling Materials.

Refer to Article 106.09, same title, of the AASHTO Sample at the end of this Chapter. All materials require considerations and precautions in their handling, if they are to retain their quality and fitness for the Work. Epoxy coated reinforcing bars for example, should not be bundled with plain steel wire nor handled with wire rope slings, as is normally done with uncoated reinforcing bars. The epoxy coating must not be broken or punctured if it is to provide full protection against corrosion.

K. Unacceptable Materials.

Refer to Article 106.10, same title, of the AASHTO Sample at the end of this Chapter. Occasionally, a contractor will procrastinate in removing rejected material from the site. When the inspector is not present, the Contractor may attempt to incorporate the rejected material in the Work if it can be covered over quickly. Defective storm drain pipe laid in a trench and covered over may go undetected, because storm drain lines do not require a leakage test. To prevent rejected material being incorporated in the Work, the referenced AASHTO Article in unmistakable language, requires the Contractor to immediately remove this material from the site.

The last sentence of the AASHTO Article reminds the Contractor that even though he may be able to correct the defects of a material previously rejected, he is not to place it in the Work until it has been inspected and approved. If this requirement were not specified, there would be nothing to prevent the Contractor from placing the rejected material, covering it over, and maintain that the defects had been corrected.

L. Owner Furnished Material.

Refer to Article 106.11, Department-Furnished Material, of the AASHTO Sample at the end of this Chapter. On some projects, the Owner will furnish certain materials for the Contractor to incorporate in the Work. This may occur for various reasons.

If a material or product is in short supply and late delivery can delay completion of the Contract, the Owner may decide to purchase the material or item beforehand and make it available to the Contractor. For example, in the design stage of a multicontract toll road project, structural steel is in short supply. The design for most of the grade separation structures utilizes structural steel. To prevent any possible delay to the completion of this income-producing facility, the Owner will award an early procurement contract for furnishing structural steel members. The steel is then stored by the Owner and later furnished to each prime Contractor.

When the Contractor takes possession of Owner-furnished material, he assumes full responsibility for it from the time of pickup at the Owner's storage yard until final acceptance of the Contract.

4.6 Legal Relations and Responsibility to the Public

A. Introduction.

Construction, particularly that involved in public works contracts, can affect the daily habits and safety of the general public. The rights of the public therefore have to be protected. Another responsibility to be considered in construction contracts is the providing of suitable working conditions on the site for the construction worker. There are many other items of legal significance to be considered, including nondiscrimination in hiring, minimum wage rates, protection of property, and damage claims. This subsection deals principally with the Contractor's legal obligations in these and other areas, plus his legal relations with the Owner.

B. Laws to be Observed.

Refer to Article 107.01, same title, of the AASHTO Sample at the end of this Chapter. There are many Federal, State, and Local laws and regulations governing construction work, particularly where it affects the life and health of the workmen and the public. The referenced AASHTO Article makes the Contractor responsible for keeping himself informed of all those laws affecting his work and the people involved with it. He is also to protect the Owner and Engineer and hold them harmless, should he violate the provisions of any of these laws. Many of the laws to be observed are discussed in the following Articles.

C. Permits, Licenses, and Taxes.

Refer to Article 107.02, same title, of the AASHTO Sample at the end of this Chapter. Permits for permanent construction or permanent changes to existing facilities are generally obtained and paid for by the Owner. For example, if the project involves work in navigable waters, the Owner will obtain permits from jurisdictional agencies such as the U.S. Coast Guard or U.S. Army Corps of Engineers. Should the Contractor find it necessary to do work of a temporary nature like dredging to facilitate his operations, he would have to obtain permits governing his dredging and disposal of the dredged material.

One of the more common types of permits is the Street Opening Permit, issued by the local jurisdictional agency to a contractor before he may excavate in a public street or roadway. The Contractor would also have to obtain a permit to close a street or to demolish a building. The Contractor is generally responsible for obtaining permits required for the Work, except for those permits that may have already been obtained by the Owner.

Payment of taxes applies to all types of endeavor, and construction contracts are no exception. The Contractor is made responsible to ascertain and pay all Federal, State, and Local taxes applicable to his work. When the work is being performed for a Government agency, certain taxes may be exempt. This would include State sales tax on work being performed for a State agency.

D. Patented Devices, Materials, and Processes.

Refer to Article 107.03, same title, of the AASHTO Sample at the end of this Chapter. Certain materials, processes and items of equipment, are protected by copyright or patent. The referenced AASHTO Article makes it the responsibility of the Contractor to pay all license fees and royalties on any patented device, material, equipment, or process used by him in the performance of the Work. He is also required to protect the Owner and specified others against claims, should he fail to pay these fees and royalties.

E. Restoration of Surfaces Opened by Permit.

Refer to Article 107.04, same title, of the AASHTO Sample at the end of this Chapter. When work affects existing streets or roadways, it sometimes becomes necessary for Utility Owners to perform work on their underground lines. In doing so, they have to cut through the pavement. It would be impractical to have each Utility Owner restore the portion of permanent pavement that it disturbed. In the interest of uniform pavement construction, this work is assigned to the construction Contractor. The Utility Owners generally have the responsibility for providing and maintaining the temporary pavement surface.

Referenced AASHTO Article 107.04 advises the Contractor that he does not have exclusive rights to the work area. It also informs him that, should he experience any delays resulting from work by the Utility Owners, he will not be compensated for any damages resulting therefrom. There are provisions however, to compensate the Contractor for his costs in reconstructing the roadway pavement.

F. Federal Aid Participation.

Refer to Article 107.05, same title, of the AASHTO Sample at the end of this Chapter. Many projects, particularly those dealing with transportation, the environment, and water resources, receive federal aid. In order to satisfy itself that its funds are being properly spent, the Federal Government will reserve the right to send its representatives out to the project to inspect the Work. These representatives are not authorized to communicate with or issue any instructions to the Contractor. All of their dealings are with the Owner and his site representatives.

Referenced AASHTO Article 107.05 advises the Contractor on a federal aid project that his work will be subject to inspections by Federal representatives. It also makes it clear that these representatives will have no direct authority over the Contractor's operations.

G. Sanitary, Health, and Safety Provisions.

Refer to Article 107.06, same title, of the AASHTO Sample at the end of this Chapter. The Contractor is responsible for providing suitable sanitary and health facilities and safe working conditions on the job site, as required by Federal, State, and Local laws. Many of these laws came into being as a result of the unsatisfactory conditions that previously existed on the job site, particularly those affecting safety.

Some of the health requirements are common to all jobs. To mention a few: drinking water is to be provided in closed containers with individual paper cups for drinkers; first aid facilities to be provided, which may vary from a first aid kit to a completely equipped first aid station having a full time nurse in attendance depending on the size of project; specified allowable maximum noise levels to be maintained on construction equipment.

Other health requirements are peculiar to the particular type of work involved. Two such examples are face masks, when spray painting or when working in underground enclosed areas, and protective clothing, when handling toxic materials such as creosoted timbers and wood piles.

Preventing physical injury by providing safe working conditions is the aim of the Construction Safety and Health Standards promulgated by the U.S. Secretary of Labor in accordance with the Williams-Steiger Occupational Safety and Health Act of 1970 (OSHA). In addition, most States have prepared their own Construction Safety Standards patterned after the Federal Regulations.

On many large projects, the Specifications will require that the Contractor provide a full-time safety supervisor. Duties of the supervisor are to set up and effectively maintain a safety program covering all phases of the Work, including weekly safety meetings.

Common examples of unsafe conditions on the job include:

1. Trench excavations over four or five feet in depth which are not braced or sheeted.
2. Openings in roof and floor areas that are not protected.
3. Scaffolds with insufficient or no guard railing.
4. Not wearing a "hard hat."

5. Not wearing safety goggles when using an acetylene torch or when breaking concrete.

H. Public Convenience and Safety.

Refer to Article 107.07, same title, of the AASHTO Sample at the end of this Chapter. Construction performed in or through populated areas can subject the public to a great deal of inconvenience and exposure to possible injury. Public roadways and streets used as haul routes may often exhibit spillage of materials and trash. Temporary detour roads may be poorly maintained. Dust in work areas and on roadway surfaces is often not controlled. When streets are closed to traffic with insufficient notice to the public, and if bus routes are involved, this can make a lot of people angry. Street closings can also have a disastrous effect on small private businesses. When sidewalks are blocked, pedestrians often have to make their way as best they can in the street. The excessive noise of construction equipment can be a problem in residential areas.

The problem of construction noise brings to mind the following case history illustrating the effect of excessive noise in a residential area. The incident occurred during construction of the sanitary sewer system described in a case history presented in Article 11.3F, Maintenance and Protection of Traffic. Sewer pipe was being laid in a part of town that had a high groundwater table. The Contractor installed a temporary wellpoint system to lower the groundwater level so that pipe could be laid in a dry trench. In this system, a pump lowers the groundwater level by withdrawing the water through wellpoints and discharging it to the nearest catch basin or stream. The pump was being driven by a gasoline engine and was positioned in the sidewalk area. In order to maintain a dry condition at the bottom of the trench, the pump had to operate continuously around the clock. Early one morning at about 2:00 A.M., the author was awakened by a ringing telephone. An irate Contractor was at the other end angrily exclaiming that someone had shut off his pump, and as a result his pipe trench was flooded. It was later established that one of the residents was responsible. The man of the house where the pump had been positioned slept in a bedroom facing the street. Being unable to sleep because of the noise from the pump, he simply went outside and shut it off.

It is possible to minimize public inconvenience and exposure to injury in a business area by requiring a contractor to follow specific precautionary procedures and provide the necessary safeguards. Necessary safeguards include warning and detour signs, barricades and guard rails, overhead lighting, warning lights, pavement markings, temporary sidewalks, and flagmen. The Contractor would not be permitted to make any changes to an established traffic plan without first obtaining approval from the Engineer. The Contractor may also be required to provide a traffic coordinator. One of the duties of the coordinator would be to keep business owners advised of activities by the Contractor which may affect them.

I. Railway-Highway Provisions.

Refer to Article 107.08, same title, of the AASHTO Sample at the end of this Chapter. When a project involves work in or adjacent to the right-of-way of an operating railroad, consideration must be given to special requirements for protecting the rights of the Railroad Company. These requirements are initially presented to the Owner by the Railroad Company during the design phase, and are then incorporated in the Contract Specifications.

Some typical railroad requirements that apply to the Contractor's operations are listed below:

1. The Contractor shall protect and safeguard the Railroad's tracks, trains, and property.
2. If the Contractor desires to cross the tracks at a location other than that provided for in the Contract, he is to make his own arrangements with the Railroad for a private crossing.
3. Work required within the railroad right- of-way shall be scheduled so as not to unnecessarily interfere with railroad traffic.
4. The Contractor's proposed construction methods for work within the railroad right-of-way must also receive prior approval of the Chief Engineer of the Railroad.
5. The Contractor shall, in addition to his normal insurance requirements, carry railroad protective liability insurance of the kinds and amounts required by the Railroad.
6. In performing his work, the Contractor shall not foul an operating track or wire line without the written permission of the Railroad Company. An operating track is considered to be fouled when any object is brought closer than a specified distance (usually 10 or 12 feet) from the centerline of the track. Cranes, shovels, and other equipment are considered to be fouling a track when they are located in such a position that failure of the equipment will bring it within the fouling limit. A power line is considered to be fouled when any object is brought within six feet of the line.
7. Concerning labor charges:

   a. When work required by the Contract is done within the railroad right-of-way, the Railroad will furnish its own flagmen. Costs for these flagmen will be borne by the Owner.

   b. The costs for flagmen or other personnel provided by the Railroad to cover work that the Contractor is doing for his own convenience are to be paid by the Contractor.

J. Construction Over, In, or Adjacent to Navigable Waters.

Refer to Article 107.09, Construction Over or Adjacent to Navigable Waters, of the AASHTO Sample at the end of this Chapter. Projects involving construction of tunnels and long bridges usually traverse navigable waterways. In these situations, the Owner is required to obtain a construction permit from the jurisdictional Federal Agency; the U.S. Coast Guard or the U.S. Army Corps of Engineers. The permit will contain specific requirements for protecting navigation, as they relate to the design and construction of the project.

Some typical requirements would be:

1. The Contractor shall provide and maintain temporary navigation lights and signals as required for the protection of navigation.
2. Dredges are to be held in a stationary position. They are to be identified by color markings in the daytime and by colored lights at night, which can be seen from all directions. Pipelines attached to dredges are to be similarly marked.
3. Anchors are to be marked by buoys. The buoys are to be lighted at night.
4. Regarding misplaced material, if the Contractor should lose any material, plant, or equipment overboard, he is to promptly recover and remove it. If it cannot be promptly removed and it is considered dangerous to navigation, he is to notify the U.S. Coast Guard and mark the obstruction with a buoy.

K. Use of Explosives.

Refer to Article 107.11, same title, of the AASHTO Sample at the end of this Chapter. The most common use for explosives in construction is in the removal of rock or large boulders encountered in excavations for highways, tunnels, dams, and deep foundations; and in the removal of existing structures. Because of the element of danger to life and property whenever explosives are involved, Federal, State, and Local laws and ordinances governing the handling, use, and storage of explosives have come into being.

Referenced AASHTO Article 107.11 covers general requirements on the use of explosives. Detailed requirements governing the use of explosives for a particular operation are specified in the Technical Section dealing with the operation. Examples of some general requirements are:

1. The Contractor shall comply with all governing laws and ordinances in the handling and storage of explosives. Explosives shall be stored in a secure manner. All such storage places shall be clearly marked "DAN-

GEROUS—EXPLOSIVES" and shall at all times be in the care of competent watchmen. Some projects may prohibit the overnight storage of explosives on the site.

2. The Contractor is reminded of the necessity for safeguarding the traveling public during blasting operations. He shall provide a sufficient number of watchmen, flagmen, and signs to warn motorists during periods of blasting.
3. The Contractor shall notify each property owner and Public Utility having structures near the site of blasting, of his intention to use explosives. All such owners within 200 feet shall be given sufficient notice before each blast, to allow them time to take such steps as necessary to protect their property from damage.
4. Such prior notice however, shall not relieve the Contractor of full responsibility for any damage resulting from his blasting operations.

Some projects may require the services of a blasting consultant for advice and assistance to the Designer in the presentation of blasting requirements.

L. Protection and Restoration of Property and Landscape.

Refer to Article 107.12, same title, of the AASHTO Sample at the end of this Chapter. Irresponsible contractors will not hesitate to cause damage to existing property if, by so doing, it will help them complete an operation in less time or at less cost. Existing property and facilities vulnerable to damage include trees, shrubbery, lawns, fences, ground signs, drainage facilities, and survey and property monuments. When work is performed in street areas, a common violation is the storing of materials or the stockpiling of excavated material, on the front lawns of private property. It becomes necessary therefore, to remind the Contractor that he is responsible for protecting property, structures, and facilities not planned to be removed or modified.

Another area for consideration is unintentional damage caused by the Contractor's operations. One such example is the damage resulting from settlement of a building adjacent to an excavation, where the groundwater level was being lowered by pumping.

M. Forest Protection.

Refer to Article 107.13, same title, of the AASHTO Sample at the end of this Chapter. There is not much to add to this subject, since not very many Contracts impinge on State or National forests. The major message to the Contractor, in addition to that of maintaining the work area in a clean, orderly condition, is the prevention of fires.

N. Responsibility for Damage Claims.

Refer to Article 107.14, same title, of the AASHTO Sample at the end of this Chapter. A pedestrian who is injured as a result of the Contractor's negligence in maintaining a temporary sidewalk brings suit against the Owner for recovery of damages. An owner of property adjacent to the construction area brings suit against the Owner to pay for damage to his property caused by the Contractor's operations. An injured workman, who is covered under Workmen's Compensation insurance and cannot sue the Contractor, but who believes he was not adequately compensated, files a claim against the Owner.

In the above examples, we see where the Owner is forced to defend himself against claims for damages that are the result of the Contractor's negligence. As a protection against these kinds of lawsuits, the Owner has provided an indemnity or hold-harmless clause in the Contract. Basically, its purpose is to have the Contractor indemnify the Owner and hold him harmless against the consequences of a claim or suit brought against the Owner by a third party who has been damaged as a result of the actions of the Contractor. Simply stated, the Contractor is held responsible for his own negligence.

When a portion of the project is accepted by the Owner before its total completion, the Contractor is generally relieved of responsibility for claims of any related damage occurring after it has been accepted.

The hazards and risks involved in construction are innumerable. In addition to the Contractor's risk of financial loss resulting from damage to persons or property by his construction activities, there is the matter of possible loss due to fire, storms, floods, thefts, and similar occurrences during construction. Many of the liabilities which may arise out of a construction project can fall on the Owner if he is not adequately protected in the Contract. It is therefore, common practice for a construction contract to require the Contractor to obtain insurance coverage against known risks. To ensure that proper coverage is furnished, the Contractor is required to submit his certificates of insurance to the Owner before he may begin work.

Most contracts will require the Contractor to carry the insurance necessary to provide adequate protection for himself, the Owner, and the Engineer against all claims, liabilities and damages, resulting from his performance of the Work. The types of policies and amounts of insurance required to protect the parties to the Contract are determined after full consideration of the risks involved. The limits will vary with the character, size, and scope of the work involved.

Protection against possible loss from such risks as fire, storms, explosions, and water leakage is provided separately by insurance policies, under the general classification of Casualty Insurance. Risk of loss due to injuries to workmen on the job, is covered by Workmen's Compensation Insurance. Burglary, robbery and theft insurance, provides protection against losses from these causes. Public Liability Insurance covers loss from damage claims arising out of injuries to persons not connected with the Work or to property other than that owned by the Contractor.

Workmen's Compensation Insurance is compulsory on employers. An employee injured and covered under Workmen's Compensation cannot sue his employer for damages in excess of the amount awarded him under the Plan. A Federal law called the Longshoremen's and Harbor Workers' Compensation Act protects employees for injuries in work involving marine exposure. This coverage can be included in the Workmen's Compensation Insurance policy.

Marine or Inland Marine Insurance provides protection against loss incurred through transportation in navigable waters. It includes loss or damage to ships, ship's cargoes, and floating equipment that is being transported.

Comprehensive Builder's Risk Insurance provides insurance protection for the increased value of completed work, as construction progresses.

On large multicontract projects such as those involving urban rapid transit construction, it may be in the Owner's interests to provide and administer an overall, coordinated insurance program with respect to all construction contracts awarded for the project. This form of arrangement is commonly referred to as "wrap-up" insurance. Under such a program, the Owner provides various insurance coverages. For the Contractor: workmen's compensation and employer's liability insurance; personal injury, bodily injury, and property damage liability insurance; all-risks construction insurance; and railroad protective insurance in amounts and on a form as required by the railroad involved. For the Engineer, professional indemnity insurance is provided. Coverages not provided are Contract bonds, Contractor's equipment and automobile liability, and property damage insurance. Coverages provided are for the protection of the Owner, Engineer, and all prime contractors and their subcontractors of any tier who perform work on the job site.

Features of such "wrap-up" insurance attractive to the Owner include:

1. Tax advantages.
2. Simplification of audit and claims handling by a single insurance group.
3. Coordination and implementation of safety programs by one insurer involving the cooperation of the Owner, Engineers, and all Contractors.
4. Elimination of confusion and overlapping in the hold harmless provisions of Contracts and the insurance therefor.
5. Discount allowable on premium because of the size of the insurance.
6. Adequate medical and first aid facilities which can be set up for all employees, regardless of their employer.

Objections to such a wrap-up arrangement usually come from Contractors who feel that, because of the loss of their own premium volume, they lose negotiating capability with their insurance carrier on other projects. They sometimes consider the liability program offered by the Owner to be less broad or complete than they usually carry for themselves. This results in extra insurance costs to them for the purchase of higher limits, completed operations, and various fringe coverages. A safe, efficient contractor may not benefit. He may lose his future bidding advantage which accrues from a good safety record. Contractors with poor safety records on the other hand, will do better under wrap-up insurance.

O. Third Party Beneficiary Clause.

Refer to Article 107.15, same title, of the AASHTO Sample at the end of this Chapter. This Article is intended to discourage anyone not a party to the Contract from filing a suit against the Contractor or Owner for damages resulting indirectly from failure to comply with the terms of the Contract. The following illustration presents a simple example of the type of legal action this Article is intended to discourage:

Mr. A. owns a piece of property adjacent to the site of a proposed sports stadium. The contract for construction of the stadium has been awarded. Mr. A. proceeds to construct a fast food restaurant in anticipation of the stadium business. Shortly after sitework begins for construction of the stadium, the project is abandoned. As a result, Mr. A. sustains a loss of thousands of dollars. In an attempt to recoup his losses he files suit against the Owner for not following through on his construction contract.

P. Possession and Use Prior to Completion.

Refer to Article 107.16, Opening Sections of Project to Traffic, of the AASHTO Sample at the end of this Chapter. On many projects, it may be desirable for the Owner to take possession and use a portion of the Work before total completion of the Contract. A common example, as illustrated in the referenced AASHTO Article, is the opening of a portion of new roadway in order to restore the original traffic pattern. Again, a contract may involve stage construction in which different completion dates have been established for specified stages of the Work.

Once the Owner takes possession of a portion of the Work, the Contractor is relieved of responsibility for its maintenance and any damage that may result after being placed in use. If possession of a portion of the Work could not have been anticipated in the Contract Documents, the Contractor is given consideration for any added costs and time delays caused thereby. The procedure for arriving at these costs is spelled out in Article 4.8F, Payment for Extra and Force Account Work.

Q. Contractor's Responsibility for Work.

Refer to Article 107.17, same title, of the AASHTO Sample at the end of this Chapter. During the life of a Contract, payments are made to the Contractor by the Owner, generally on a monthly basis, for work performed during the previous month (see Article 4.8H, Progress Payments). Some contractors may come to believe that when they have been paid for work performed, they are no longer responsible for it. The referenced AASHTO Article eliminates any doubt that may exist of a contractor's responsibility for work performed and paid for prior to completion of the Contract. The Contractor is responsible for protecting all phases of the Work, regardless of its stage of completion or of any payments therefor, until the Contract has been completed in its entirety and accepted.

There are some exceptions to holding the Contractor responsible until final acceptance, for work performed. Responsibility would end for those portions of the Work that the Owner may have taken possession of and placed in use prior to final acceptance (see Article 4.6P). Repair of damage to the Work by unforseeable causes beyond the control of the Contractor, as enumerated in referenced AASHTO Article 107.17, would also not be his responsibility.

R. Contractor's Responsibility for Utility Property and Services.

Refer to Article 107.18, same title, of the AASHTO Sample at the end of this Chapter. Utilities are everywhere; beneath the ground, on the surface of the ground, and overhead. It is rare when the work of a project does not affect some utility. In addition to the loss of service, some utilities damaged by construction operations can have serious consequences. A ruptured gas line may trigger an explosion, or accidental contact with an electric power line may result in an electrocution.

The greatest possibility for damage lies with utilities that are buried and hidden from view. Many States now have a law on the books that requires a contractor to give prior notice by calling a designated telephone number, before beginning an excavation or driving piles. This alerts Utility Owners. Those Utility Owners affected will then send their representatives out to the site to indicate on the ground surface, the horizontal location of their underground facilities. The Contractor is then required to verify the location by digging a test pit to uncover and expose the facility. Having accomplished this, he can then proceed with his below-surface operation.

The proposed construction may make it necessary for some Utility Owners to relocate their facilities or provide for their temporary protection.

Referenced AASHTO Article 107.18 forbids the Contractor from beginning any work that may disrupt a utility service until the Utility Owner has completed its relocation or the arrangements necessary for its protection. It also requires the Contractor to cooperate with the Utility Owners when necessary, either in their preparations or in the restoration of service interrupted by the Contractor's operation. The Contractor is responsible for any damage or disruption of service resulting from his negligence. However, if an underground utility not shown on the Plans or otherwise indicated, is damaged by the Contractor and this was not due to his negligence, he will generally be reimbursed for any related costs of repair.

S. Furnishing Right-of-Way.

Refer to Article 107.19, same title, of the AASHTO Sample at the end of this Chapter. The Contractor has a right to expect that, upon award of the Contract, the site of the Work including access to it will be made available, unless there are exceptions noted.

Projects such as new highway construction involve the acquisition of numerous parcels of private property. Some of these properties may have to be acquired by legal condemnation, which is time consuming. Also, some buildings or dwellings scheduled for demolition may not have been vacated by the time the Contract has been awarded. Those structures not immediately available to the Contractor are identified in the Contract Documents, with the dates of their availability specified.

On a multicontract project in which adjacent contracts are interrelated, portions of the work area for one contractor may not be available until the work of another contractor is completed. The dates on which these areas are scheduled to be made available to the Contractor are presented in the Specifications.

Should an area, parcel, or building not be available to the Contractor on the date specified, the Contractor will usually be granted an extension of time to his Contract.

T. Personal Liability of Public Officials.

Refer to Article 107.20, same title, of the AASHTO Sample at the end of this Chapter. The purpose of the referenced AASHTO Article is to provide liability protection for individual representatives of Owners including representatives of the Engineer, from being held personally responsible by those who may have been damaged because of actions by the representatives or by their failure to take action.

A simple example would be that of an injured workman bringing suit against an inspector who failed to notify the Contractor of the unsafe condition that caused the workman's injury.

U. No Waiver of Legal Rights.

Refer to Article 107.21, same title, of the AASHTO Sample at the end of this Chapter. Payment items in the majority of engineering construction contracts are of the unit price type. Payment of these items is based on the number of completed units of work, as determined by field measurements. When efforts of individuals are involved, the human error factor is ever present. Consequently, overpayments can be made to contractors when errors occur in field measurements or in computing the final pay quantities. If these errors are discovered before final payment is made, the adjustment is easily accomplished. If however, the errors are discovered after final payment, the return of an overpayment may require the Owner to go to litigation. With the provisions of the referenced AASHTO Article, return of an overpayment to the Contractor should present no problem.

V. Environmental Protection.

Refer to Article 107.22, same title, of the AASHTO Sample at the end of this Chapter. Protection of the environment is becoming more critical with each passing year. At one time there were very few, if any, environmental restrictions placed upon a contractor. Burnable trash and timber was consumed on the site; pumps dewatering excavations could discharge directly into streams or existing storm drainage systems; and pile drivers banged away full blast.

Today, this is all changed. Burning on the site is prohibited; the polluting or muddying of streams is forbidden; the levels of construction noise in inhabited areas are limited to maximum sound levels during specified hours; dust in the atmosphere shall be controlled; and siltation of streams from run-off of "raw" slopes shall be prevented.

Measures for controlling soil erosion, particularly of newly-formed earth slopes in deep cuts and high embankments as in road construction, can take many forms, depending on the particular situation. Because the limits of these measures generally cannot be established in advance, they are frequently presented as unit price bid items in the Contract. Their requirements are presented in a Technical Section.

W. Minimum Wage Rates.

The Davis-Bacon Act requires that contractors on federally funded or federally assisted projects over a stated amount pay workers no less than the prevailing wage in the project area. Wage rates for federally funded projects are predetermined for each contract by the Secretary of Labor and inserted in the Specifications. Minimum wage rates for federally assisted projects are also established by the particular State Department of Labor, and made a part of the Contract.

When two sets of wage rates (Federal and State) are included in a contract, two different rates may be specified for the same labor classification. To resolve these differences, the Specifications will usually include a statement to the effect that where there is a conflict between the State and Federal wage rate, the higher rate will prevail. The prevailing wage rates in an area are generally those rates that are being paid the union workmen. The requirements of the Davis-Bacon Act will sometimes force nonunion contractors to pay higher wages than they could offer on the open market.

Some of the regulations and requirements governing working hours and wages, read as follows:

1. Eight hours shall constitute a regular work day for every laborer, mechanic, and apprentice working on the site of the Work.
2. All hours worked on Saturdays and Sundays, all hours worked in excess of eight hours per day on Monday through Friday, and all hours worked on legal holidays constitute overtime hours.
3. Every mechanic, laborer, and apprentice shall be paid not less often than once a week.
4. A copy of the minimum hourly wage rates shall be kept posted by the Contractor at the site of the Work, in a prominent place where it can be easily seen and read by the workmen.
5. The Contractor shall maintain payroll records and preserve them for a period of three years thereafter for all laborers, mechanics, and apprentices.

X. Equal Employment Opportunity.

Contractors are required by law to adhere to fair employment practices in connection with the performance of their work under the Contract. Basically, they are forbidden to discriminate against any employee or applicant for employment because of age, race, color, ancestry, marital status, sex, or national origin. These requirements also apply to subcontractors. Additional requirements, depending on the funding for the project and applicable State and Local laws, may call for the employment of a specified minimum number or percentage of minority representatives, training programs, and the soliciting of bids for subcontracts from available minority businesses. A minority business is generally defined as one in which at least 50 percent is owned by minority group members. Minority group members are Blacks, Spanish speaking American persons, American Orientals, American Indians, American Eskimos, and American Aleuts.

Detailed procedures for ensuring contractor compliance are spelled out in the Contract Documents. Failure to comply with the requirements can result in penalties.

4.7 Prosecution and Progress

A. Introduction.

Progress is the key to successful completion of a construction contract; but without prosecution there can be no progress. Time is of primary importance in a construction project, particularly when the Contractor has costly construction equipment and extensive labor forces on the site. A contractor will maintain that the difference between a profitable job and a losing job can be measured by the factor of time in completing his work. Extended equipment rentals, labor rate escalations, material cost increases, and prolonged supervisory costs, are all consequences of delays in the progress of a construction project.

Many factors influence progress. There are those that are controllable by the Contractor and others that cannot be controlled by the Contractor. Factors like labor strikes, acts of God, actions of the Owner and Engineer, and site conditions, generally cannot be controlled by the Contractor. On the other hand, factors like scheduling, material deliveries, equipment selectivity, labor distribution, construction methods, and the capability of supervisory personnel, are within the control of the Contractor.

B. Subletting of Contract.

Refer to Article 108.01, same title, of the AASHTO Sample at the end of this Chapter. Contractors find it necessary to sublet a portion of their work, for various reasons. It is rare that any one construction contractor will be equally skilled in all the trades and techniques required for a major construction contract. The Contractor will therefore, utilize the services of specialty contractors under subcontract agreements. The Contract between the Owner and the Contractor is generally known as the prime Contract, in which the Contractor is responsible to the Owner for all the work of the Contract. This includes work performed by his subcontractors as well as that accomplished by his own forces. The Contractor is also held responsible for the management, scheduling, and coordination of his subcontractors' work, as well as that of his own.

Items of work commonly let out to subcontractors include foundation bearing piles, structural steel fabrication and erection, furnishing concrete, electrical work, fencing, painting, and landscaping. The subcontract will include terms of the prime Contract and the Plans and Specifications, insofar as they apply to the subcontract. The Owner generally reserves the right to approve or disapprove the selection of a subcontractor. The Owner however, does not recognize subcontractors. In all matters concerning the work of subcontractors, the Owner will deal directly with the prime Contractor, whom he holds responsible.

Some of the requirements governing subcontracts are:

1. Before making any subcontracts, the Contractor must submit a written statement to the Engineer giving the name and address of the proposed subcontractor, the portion of the work which he is to perform and materials which he is to supply, and furnish any other information tending to prove that the proposed subcontractor has the necessary facilities, skill, integrity, past experience, and financial resources to perform the work in accordance with the terms and conditions of the Contract.
2. If the Engineer finds that the proposed subcontractor is qualified, he will notify the Contractor within ten days. If the determination is to the contrary, the Engineer will notify the Contractor, who may submit another proposed subcontractor, unless he decides to do the work himself.
3. The Engineer's approval of a subcontractor does not relieve the Contractor of any of his responsibilities, duties, and liabilities under the Contract. The Contractor is solely responsible for the acts or defaults of his subcontractor.
4. No subcontractor will be permitted to perform work at the site until he has furnished satisfactory evidence of insurance covering Workmen's Compensation, Public Liability, and Property Damages, as required.
5. The Contractor shall upon request, file with the Engineer a conformed copy of the subcontract, with prices and terms of payment deleted.

The requirement presented in referenced AASHTO Article 108.01, placing a limitation on the value of work that can be subcontracted, serves a good purpose. Allowing contractors to sublet their work without imposing any limit can bring on problems. Given this opportunity, some contractors may take full advantage and sublet the major portion of the Work. This would in essence place the Contractor in the role of a broker, in which he would provide the general superintendence, and not much more.

It is recognized that a prime contractor of heavy construction cannot be equipped to perform minor items of specialized work, such as electrical work, painting, or landscaping. These items of work are referred to as Specialty Items (see the definition in Article 4.2C). As specified in the referenced AASHTO Article, when computing the dollar value of work that must be performed by the Contractor's own organization, the dollar value of the specialty items can be deducted from the total Contract amount.

C. Preconstruction Conference.

This is a meeting conducted by the Owner or his representative shortly after the award of the Contract and before the Contractor begins any work. Attending this conference are representatives of the Owner, Contractor, Designer, and Engineer or Construction Manager. Among items for discussion are:

1. Scope of Work.
2. Establishing the lines of communication.
3. Authority and responsibilities of the Owner's site representatives.
4. Procedure for submittals by the Contractor.
5. Job meetings.
6. Preliminary progress schedule.
7. Unusual items of construction or unusual requirements in the Contract Documents that should be brought to the attention of the Contractor.
8. Questions by the Contractor concerning clarification of the Plans and Specifications.

D. Notice to Proceed.

Refer to Article 108.02, same title, of the AASHTO Sample at the end of this Chapter. After the Contract has been executed by the successful Bidder (See Article 5.3F, Execution and Approval of Contract), and the Contractor has submitted his certificates of insurance, the Owner will, within a specified time limit, issue to the Contractor the Notice to Proceed. This is the official notice to the Contractor to begin work on the Contract. As stated in the referenced AASHTO Article, the date stipulated in the Notice for beginning the Work also constitutes the official date of the beginning of the Contract. This is an important date because it will be used in determining the extent of any liquidated damages to be assessed the Contractor, should he fail to complete the Work on time, as specified in Article 4.7K, Failure to Complete on Time. The date established for beginning the Work takes into account the time required for the Contractor to assemble his force and equipment.

E. Prosecution and Progress.

Refer to Article 108.03, same title, of the AASHTO Sample at the end of this Chapter. Having received his Notice to Proceed, the Contractor now has the "green light" to begin work. Each construction contract has a specified time for completion. One of the first submittals required from the Contractor is a preliminary progress schedule, indicating his proposed programs of operation. In preparing his schedule, the Contractor will take into account those conditions described in Article 4.7F, Limitation of Operations, that will affect his progress. Submittal of this preliminary schedule for the Engineer's review is generally required within 15 days after Award of Contract. It is one of the items for discussion at the preconstruction conference. The Contractor indicates in this schedule his proposed sequence of operations for performing the Work within the specified time, showing the dates of start and finish for each major item of work. The schedule is updated each month, showing actual progress against scheduled progress. This affords both the Contractor and Engineer a means of monitoring the progress of the Work.

The Contractor is expected to begin work promptly within the specified time and prosecute it continuously on all suitable working days until the Contract is completed within the allotted time. Sometimes the Contractor's progress will fall behind that indicated on his submitted schedule. If the Engineer believes the Contractor's rate of progress is insufficient to complete the Work within the Contract time, he may order the Contractor to submit a revised schedule and modify his operations by either increasing his work force and equipment, or increasing his working hours, or both. Two common types of progress schedules are the bar chart and the critical path-type analysis (CPM).

The bar chart schedule is suitable for the relatively simple and straightforward type of contract. Projects like the construction of a new highway or a new pipeline would be more likely to utilize a bar chart. In this type of schedule, major items of work like roadway excavation, paving, and sanitary sewer pipe, would each be represented by a horizontal bar drawn on the chart to a convenient scale. The scale would equate a unit of length to a unit of time, such as: one inch equals 20 days. The relative location of a bar on the chart would indicate when the activity represented by that bar will begin.

The CPM type of schedule is applicable for the more complicated construction projects where stage construction or overlapping contracts are involved. The individual construction activities are integrated within a logic tied network diagram. Each activity is linked in a network with durations, start/finish dates (early and late), and criticality/float defined. A CPM type of schedule has the capacity to monitor more activities than the construction activities shown on a bar chart. It can also keep track of such activities as forecasting manpower requirements by principal trades, shop drawing submittals, and cash flow projections for progress payments. A CPM schedule can show the status of the project at any time and highlight potential trouble spots early enough for corrective action to be taken. Preparing and monitoring the CPM schedule requires specially trained personnel utilizing electronic computers.

F. Limitation of Operations.

Refer to Article 108.04, same title, of the AASHTO Sample at the end of this Chapter. There are few construction sites on which a contractor can perform his work with no restrictions or limitations. Some of the conditions or requirements that can limit a contractor's operations are:

1. Maintaining and protecting vehicular, railroad, marine, or pedestrian traffic.
2. Maintaining ingress to and egress from, adjacent businesses and properties.

3. Stage construction, where one phase of the Work must be completed before a subsequent phase may begin.
4. Certain parcels within the right-of-way or certain structures scheduled for demolition, may not be available to the Contractor upon Notice to Proceed.
5. Pile driving operations may be limited to specific hours of the day.
6. Limitation on blasting operations because of the presence of sensitive instruments in nearby buildings.
7. Other contractors working in the area.

The bidder is advised beforehand of the limitations that may affect his operations.

G. Character of Workmen; Methods and Equipment.

Refer to Articles 108.05, Character of Workmen, and 108.06, Methods and Equipment, of the AASHTO Sample at the end of this Chapter. The referenced AASHTO Articles have been combined in order to be consistent with the arrangement followed in the FHA and in most State DOT Standard Specifications. Workmen play an important role in construction. When large numbers are involved, inexperienced or undesirable individuals can sometimes be included among those employed. In addition to performing substandard work, an inexperienced workman represents a potential hazard to his fellow workers. This would be particularly true in the erection of structural steel or in blasting work. Similar problems can develop if a workman drinks on the job. When construction is taking place in a populated area, a workman using foul language can create additional problems.

Designers generally do not specify the construction methods or equipment to be used in performing the work. These decisions are left to the Contractor to allow him to utilize his expertise in these matters. It then becomes his responsibility to select the methods and equipment that will produce the desired results.

In some instances however, the method and equipment will be specified for a particular item of construction. For example, when the same subbase material has been used through the years for roadway construction such as in State highway work, the method and equipment can be specified with assurance that the desired result will be obtained. However, by giving the Contractor an opportunity to propose an alternate method or equipment, the Owner may benefit from a betterment. This also enables the Contractor to use a piece of equipment that he may have available and which is capable of producing the desired result.

H. Progress Photographs.

It is said that a picture is worth a thousand words. This is particularly true in construction. Words are incapable of adequately describing a situation or condition at a construction site, that no longer exists. In addition to providing a pictorial history of the progress of a project, photographs are invaluable in helping to resolve disputes. Photographs constitute an important part of the construction records of a project.

Furnishing progress photographs is generally made a responsibility of the Contractor. Depending on the size and complexity of the project, a specified number of photographs are to be taken before start of the Work, every month thereafter during the life of the Contract, and upon completion. The Contractor is required to use the services of a professional photographer. Upon completion of the Contract, the Contractor delivers all negatives to the Engineer.

I. Suspension of Work by Owner.

There are times when an Owner may, because of conditions unsuitable for prosecuting the Work or for other reasons, wish to suspend, delay, or interrupt the Work, wholly or in part, for such periods of time that he finds necessary. Under this Article in the Specifications, the Owner reserves the right to proceed with this action. By complying with the order of the Owner, the Contractor may sustain financial damage in addition to the loss of time. If the Contractor is unreasonably delayed in the performance of his work by acts of the Owner, it is generally accepted that the Contractor may recover damages arising from the delay. Actual loss for which the Contractor would be compensated includes the idle time of equipment, necessary payments for the idle time of labor, cost of the extra moving of equipment necessitated by the stop order, and the cost of overhead items such as the Contractor's jobsite office, insurance, utility services, and wages to supervisory and office personnel. Compensation would be for actual loss only; no markup for profit. The suspension, delay, or interruption of the Contractor's operation would also qualify for an extension of time to the Contract, as described in Article 4.7J.

Some Specifications may provide for the Contractor to completely withdraw from the Contract if a job shutdown initiated for the Owner's convenience continues beyond an unreasonable length of time, say 60 days or more.

J. Determination and Extension of Contract Time.

Refer to Article 108.07, same title, of the AASHTO Sample at the end of this Chapter. Basically, all construction contracts provide for a completion date. When the Contract states that "time is of the essence," it generally means that should a delay to the completion of the project occur, the party causing such delay can be held responsible for costs and damages arising from such delay.

Delays can be classified as excusable or not excusable. Delays beyond the control of the Contractor are generally considered excusable, and qualify for an extension of time. A time extension is an adjustment to the Contract completion date by the allowance of an additional period of time. Excusable delays include:

1. Delays caused by unforseeable circumstances beyond the control and without the fault or negligence of the Contractor, including such causes as "Acts of God", acts of public enemy, fire, strikes, labor problems, unusually severe weather, and freight embargoes. "Acts of God," are generally construed to mean earthquakes, floods, cyclones, or other cataclysmic phenomena of Nature beyond the power of the Contractor to foresee. Unusually severe weather can generally be verified by checking the meteorological records for the past five or ten years.
2. Acts of other contractors over whom the Contractor has no control.
3. When final Contract quantities of work are greater than the estimated quantities listed in the Unit Price Schedule, resulting in the final Contract price being greater than the original Contract price. In this situation, the Contract time can be increased in the same ratio that the final Contract price bears to the original Contract price.
4. Differing subsurface conditions which could not be anticipated. (see Article 4.3C, Differing Subsurface Conditions).
5. Extra work orders or change orders by the Owner or Engineer, which add to the Work. (see Article 4.3B, Changes)
6. Other acts of the Owner or Engineer, such as inadequate or faulty Plans and Specifications, delayed decisions, delay in the review of shop drawings, suspension of work (except suspensions because of a failure of the Contractor to conform to Contract requirements), or delay in making the site available; all of which actions may delay the work of the Contractor.

Granting extensions of time for the reasons described in the first and third of the points listed above will generally not be a basis of compensation for additional costs incurred by the Contractor during the period of delay.

Delays that are considered inexcusable and for which the Contractor would not be granted an extension of time include:

1. Those due to the Contractor's negligence, such as failure to properly protect the Work, and in correcting work that did not conform to the requirements.
2. Poor management performance resulting in understaffed work crews, lack of material, and insufficient equipment.
3. Delays resulting from acts of the Contractor's subcontractors.

Many Specifications will require that when request is made for an extension of time, the Contractor must notify the Engineer in writing within a specified time (generally 15 days) from the beginning of any such delay. This is to allow the Engineer time to make an adequate evaluation of the Contractor's request.

Additional discussion on the subject of Contract time is presented in Article 11.7J, Determination and Extension of Contract Time.

K. Failure to Complete on Time.

Refer to Article 108.08, same title, of the AASHTO Sample at the end of this Chapter. As indicated in the referenced AASHTO Article, when the Contractor fails to complete the Work within the Contract time, the Owner may suffer financial losses and other hardships. In the case of income-producing projects such as rapid transit projects and toll roads, the loss of anticipated revenue because of late completion can be significant. Added to this loss are the additional interest charges the Owner would have to pay on his construction loans.

The assessment of liquidated damages is to compensate the Owner for the damages he will suffer because of a delay in the completion of the Work. Without a liquidated damages clause, it might be difficult for the Owner to enforce timely completion of his project without resorting to a full scale lawsuit for actual damages. The inclusion of a liquidated damages clause provides one more incentive for the Contractor to complete the Work on time.

L. Default of Contract.

Refer to Article 108.09, same title, of the AASHTO Sample at the end of this Chapter. There are many areas, as listed in the referenced AASHTO Article, in which the Contractor may fail to fulfill his obligations under the Contract. The Owner has to be protected against the consequences of such failures on the part of the Contractor. This is handled in the Specifications by giving the Engineer the authority to notify the Contractor in writing of the Contractor's failure to carry on the Work in an acceptable manner. Should no action be taken by the Contractor, the Owner then has the authority to terminate the Contractor's right to continue with the Work. Completion of the Work will then be awarded to others. The authority of the Owner in this situation includes the right to make use of any materials or equipment remaining on the site. The additional expense incurred by the Owner in taking this action plus the cost of completing the Work, are deducted from the monies due or to become due the Contractor.

M. Termination of Contract.

Refer to Article 108.10, same title, of the AASHTO Sample at the end of this Chapter. As the referenced AASHTO Article illustrates, a contract may have to be terminated for reasons that are beyond the control of either the Owner or Contractor. This AASHTO Article outlines a procedure to be followed for arriving at an equitable settlement with the Contractor for work performed up to the time of termination and for additional costs attributed to the termination. It should be noted that the Contractor is still held to his contractual obligations for that portion of the Work which is completed.

Some contracts also contain a provision that the Owner may terminate the

Contract when it is in his interest to do so. Under such action, the Contractor is entitled to be compensated for the work he did prior to the time of termination and for all other unpaid expenses resulting from the abandonment of the Work. The Contractor's reimbursible costs would be determined as described in referenced AASHTO Article 108.10.

The Contract may also be terminated by the Owner because of failure of the Contractor to perform in accordance with the requirements of the Contract, as outlined in Article 4.7L, Default of Contract.

N. Disputes.

Disputes on a construction project are inevitable (see Article 4.4S, Claims for Adjustment). Disputes which cannot be satisfactorily settled by the parties themselves can be resolved by resorting to mediation, arbitration, or court action. Unresolved disputes on public works contracts may be referred to a Contract Board of Appeals or State Court.

1. Mediation. Mediation, a relatively new procedure, is a nonbinding voluntary approach for settling disputed issues before the positions of both parties harden into an adversarial posture. In this process, a neutral mediator acts as a go-between and meets separately with each party, pointing out the weaknesses in their case and encouraging them to work out their own settlement by direct negotiations. The mediator cannot impose a settlement. Mediation appears to be confined to disputes which are relatively small compared to the overall contractual relationship of the parties. Mediation is administered by the American Arbitration Association (AAA) in accordance with the Construction Industry Mediation Rules.

2. Arbitration. Many prefer the course of arbitration. It is a legal alternative for resolving disputes without having to go to court. Arbitration clauses may be more common in private than in public construction contracts. In arbitration, a dispute is submitted to three disinterested arbitrators and settled in accordance with the Construction Industry Arbitration Rules of the AAA. The AAA is a not-for-profit public service organization which provides private dispute settlement services. The Association maintains throughout the United States a national panel of impartial arbitrators consisting of experts in all the trades and professions. The AAA does not itself get involved in deciding cases. It supplies lists from which the parties of the dispute mutually select arbitrators.

When an arbitration clause is in the Contract, an unresolved dispute can be submitted to arbitration on demand of either party. After the filing of a demand for arbitration, the AAA will submit simultaneously to each party to the dispute an identical list of names of qualified arbitrators. These names may include Builders, Contractors, Engineers, Architects, other business people familiar with the construction industry, and Attorneys who represent such clients; all competent to hear and determine disputes administered under the Construction Industry Arbitration Rules. Each party selects the name of an arbitrator to represent it.

The two chosen arbitrators will then select a third neutral arbitrator to complete the three-man panel that will hear and resolve the dispute.

Arbitration seems to be preferable to litigation principally where construction claims under one million dollars are involved. Some of the reasons for this preference are:

a. Arbitrators are selected who know the industry and understand the technical issues, thus being knowledgeable in construction.
b. Arbitration takes place in an informal atmosphere which is less hostile than the atmosphere in court proceedings.
c. Less time is required to prepare and present a case for arbitration than for a court of law, because of the less formal procedural rules and the informal atmosphere. This usually results in lower legal costs.
d. The proceedings take place in an atmosphere of privacy; there is no press coverage or public record as in most court cases.

During the arbitration proceeding the Contractor does not slow down; he continues with the Work and maintains his construction schedule. Arbitration does have some shortcomings. There is a loss of time where major construction claims are involved. Arbitrators are highly capable individuals who volunteer to serve. They have busy schedules in their own private practice and are unable to devote the time to hear a case continuously. Arbitration has to suit the convenience of the arbitrators, the litigants, and their counsel. On the other hand, when a trial begins in court it continues to conclusion without interruption. Also, there is lack of a legal safeguard. A court ruling can be appealed, whereas an arbitration award is generally binding upon both parties and is not subject to appeal.

4.8 Measurement and Payment

A. Introduction.

Measurement of the Work and the payment for it represents an important feature of the construction Contract. The average contractor could not continue very long on the job if he were not being paid periodically for the work he performed. Unless a contract is of very short duration, few contractors have the resources to finance a job to its completion. The following Articles cover the various areas that may be involved in determining the number of completed units of work, the amount of payments, and establishing when they will be made to the Contractor.

B. Measurement of Quantities.

Refer to Article 109.01, same title, of the AASHTO Sample at the end of this Chapter. Payment quantities for unit price items are determined through a combination of field measurements and mathematical calculations. Detailed procedures for measuring individual items of work are specified in the Method of Measurement subsection of each Technical Section.

Referenced AASHTO Article 109.01 presents general guidelines on the measurement and determination of quantities. The following explanations and comments refer to the contents of this AASHTO Article:

1. Referring to the first paragraph; note that all work "will be measured by the Engineer," for obvious reasons.

2. Referring to the fourth paragraph; when taking measurements for computing the areas of constructed sidewalks and pavements, it is the practice to make no deductions for relatively small penetrations through the slab, like those made by manholes, valve boxes, hydrants, and lightpole foundations. The engineering labor and time expended in taking the additional measurements and computing these small areas would be more costly than the small savings resulting from these deductions.

3. In the seventh paragraph, the average end area method for determining volumes of earthwork is commonly used in computing roadway excavation and embankment, when measurements are based on cross-sections. This method would not be appropriate for determining the volume of excavation for such work as structure foundations, drainage structures, and test pits.

4. Referring to the ninth paragraph, another designation of weight is the "pound", which means 16 ounces avoirdupois.

5. Referring to the thirteenth paragraph, on volume measurement; in the second line it is believed that "6° F" should read "60° F."

6. Referring to the sixteenth paragraph; cement is also measured by the bag or barrel. A bag of cement contains 94 pounds net and is considered equal to one cubic foot in volume. A barrel of cement is equal to four bags and should weigh 376 pounds.

When a payment item in a contract is designated as a Lump Sum item or when an estimated quantity is designated as a Fixed (Plan) Quantity (see following Article), field measurements are not required for payment purposes. Field measurements would be required if changes to the Plan dimensions have been authorized by the Engineer.

C. Fixed (Plan) Quantities.

Some Contract Specifications may specify that certain unit price items will be paid for on the basis of their estimated quantities listed in the Unit Price Schedule. This arrangement is mentioned in the last paragraph of AASHTO Article 109.01, referred to in Article 4.8B. These quantities have been determined beforehand during the design phase, from information shown on the Plans. Under this arrangement, no field measurements have to be taken nor do computations have to be made to determine final payment quantities. This is particularly beneficial in highway contracts, by eliminating the tedious field work of taking survey cross-sections plus the attendant office computations in determining earthwork quantities. This is also beneficial in determining excavation quantities for a trench type underwater vehicular or rapid transit tunnel.

As mentioned in the last paragraph of AASHTO Article 109.01, the Contractor is paid on the basis of the Fixed Quantities listed in the Unit Price Schedule. If a change to the Plans is authorized by the Engineer and it affects the work of a fixed quantity item, the fixed quantity is increased or decreased by the amount represented by the authorized change.

D. Scope of Payment.

Refer to Article 109.02, same title, of the AASHTO Sample at the end of this Chapter. The intent of the second paragraph of the referenced AASHTO Article is to prevent the Owner from having to pay the Contractor twice for the same work or material. This can happen if similar work is specified in two different Sections of the Specifications. For example, in a Section specifying Storm Drainage, the Basis of Payment may state that payment for Storm Drain Manholes will be made at the Contract unit price per each manhole acceptably constructed, which price shall include the costs of excavation and backfill, concrete, reinforcing steel, ladder rungs, and manhole frame and cover. In another Section specifying Reinforcing Steel, the Basis of Payment would state that payment will be made at the Contract unit price per pound for reinforcing steel acceptably incorporated in the Work. The Contractor may attempt to get duplicate payment for the same reinforcing steel in storm drain manholes, by requesting that payment be made on the unit price basis as specified in the Section on Reinforcing Steel.

E. Compensation for Altered Quantities.

Refer to Article 109.03, same title, of the AASHTO Sample at the end of this Chapter. Referenced AASHTO Article 109.03 permits no normal renegotiation of the Contract Unit Price for an item of work when its final quantity varies from the estimated quantity listed in the Bid Schedule. However, many Contract Specifications now include a provision for renegotiation of Contract Unit Prices for major items, when their final quantities vary from the estimated quantities by more than a specified percentage. This is described in Article 4.3D, Variations in Estimated Quantities.

F. Payment for Extra and Force Account Work.

Refer to Article 109.04, Extra and Force Account Work, of the AASHTO Sample at the end of this Chapter. When no prior agreement can be reached on a price to be paid to the Contractor for Extra Work (see Article 4.3E, Extra Work), the

Owner may require the work to be done on a force account basis, or he may have it done by others, if it suits his purposes. Work done by force account is reimbursible for the actual costs, plus an allowance for overhead and profit, as illustrated in referenced AASHTO Article 109.04. Overhead includes the costs of superintendence, field office, and home office.

Generally, the inspector assigned to the force account work and the Contractor's representative compare at the end of each day's work their records for labor, equipment, and material used. Agreement is signified by their signatures. This eliminates any later disagreements that may arise on field records, when the final cost is being determined. Payment for force account work is normally made with the regular progress payments.

G. Eliminated Items.

Refer to Article 109.05, same title, of the AASHTO Sample at the end of this Chapter. The following examples will illustrate the need for this provision in the Specifications.

1. In the construction of a new highway, it may be necessary to demolish some existing houses and buildings which lie within the right-of-way. Generally, the demolition of structures is paid for at a separate lump sum price for each structure to be demolished. Shortly after the Contract is awarded, the Owner may decide that it would be in his best interests to occupy one of the buildings scheduled for demolition; a building that is situated outside the area of construction. The payment item for the demolition of this building would therefore be eliminated from the Contract.

2. After work has begun on a highway project, field conditions may reveal that a specified size of storm drain pipe is not required. This pipe item would then be eliminated from the Contract.

H. Progress Payments.

Refer to Article 109.06, same title, of the AASHTO Sample at the end of this Chapter. As work proceeds, progress payments are made to the Contractor by the Owner. Progress payment estimates prepared by the Engineer are based on his determination of the approximate quantities of work satisfactorily performed during the period covered by the payment estimate. Since these estimates are approximations they are subject to correction in subsequent progress payments and in the final payment. Progress payments may also include advance payment on materials not yet incorporated in the Work (see Article 4.8I, Payment for Material on Hand).

On lump sum contracts, the progress payment estimate is expressed as a percentage of the total work of the Contract. On unit price contracts, the estimate is based on the number of units satisfactorily completed for each item. Work performed on a lump sum item is expressed as a percentage of the total work of that item.

Progress payments are normally made each month. Large multimillion dollar contracts may provide for a payment in the middle of the month. A specified minimum value of work to be completed in the midmonth period would be required before the Contractor could receive a midmonth payment.

As specified in referenced AASHTO Article 109.06, the Contractor does not receive the entire sum of a progress payment estimate. A specified percentage of the amount due is retained by the Owner, until the Work is at least 95 percent complete. Various reasons are given for imposing this "retainage," as it is commonly called. Some of them are:

1. It is a protection for the Owner against possible claims or charges against him for some act or omission on the part of the Contractor, such as failure to reimburse subcontractors, damage to adjacent property or to another contractor, or defective work that has not been corrected.
2. It provides an incentive for the Contractor to complete his work, so that he may collect his retainage.
3. It provides a degree of protection for the Owner against an overpayment made for work accomplished.

A common arrangement is retention of ten percent of the payment estimate until the Contract is 50 percent complete. Subsequent payments are then made in full if the Contractor continues to maintain his work on schedule. Should he fall behind schedule, the retainage is then reimposed until such time that progress gets back on schedule.

The subject of retainage is a controversial matter. Construction Contractor Associations would like to see the retainage percentage reduced, since it represents money that has already been earned by the Contractor. It is argued by some that the retainage duplicates protection the Owner already receives from the Contractor's performance and payment bonds (see Article 5.3E, Requirements of Contract Bonds). Some Government Agencies are now abolishing or reducing retainage requirements on their construction contracts. The current trend appears to be leaning toward relaxing the long standing requirements on retainage.

The 95 percent completion stage mentioned in the last paragraph of referenced AASHTO Article 109.06, is sometimes referred to as the time of substantial completion (see Article 4.2C, Definitions) and the issuance of a semifinal payment estimate.

The close-out of a contract and release of all monies due the Contractor, can sometimes take many months. For example, after completing the major portion of the Work, a contractor may be prevented from completing the Contract because of seasonal or weather conditions which do not permit him to work on relatively minor items such as painting, seeding and planting. In such instances, the Engineer will make a semifinal inspection of the Work and list the items of incomplete work. This list will among other things, help to determine the amount of money that is to be retained for work still to be completed.

The semifinal payment estimate is based upon those quantities the Engineer has computed and set up as proposed final quantities, plus a reasonably accurate estimate of those items for which the Engineer has not yet computed the final quantities. To arrive at the amount of the semifinal payment, there is deducted from the estimated value of the Contract the total of all amounts previously paid the Contractor in progress payments, sums deemed properly chargeable against the Contractor including liquidated damages, and a sum retained to cover the work still to be completed.

I. Payment for Material on Hand.

Refer to Article 109.07, same title, of the AASHTO Sample at the end of this Chapter. Because of a scarcity or limited production, it may be necessary for the Contractor to purchase certain materials far in advance of their incorporation into the Work, to ensure their availability when needed. Materials like structural steel, prestressed concrete beams, reinforcing steel, certain aggregates, and certain pipe, would be likely candidates. Certain items of equipment might also be included.

Making these advance purchases can tie up part of a contractor's working capital and increase his costs. To ease this financial burden on the Contractor, advance payment for material on hand is made, subject to conditions outlined in referenced AASHTO Article 109.07 and as follows:

1. Advance payment will apply only to designated materials and equipment.
2. Request for advance payment will have to be initiated by the Contractor.
3. Proof of ownership, such as receipted invoices, will have to be submitted.
4. Proof of acceptability of the material will have to be furnished.
5. Material will have to be stored in an approved manner and in an area where damage is not likely to occur.

Advance payments for material are included in, and made a part of, the normal progress payments.

J. Lump Sum Breakdown.

In order for the Engineer to better determine the value of progress payments on a lump sum contract or on a contract containing lump sum pay items, he requires back-up information from the Contractor.

On a lump sum contract (see Article 1.3A, Fixed Price Contract), the Contractor would be required to submit a breakdown of his bid price to the Engineer for approval. Why the Engineer's approval? Without this requirement, a contractor could take advantage of the opportunity to submit an unbalanced breakdown that would allow him to receive larger progress payments in the early stages of the Contract; more than he actually earned. An acceptable breakdown would present the Contractor's estimated cost of the operations indicated in his Progress Schedule. The total of the values for these operations would equal the lump sum price bid.

For a contract containing individual lump sum pay items, the Contractor would be required to submit a similar breakdown for each lump sum pay item, when specified. Guidelines for establishing individual lump sum pay items are basically the same as those for a lump sum contract; namely that the limits of the work must be accurately defined on the Plans. The breakdown for each item would show the Contractor's estimated cost for each principal operation of work in that item. To illustrate; a contract involving construction of a sanitary sewer system included a lump sum pay item titled Thruway Force Main Crossing. This item consisted of the installation of a sewer force main (pressure line) under a Thruway. The 12-inch diameter force main was to be installed inside a 20-inch diameter steel casing pipe which was to be jacked under the roadway. Principal operations of work accounted for in the lump sum breakdown were:

1. Construction of two work pits; one at each end of the crossing. This involved excavation, sheeting and bracing, and dewatering.
2. Jacking the sections of steel casing pipe and removing the soil from within the casing. The casing was to serve as a pipe sleeve under the roadway.
3. Installing the carrier pipe within the sleeve and performing pressure and leakage tests.
4. Backfilling the work pits and restoring the ground surface.

K. Payment of Withheld Funds.

Refer to Article 109.08, same title, of the AASHTO Sample at the end of this Chapter. To ease the burden represented by the retainage of a portion of the Contractor's money, many governmental agencies have established procedures to enable the Contractor to substitute approved securities for the retained money. Under this arrangement the Contractor receives the retained money and the interest from the securities that he substituted for the retained monies.

L. Acceptance and Final Payment.

Refer to Article 109.09, same title, of the AASHTO Sample at the end of this Chapter. A contract is considered complete when all work has been finished, the final inspection made, and the Work accepted (see Article 4.4R, Acceptance of the Work).

The date of final acceptance of the Work constitutes the official date of completion of the Contract. Following this, the next function of the Engineer is to complete his field measurements of the finished Work and compute final quantities for the final payment estimate. Since this may involve some time, the Specifications will usually include a statement that the final quantities will be submitted to the Contractor within a stipulated period of time after the date of final acceptance. This period of time may vary from 30 to 60 days, depending on the complexity of the Contract. The Contractor is allowed a specified number of days to review the final quantities and notify the Owner in writing of his approval or disapproval. The final payment sum will reflect the return of any retained monies, deduction for liquidated damages if any, and any other deductions provided in the Contract that may be applicable.

If there are any outstanding claims or liens against the Contractor, an amount of money sufficient to cover these claims or liens may be withheld from the Contractor until these matters are settled.

The following Pages 95 through 142 have been reproduced from the AASHTO Guide Specifications for Highway Construction 1985, Sections 101 and 104 through 109. The Articles of these Sections are concerned with the General Conditions of the Contract and are referred to in this text.

SECTION 100—GENERAL PROVISIONS

Section 101—Definitions and Terms

Wherever in these specifications or in other contract documents the following terms or pronouns in place of them are used, the intent and meaning shall be interpreted as follows:

101.01 Abbreviations. Wherever the following abbreviations are used in these specifications or on the plans, they are to be construed the same as the respective expressions represented:

A.A.N.	American Association of Nurserymen
A.A.R.	Association of American Railroads
A.A.S.H.T.O.	American Association of State Highway and Transportation Officials
A.G.C.	Associated General Contractors of America
A.I.A.	American Institute of Architects
A.I.S.I.	American Iron and Steel Institute
A.N.S.I.	American National Standards Institute
A.R.A.	American Railway Association
A.R.E.A.	American Railway Engineering Association
A.S.C.E.	American Society of Civil Engineers
A.S.L.A.	American Society of Landscape Architects
A.S.T.M	American Society for Testing and Materials
A.W.P.A.	American Wood Preservers' Association
A.W.W.A.	American Water Works Association
A.W.S.	American Welding Society
F.H.W.A.	Federal Highway Administration
F.S.S.	Federal Specifications and Standards. General Services Administration
S.A.E.	Society of Automotive Engineers
U.L.	Underwriter's Laboratory
MIL	Military Specifications

101.02 Advertisement. A public announcement, inviting bids for work to be performed or materials to be furnished.

101.03 Award. The acceptance by the contracting agency of a proposal.

101.04 Bidder. An individual, partnership, firm, corporation, or any acceptable combination thereof, or joint venture, submitting a proposal.

101.05 Bridge. A single or multiple span structure, including supports, erected over a depression or an obstruction, as water, highway, or railway, and having a track or passageway for carrying traffic or other moving loads and having an opening measured along the center of the roadway of more than 20 feet between undercopings of abutments or spring lines of arches or extreme ends of openings for multiple boxes. If there are no abutment copings or fillets, the 20-foot measurement shall be between points six inches below the bridge seats or, in the case of frame structures, immediately under the top slab. A bridge may include multiple pipes where the clear distance between openings is less than half of the smaller contiguous opening. All measurements shall include the widths of intervening piers or division walls.

Bridge Length — The greater dimension of a structure measured along the center of the roadway between backs of abutment backwalls or between ends of bridge floor.

Bridge Roadway Width — The clear width of structure measured at right angles to the center of the roadway between the bottom of curbs or, if curbs are not used, between the inner faces of parapet or railing.

101.06 Calendar Day. Any day shown on the calendar, beginning and ending at midnight.

101.07 Change Order. A written order to the Contractor, covering contingencies, extra work, increases or decreases in contract quantities, and additions or alterations to the plans or specifications, within the scope of the contract, and establishing the bases of payment and time adjustments for the work affected by the changes.

101.08 Contract. The written agreement between the contracting agency and the Contractor, setting forth the obligations of the parties for the performance of the prescribed work.

The contract includes the invitation for bids, proposal, contract form and contract bond, specifications, supplemental specifications, special provisions, general and detailed plans, and notice to proceed, also any change orders and agreements that are required to complete the construction of the work in an acceptable manner, including authorized extensions thereof, all of which constitute one instrument.

101.09 Contract Bond. The approved form of security, executed by the Contractor and his Surety or Sureties, guaranteeing complete execution of the contract and all supplemental agreements pertaining thereto and the payment of all legal debts pertaining to the construction of the project.

101.10 Contract Item (Pay Item). A specific unit of work for which a price is provided in the contract.

101.11 Contract Payment Bond. The security furnished to the contracting agency to guarantee payment of prescribed debts of the Contractor covered by the bond.

101.12 Contract Performance Bond. The security furnished to the contracting agency to guarantee completion of the work in accordance with the contract.

101.13 Contract Time. The number of working days or calendar days allowed for completion of the contract.

If a calendar date of completion is shown in the proposal in lieu of a number of working or calendar days, the contract shall be completed by that date.

101.14 Contractor. The individual, partnership, firm, corporation, or any acceptable combination thereof, or joint venture, contracting with the highway agency for performance of prescribed work through its agents or employees.

101.15 County. The county, borough, or parish in which the work herein specified is to be done.

101.16 Culvert. Any structure which provides an opening under the roadway but which does not meet the classification of a bridge as defined in subsection 101.05.

101.17 Department, Commission or Agency. The State Highway or Transportation Department, Commission, or other organization as constituted under the laws of said State, or Commonwealth, for the administration of highway or transportation work.

101.18 Engineer. The Chief Engineer of the Department, Commission or Agency, acting directly or through a duly authorized representative, who is responsible for engineering supervision of the construction.

101.19 Equipment. All machinery and equipment, together with the necessary supplies for upkeep and maintenance, and also tools and apparatus necessary for the proper construction and acceptable completion of the work.

101.20 Extra Work. An item of work not provided for in the contract as awarded but found by the Engineer essential to the satisfactory completion of the contract within its intended scope.

101.21 Extra Work Order. A change order concerning the performance of work or furnishing of materials involving extra work. Such extra work may be performed at agreed prices or on a force account basis as provided elsewhere in these specifications.

101.22 Highway, Street, or Road. A general term denoting a public way for purposes of vehicular travel, including the entire area within the right-of-way. (Recommended usage: in urban areas—highway or street; in rural areas—highway or road.)

101.23 Holidays. In the State of ______________________, holidays occur on:

If any holiday listed above falls on a Sunday, the following Monday shall be considered a holiday.

101.24 Inspector. The Engineer's authorized representative assigned to make detailed inspections of contract performance.

101.25 Invitation for Bids. The advertisement for proposals for all work or materials on which bids are required. Such advertisement will indicate with reasonable accuracy the quantity and location of the work to be done or the character and quantity of the material to be furnished and the time and place of the opening of proposals.

101.26 Laboratory. The testing laboratory of the Department or any other testing laboratory which may be designated by the Engineer.

101.27 Major and Minor Contract Items. Major contract items are listed as such in the bid schedule or in the special provisions; all other original contract items shall be considered as minor items; or

In cases where the major contract items are not listed as such, the original contract item of greatest cost, computed from the original contract price and estimated quantity, and such other original contract items next in sequence of lower cost, computed in like manner, necessary to show a total cost at original prices and quantities of not less than .... percent (60 percent suggested) of the original contract cost shall be considered as a major item or items; or

Any item having an original contract value in excess of .... percent (5 to 10 percent suggested) of the original contract amount shall be considered as a major item or items.

101.28 Materials. Any substances specified for use in the construction of the project and its appurtenances.

101.29 Notice to Proceed. Written notice to the Contractor to begin with the contract work; when applicable, includes the date of beginning of contract time.

101.30 Pavement Structure. The combination of subbase, base course, and surface course placed on a subgrade to support the traffic load and distribute it to the roadbed.

Base Course — The layer or layers of specified or selected material of designed thickness on a subbase or a subgrade to support a surface course.

Subbase — The layers of specified or selected material of designed thickness placed on a subgrade to support a base course.

Subgrade — The top surface of a roadbed upon which the pavement structure and shoulders including curbs are constructed.

Subgrade Treatment — Modification of roadbed material by stabilization.

Surface Course — One or more layers of a pavement structure designed to accommodate the traffic load, the top layer of which resists skidding, traffic abrasion, and the disintegrating effects of climate. The top layer sometimes called "Wearing Course."

101.31 Plans. The contract drawings which show the location, character, and dimensions of the prescribed work, including layouts, profiles, cross sections, and other details.

Standard Plans — Drawings approved for repetitive use, showing details to be used where appropriate.

Working Drawings — Supplemental design sheets or similar data which the Contractor is required to submit to the Engineer such as stress sheets, shop drawings, erection plans, falsework plans, framework plans, cofferdam plans, and bending diagrams for reinforcing steel.

101.32 Profile Grade. The trace of a vertical plane intersecting the top surface of the proposed wearing surface, usually along the longitudinal centerline of the roadbed. Profile grade means either elevation or gradient of such trace according to the context.

101.33 Project. The specific section of the highway together with all appurtenances and construction to be performed thereon under the contract.

101.34 Proposal. The offer of a bidder, on the prescribed form, to perform stated construction work at the prices quoted.

101.35 Proposal Form. The prescribed form on which the offer of a bidder is to be submitted.

101.36 Proposal Guaranty. The security furnished with a bid to assure that the bidder will enter into the contract if his offer is accepted.

101.37 Questionnaire. The specified forms on which the Contractor shall furnish required information as to his ability to perform and finance the work.

101.38 Right-of-Way. A general term denoting land, property, or interest therein, usually in a strip, acquired for or devoted to transportation purposes.

101.39 Roadbed. The graded portion of a highway within top and side slopes, prepared as a foundation for the pavement structure and shoulders.

101.40 Roadbed Material. The material in cuts and embankments and in embankment foundations from the subgrade surface down, extending to such depth as affects the support of the pavement structure.

101.41 Roadside. A general term denoting the area adjoining the outer edge of the roadway. Extensive areas between the roadways of a divided highway may also be considered roadside.

101.42 Roadside Development. Those items necessary to the complete highway which provide for the preservation of landscape materials and features; the rehabilitation and protection against erosion of all areas disturbed by construction through seeding, sodding, mulching, and the placing of other ground covers; such suitable planting and other improvements as may increase the effectiveness and enhance the appearance of the highway.

101.43 Roadway. The portion of a highway within limits of construction.

101.44 Shoulder. The portion of the roadway contiguous with the traveled way for accommodation of stopped vehicles for emergency use, and for lateral support of base and surface courses.

101.45 Sidewalk. That portion of the roadway primarily constructed for the use of pedestrians.

101.46 Special Provisions. Additions and revisions to the standard and supplemental specifications applicable to an individual project.

101.47 Specifications. The compilation of provisions and requirements for the performance of prescribed work.

Standard Specifications — A book of specifications approved for general application and repetitive use.

Supplemental Specifications — Approved additions and revisions to the standard specifications.

Special Supplemental Specifications — Approved additions and revisions to the Standard and Supplemental Specifications applicable to certain specialty work and included in only those contracts containing applicable work.

101.48 Specified Completion Date. The date on which the contract work is specified to be completed.

101.49 Stabilization. Modification of soils or aggregates by incorporating materials that will increase load-bearing capacity, firmness, and resistance to weathering or displacement.

101.50 State or Commonwealth. The State or Commonwealth of ...., acting through its authorized representative.

101.51 Structures. Bridges, culverts, catch basins, drop inlets, retaining walls, cribbing, manholes, endwalls, buildings, sewers, service pipes, underdrains, foundation drains, and other features which may be encountered in the work and not otherwise classed herein.

101.52 Subcontractor. An individual partnership, firm, corporation, or any acceptable combination thereof, or joint venture, to which the Contractor sublets part of the contract.

101.53 Substructure. All of that part of a structure below the bearings of simple and continuous spans, skewbacks of arches and tops of footings of rigid frames, including backwalls, wingwalls, and wing protection railings.

101.54 Superintendent. The Contractor's authorized representative in responsible charge of the work.

101.55 Superstructure. All that part of a structure above the bearings of simple and continuous spans, skewbacks of arches and top of footings of rigid frames, excluding backwalls, wingwalls, and wing protection railings.

101.56 Surety. The corporation, partnership, or individual other than the Contractor, executing a bond furnished by the Contractor.

101.57 Title (or Headings). The titles or headings of the sections and subsections herein are intended for convenience of reference and shall not be considered as having any bearing on their interpretation.

101.58 City, Township, Town, or District. A subdivision of the county used to designate or identify the location of the proposed work.

101.59 Traveled Way. The portion of the roadway for the movement of vehicles, exclusive of shoulders and auxiliary lanes.

101.60 Work. The furnishing of all labor, materials, equipment, and incidentals necessary or convenient to the successful completion of the project and the carrying out of the duties and obligations imposed by the contract.

101.61 Working Day. A calendar day during which normal construction operations could proceed for a major part of a shift, normally excludes Saturdays, Sundays, and holidays.

101.62 Work Order. A written order, signed by the Engineer, of a contractual status requiring performance by the Contractor without negotiation of any sort.

101.63 In order to avoid cumbersome and confusing repetition of expressions in these specifications, it is provided that whenever anything is, or is to be, done, if, as, or, when, or where "contemplated, required, determined, directed, specified, authorized, ordered, given, designated, indicated, considered necessary, deemed necessary, permitted, reserved, suspended, established, approval, approved, disapproved, acceptable, unacceptable, suitable, accepted, satisfactory, unsatisfactory, sufficient, insufficient, rejected, or condemned," it shall be understood as if the expression were followed by the words "by the Engineer" or "to the Engineer."

SECTION 104—SCOPE OF WORK

104.01 Intent of Contract. The intent of the contract is to provide for the construction and completion in every detail of the work described. The Contractor shall furnish all labor, materials, equipment, tools, transportation, and supplies required to complete the work in reasonably close conformity with the plans, specifications, and terms of the contract.

104.02 Alteration of Plans or Character of Work. The Department reserves the right to make, at any time during the progress of the work, such increases or decreases in quantities and such alterations in the work within the general scope of the contract, including alterations in the grade or alignment of the road or structure or both, as may be found to be necessary or desirable. Such increases or decreases and alterations shall not invalidate the contract nor release the surety, and the Contractor agrees to perform the work as altered.

Under no circumstances shall alterations of plans or of the nature of the work involve work beyond the termini of the proposed construction except as necessary to satisfactorily complete the project.

Unless such alterations and increases or decreases materially change the character of the work to be performed or the cost thereof, the altered work shall be paid for at the same unit prices as other parts of the work. If, however, the character of the work or the unit costs thereof are materially changed, an allowance shall be made on the basis agreed to in advance of the performance of the work, or if a basis has not been previously agreed upon, then an allowance shall be made,either for or against the Contractor, in such amount as the Engineer may determine to be fair and equitable.

No claim shall be made by the Contractor for any loss of anticipated profits because of any alteration, or by reason of any variation between the approximate quantities and the quantities of work as done.

If the altered or added work is of sufficient magnitude as to require additional time to complete the project, such time adjustments may be made in accordance with the provisions of subsection 108.07.

Should the Contractor encounter or the Department discover during the progress of the work subsurface of latent physical conditions at the site differing materially from those indicated in this contract, or unknown physical conditions at the site of an unusual nature, differing materially from those ordinarily encountered and generally recognized as inherent in work of the character provided for in the contract, the Engineer shall be promptly notified in writing of such conditions before they are disturbed. The Engineer will thereupon promptly investigate the conditions and if the Engineer determines that they do materially differ and cause an increase or decrease in the cost of, or the time required for, the performance of the contract, an equitable adjustment will be made and the contract modified in writing accordingly.

104.03 Extra Work. The Contractor shall perform unforeseen work, for which there is no price included in the contract, whenever it is deemed necessary or desirable to complete the work as contemplated. Such work shall be performed in accordance with the specifications and as directed, and will be paid for as provided under subsection 109.04.

104.04 Maintenance of Traffic. Unless otherwise provided, the road while undergoing improvements shall be kept open to all traffic by the Contractor. Where provided on the plans, or approved by the Engineer, the Contractor may bypass traffic over an approved detour route. The Contractor shall keep the portion of the project being used by public traffic in such condition that traffic will be adequately accommodated, shall furnish, erect and maintain barricades, warning signs, delineators, flaggers, and pilot cars in accordance with the *Manual on Uniform Traffic Control Devices for Streets and Highways*, and Section 618—Traffic Control of these specifications. The Contractor shall also provide and maintain in a safe condition temporary approaches or crossings and intersections with trails, roads, streets, businesses, parking lots, residences, garages, and farms; however, snow removal will not be required of the Contractor. The Contractor shall bear all expense of maintaining the traffic over the section of road undergoing improvement and of constructing and maintaining such approaches, crossings, intersections, and other features as may be necessary, without direct compensation, except as provided below:

(a) *Special Detours* — When the proposal contains an item for "Maintenance of Detours" or "Removing Existing Structures and Maintaining Traffic," the payment for such item shall cover all cost of constructing and maintaining such detour or detours, including the construction of any and all temporary bridges and accessory features and the removal of the same, and obliteration of the detour road. Right-of-way for temporary highways or bridges will be furnished by the Department.

(b) *Maintenance of Traffic During Suspension of Work* — The Contractor shall make passable and shall open to traffic such portions of the project and temporary roadways as may be agreed upon between the Contractor and the Engineer for the temporary accommodation of necessary traffic during the anticipated period of suspension. Thereafter, and until an issuance of an order for the resumption of construction operations, the maintenance of the temporary route or line of travel agreed upon will be the responsibility of the Department. When work is resumed, the Contractor shall replace or renew any work or materials lost or damaged because of temporary use of the project; shall remove to the extent directed by the Engineer, any work or materials used in the temporary maintenance and shall complete the project as though its prosecution had been continuous and without interference. All additional work caused by such suspensions, for reasons beyond the control of the Contractor, will be paid for by the Department at contract prices or by extra work. When work is suspended due to seasonal or climatic conditions, for failure of the Contractor to correct conditions unsafe for the workmen or the general public, for failure to carry out provisions of the contract, or for failure to carry out orders of the Engineer, all costs for maintenance of traffic during the suspended period shall be borne by the Contractor.

(c) *Maintenance Directed by the Engineer* — If the Engineer directs special maintenance for the benefit of the traveling public not otherwise included in the contract, then the Contractor will be paid on the basis of unit prices or under subsection 104.03: Extra Work. The Engineer will determine the work to be classed as special maintenance.

104.05 Rights In and Use of Materials Found on the Work. The Contractor, with the approval of the Engineer, may use on the project such stone, gravel, sand, or other material determined suitable by the Engineer, as may be found in the excavation and will be paid both for the excavation of such materials at the corresponding contract unit price and for the pay item for which the excavated materials is used. He shall replace at his own expense with other acceptable material all of that portion of the excavation material so removed and used which was needed for use in the embankments, backfills, approaches, or otherwise. No charge for the materials so used will be made against the Contractor. The Contractor shall not excavate or remove any material from within the highway location which is not within the grading limits without written authorization from the Engineer.

Unless otherwise provided, the material from any existing old structure may be used temporarily by the Contractor in the erection of the new structure. Modification of such material will not be permitted except with the approval of the Engineer.

104.06 Final Cleaning Up. Before final acceptance, the highway, borrow, and local material sources and all areas occupied by the Contractor in connection with the work shall be cleaned of all rubbish, excess materials, temporary structures, and equipment, and all parts of the work shall be left in an acceptable condition. The cost of final cleanup shall be included in other items and no separate payment will be made therefor.

104.07 Value Engineering Proposals by Contractor. This subsection will apply only when it is included or is designated in the contract documents.

A. *Purpose and Scope* — The intent of this provision is to share with the Contractor any cost savings generated on this contract as a result of a proposal or proposals offered by the Contractor and approved by the Department. The purpose is to encourage the use of Contractor's ingenuity and experience in arriving at alternative, lower cost construction methods than those reflected in the contract documents by the sharing of savings resulting therefrom.

The value engineering proposals contemplated are those that could produce a savings to the Department without, in the sole judgment of the Department, impairing essential functions and characteristics of the facility including but not limited to, service life, economy of operation, ease of maintenance, desired appearance, and safety.

B. *Submittal of Proposal* — As a minimum, the following materials and information shal be submitted with each proposal, plus any additional information requested by the Department.

1. A statement that the proposal is submitted as a Value Engineering Proposal.
2. A description of the difference between the existing contract requirements and the proposed change, and the comparative advantages and disadvantages of each, including considerations of service life, economy of operations, ease of maintenance, desired appearance, and safety.
3. Complete plans and specifications showing the proposed revisions relative to the original contract features and requirements.
4. A complete cost analysis indicating the final estimate costs and quantities to be replaced by the proposal, the new costs and quantities generated by the proposal, and the cost effects of the proposed changes on operational, maintenance, and other considerations.
5. A statement of the time by which a change order adopting the proposal must be executed so as to obtain the maximum cost reduction during the remainder of the contract. This date must be selected to allow the Department ample time for review and processing a change order but without affecting the Contractor's schedule. Should the Department find that insufficient time is available for review and processing, it may reject the proposal solely on such basis. If the Department fails to respond to the proposal by the date specified, the Contractor shall consider the proposal rejected and shall have no claim against the State as a result thereof.
6. A statement as to the effect the proposal will have on the time for completion of the contract.
7. A description of any previous use or testing of the proposal on another Department project or elsewhere and the conditions and results therewith. If the proposal was previously submitted on another Department project, indicate the date, contract number, and the action taken by the Department.

C. *Conditions* — Value Engineering Proposals will be considered only when all of the following conditions are met:

1. The Contractor is cautioned not to base any bid prices on the anticipated approval of a Value Engineering Proposal and to recognize that such proposal may be rejected. In the event of rejection, the Contractor will be required to complete the contract in accordance with the plans and specifications at the prices bid.

2. All proposals, approved or not approved by the Department for use in this contract, apply only to the ongoing contracts or contracts referenced in the proposal, become the property of the Department, and shall contain no restrictions imposed by the Contractor on their use or disclosure. The Department shall have the right to use, duplicate and disclose in whole or in part any data necessary for the utilization of the proposal. The Department retains the right to utilize any accepted proposal or part thereof on any other or subsequent projects without any obligation to the Contractor. This provision is not intended to deny rights provided by law with respect to patented materials or processes.
3. If the Department already has under consideration certain revisions to the contract or has approved certain changes in specifications or standards for general use which are subsequently incorporated in a Value Engineering Proposal, the Department shall reject the Contractor's proposal and may proceed with such revisions without any obligation to the Contractor.
4. The Contractor shall have no claim against the Department for any costs or delays due to the Department's rejection of a Value Engineering Proposal, including but not limited to, development costs, anticipated profits, or increased material or labor costs resulting from delays in the review of such proposal.
5. The Department shall be the sole judge as to whether a proposal qualifies for consideration and evaluation. It may reject any proposal that requires excessive time or costs for review, evaluation, and/or investigations, or which is not consistent with the Department's design policies and basic design criteria for the project.
6. The Engineer may reject all or any portion of work performed pursuant to an approved Value Engineering Proposal if he determines that unsatisfactory results are being obtained. The Engineer may direct the removal of such rejected work and require the Contractor to proceed in accordance with the original contract requirements without reimbursement for any work performed under the proposal, or for its removal. Where modifications to the Value Engineering Proposal are approved in order to adjust to field or other conditions, reimbursement will be limited to the total amount payable for the work at the contract bid prices as if it were constructed in accordance with the original contract requirements. Such rejection or limitation

of reimbursement shall not constitute the basis of any claim against the State for delay or for any other costs.

7. The proposal shall not be experimental in nature but shall have been proven to the Department's satisfaction under similar or acceptable conditions on another Department project or at another location acceptable to the Department.
8. Proposals shall be considered only if equivalent options are not already provided in the contract documents.
9. The savings generated by the proposal must be of sufficient significance, in the sole judgment of the Department, to warrant review and processing.
10. A proposal changing the types and/or thickness of the pavement structure will not be considered.
11. If additional information is needed to evaluate proposals, this information must be provided in a timely manner. Failure to do so will result in rejection of the proposal. Such additional information could include, where design changes are proposed, results of field investigations and surveys, design computations, and field change sheets.

D. *Payment* — If the Value Engineering Proposal is accepted by the Department, the changes and payment therefor will be authorized via a change order. Reimbursement to the Contractor will be made as follows:

1. The changes will be incorporated into the contract via changes in the quantities of unit bid items, new agreed price items or by force account, as appropriate, in accordance with the specifications.
2. The cost of the revised work as determined from the aforementioned changes in quantities, new items, or force account will be paid directly. In addition to such payment, the Department will pay to the Contractor, via a separate item, 50 percent of the savings to the Department as reflected by the difference between the above payment and the cost of the related construction required by the original contract plans and specifications computed at contract bid prices.
3. The Contractor's costs for development, design, and implementation of the Value Engineering Proposal are not eligible for reimbursement.
4. The Contractor may submit Value Engineering Proposals for an approved subcontractor, provided that reimbursement is made by the Department to the Contractor and that the terms of the

passthrough to the subcontractor are satisfactorily negotiated and accepted before the proposal is submitted to the Department. Subcontractors may not submit a proposal except through the prime Contractor.

SECTION 105—CONTROL OF WORK

105.01 Authority of the Engineer. The Engineer will decide all questions that may arise as to the quality and acceptability of materials furnished and work performed and as to the rate of progress of the work; all questions regarding the interpretation of the plans and specifications; all questions as to the acceptable fulfillment of the contract on the part of the Contractor.

The Engineer will have the authority to suspend the work wholly or in part due to the failure of the Contractor to correct conditions unsafe for the workmen or the general public; for failure to carry out provisions of the contract; for failure to carry out orders; for such periods as he may deem necessary due to unsuitable weather; for conditions considered unsuitable for the prosecution of the work, or for any other condition or reason deemed to be in the public interest.

105.02 Plans and Working Drawings. Plans will show details of all structures, lines, grades, typical cross sections of the roadway, location, and design of all structures and a summary of items appearing on the proposal. Only general features will be shown for steel bridges. The Contractor shall keep one set of plans available on the work at all times.

The plans will be supplemented by such working drawings as are necessary to adequately control the work. Working drawings for structures shall be furnished by the Contractor and shall consist of such detailed plans as may be required to adequately control the work and are not included in the plans furnished by the Department. They shall include stress sheets, shop drawings, erection plans, falsework plans, cofferdam plans, bending diagrams for reinforcing steel, computations, or any other supplementary plans or similar data required of the Contractor. All working drawings must be approved by the Engineer and such approval shall not operate to relieve the Contractor of any of his responsibility under the contract for the successful completion of the work.

The contract price will include the cost of furnishing all working drawings.

105.03 Conformity with Plans and Specifications. All work performed and all materials furnished shall be in reasonably close conformity with the lines, grades, cross sections, dimensions, and material requirements, including tolerances, shown on the plans or indicated in the specifications.

Plan dimensions and contract specification values are to be considered as the target values to be strived for and complied with as the design values from which any deviations are allowed. It is the intent of the specifications that the materials and workmanship shall be uniform in character and shall conform as nearly as realistically possible to the prescribed target value or to the middle portion of the tolerance range. The purpose of the tolerance range is to accommodate occasional minor variations from the median zone that are unavoidable for practical reasons. When either a maximum and minimum value or both are specified, the production and processing of the material and the performance of the work shall be so controlled that material or work will not be preponderantly of borderline quality or dimension.

In the event the Engineer finds the materials furnished, work performed, or the finished product not within reasonably close conformity with the plans and specifications but that reasonably acceptable work has been produced, the Engineer shall then make a determination as to whether the work shall be accepted and remain in place. In this event, the Engineer will document the basis of acceptance by contract modification which will provide for an appropriate adjustment in the contract price for such work or materials necessary to conform to his determination.

In the event the Engineer finds the materials or the finished product in which the materials are used not within reasonably close conformity with the plans and specifications, but that reasonably acceptable work has been produced, the Engineer with make a determination as to whether the work will be accepted and remain in place. In such event, the Engineer will document the basis of acceptance by a contract modification which will provide for an appropriate adjustment in the contract price for such work or materials as the Engineer determines to be appropriate, based on engineering judgment.

In the event the Engineer finds the materials or the finished product in which the materials are used are not in reasonably close conformity with the plans and specifications and have resulted in an unsatisfactory or unacceptable product, the work or materials shall be removed and replaced or otherwise corrected by and at the expense of the Contractor in a manner satisfactory to the Engineer.

105.04 Coordination of Plans, Specifications, Supplemental Specifications, and Special Provisions. These specifications, the supplemental specifications, the plans, special provisions, and all supplementary documents are essential parts of the contract, and a requirement occurring in one is as binding as though occurring in all. They are intended to be complementary and to describe and provide for a complete work. In case of discrepancy, calculated dimensions will govern over scaled dimensions; plans will govern over standard and supplemental specifications; supplemental specifications will govern over standard specifications; special provisions will govern over standard specifications, supplemental specifications, and plans.

The Contractor shall take no advantage of any apparent error or omission in the plans or specifications. In the event the Contractor discovers such an error or omission, the Engineer shall be promptly notified. The Engineer will then make such corrections and interpretations as may be deemed necessary for fulfilling the intent of the plans and specifications.

105.05 Cooperation by Contractor. The Contractor will be supplied with a minimum of two sets of approved plans and contract assemblies, including special provisions. The Contractor shall keep one set available on the work at all times.

The Contractor shall give the work the constant attention necessary to facilitate the progress thereof, and shall cooperate with the Engineer, his inspectors, and other contractors in every way possible.

The Contractor shall have on the work at all times, as his agent, a competent superintendent capable of reading and thoroughly understanding the plans and specifications and thoroughly experienced in the type of work being performed, who shall receive instructions from the Engineer or authorized representatives. The superintendent shall have full authority to execute orders or directions of the Engineer without delay, and to promptly supply such materials, equipment, tools, labor, and incidentals as may be required. Such superintendence shall be furnished irrespective of the amount of work sublet.

105.06 Cooperation with Utilities. The Department will notify all utility companies, all pipeline owners, or other parties affected, and endeavor to have all necessary adjustments of the public or private utility fixtures, pipelines, and other appurtenances within or adjacent to the limits of construction, made as soon as practicable.

Water lines, gas lines, wire lines, service connections, water and gas meter boxes, water and gas valve boxes, light standards, cableways, signals, and all other utility appurtenances within the limits of the proposed construction that are to be relocated or adjusted are to be moved by the owners at their expense, unless otherwise provided in the contract.

It is understood and agreed that the Contractor has considered in the bid all of the permanent and temporary utility appurtenances in their present or relocated positions as shown on the plans and that no additional compensation will be allowed for any delays, inconvenience, or damage sustained due to any interference from the said utility appurtenances or the operation of moving them.

Note: The contract will indicate various utility items, certain of which are to be relocated or adjusted by the utility owner and others that are to be relocated or adjusted by the Contractor. The special provision shall indicate the means of adjudication, if any, in case of failure by the utility owners to comply with their responsibility in relocating or adjusting their facility.

105.07 Cooperation Between Contractors. The Department reserves the right at any time to contract for and perform other or additional work on or near the work covered by the contract.

When separate contracts are let within the limits of any one project, each Contractor shall conduct the work so as not to interfere with or hinder the progress or completion of the work being performed by other contractors. Contractors working on the same project shall cooperate with each other as directed.

Each Contractor involved shall assume all liability, financial or otherwise, in connection with his contract and shall protect and save harmless the Department from any and all damages or claims that may arise because of inconvenience, delays, or loss experienced because of the presence and operations of other contractors working within the limits of the same project.

The Contractor shall arrange the work and shall place and dispose of the materials being used so as not to interfere with the operations of the other contractors within the limits of the same project. The work shall be coordinated with that of the others in an acceptable manner and shall be performed in proper sequence to that of the others.

105.08 Construction Stakes, Lines, and Grades. Except where there are specific provisions in the contract for survey and stakeout by the Contractor, the Engineer will set construction stakes establishing lines, slopes, and continuous profile grade in road work, and centerline and benchmarks for bridge work, culvert work, protective and accessory structures and appurtenances as deemed necessary, and will furnish the Contractor with all necessary information relating to lines, slopes, and grades. These stakes and marks shall constitute the field control by which the Contractor shall establish other necessary controls and perform the work.

The Contractor shall be held responsible for the preservation of all stakes and marks, and if any of the construction stakes or marks have been carelessly or willfully destroyed or disturbed by the Contractor, the cost of replacing them will be charged against and deducted from the payment for the work.

The Department will be responsible for the accuracy of lines, slopes, grades, and other engineering work set forth under this section.

105.09 Authority and Duties of the ...... Engineer. As the direct representative of the Engineer, the ...... Engineer has immediate charge of the engineering details of each construction project and is responsible for the administration and satisfactory completion of the project. The ...... Engineer has the authority to reject defective material and to suspend any work that is being improperly performed.

105.10 Duties of the Inspector. Inspectors employed by the Department will be authorized to inspect all work done and materials furnished. Such inspection may extend to all or any part of the work and to the preparation, fabrication, or manufacture of the materials to be used. The inspector will not be authorized to alter or waive the provisions of the contract. The inspector will not be authorized to issue instructions contrary to the plans and specifications, or to act as foreman for the Contractor; however, the inspector shall have the authority to reject work or materials until any questions at issue can be referred to and decided by the Engineer.

105.11 Inspection of Work. All materials and each part or detail of the work shall be subject to inspection by the Engineer. The Engineer shall be allowed access to all parts of the work and shall be furnished with such information and assistance by the Contractor as is required to make a complete and detailed inspection.

The Contractor, at any time before acceptance of the work, shall remove or uncover such portions of the finished work as may be directed by the Engineer. After examination, the Contractor shall restore said portions of the work to the standard required by the specifications. Should the work thus exposed or examined prove acceptable, the uncovering, or removing, and the replacing of the covering or making good of the parts removed will be paid for as extra work. Should the work so exposed or examined prove unacceptable, the uncovering, or removing, and the replacing of the covering or making good of the parts removed, will be at the Contractor's expense.

Any work done or materials used without supervision or inspection by an authorized Department representative may be ordered removed and replaced at the Contractor's expense, unless the Department representative failed to inspect the work after having been given reasonable notice in writing that the work was to be performed.

When any unit of government or political subdivision or any railroad corporation is to pay a portion of the cost of the work covered by this contract, its respective representatives shall have the right to inspect the work. Such inspection shall not make any unit of government or political subdivision or any railroad corporation a party to this contract, and shall in no way interfere with the rights of either party hereunder.

105.12 Removal of Unacceptable and Unauthorized Work. All work that does not conform to the requirements of the contract will be considered unacceptable, unless otherwise determined acceptable under the provisions in subsection 105.03.

Unacceptable work, whether the result of poor workmanship, use of defective materials, damage through carelessness, or any other cause, found to exist prior to the final acceptance of the work, shall be removed immediately and replaced in an acceptable manner.

No work shall be done without lines and grades having been given by the Engineer. Work done contrary to the instructions of the Engineer, work done beyond the lines shown on the plans, or as given, except as herein specified, or any extra work done without authority, will be considered as unauthorized and will not be paid for under the provisions of the contract. Work so done may be ordered removed or replaced at the Contractor's expense.

Upon failure on the part of the Contractor to comply forthwith with any order of the Engineer made under the provisions of this article, the Engineer will have authority to cause unacceptable work to be remedied or removed and replaced and unauthorized work to be removed and to deduct the costs from any monies due or to become due the Contractor.

105.13 Load Restrictions. The Contractor shall comply with all legal load restrictions in the hauling of materials on public roads beyond the limits of the project. A special permit will not relieve the Contractor of liability for damage that may result from the moving of material or equipment.

The operation of equipment of such weight or so loaded as to cause damage to structures or the roadway or to any other type of construction will not be permitted. Hauling of materials over the base course or surface course under construction shall be limited as directed by the Engineer to prevent damage to any portion of the pavement structure. No loads will be permitted on a concrete pavement, base, or structure before the expiration of the curing period. In no case shall legal load limits be exceeded unless permitted in writing. The Contractor shall be responsible for all damage done by hauling equipment.

105.14 Maintenance During Construction. The Contractor shall maintain the work during construction and until the project is accepted. This maintenance shall constitute continuous and effective work prosecuted day-by-day, with adequate equipment and forces so the roadway or structures are kept in satisfactory condition at all times.

In the case of a contract for the placing of a course upon a course or subgrade previously constructed, the Contractor shall maintain the previous course or subgrade during all construction operations.

All cost of maintenance work during construction and before the project is accepted shall be included in the unit prices bid on the various pay items and the Contractor will not be paid an additional amount for such work.

105.15 Failure to Maintain Roadway or Structure. If the Contractor, at any time, fails to comply with the provisions of subsection 105.14, the Engineer will immediately notify the Contractor of such noncompliance. If the Contractor fails to remedy unsatisfactory maintenance within 24 hours after receipt of such notice, the Engineer may immediately proceed to maintain the project, and the entire cost of this maintenance will be deducted from monies due or to become due the Contractor.

105.16 Acceptance.

(a) *Partial Acceptance* — If at any time during the prosecution of the project, the Contractor substantially completes a unit or portion of the project, such as a structure, an interchange, or a section of road or pavement, the Contractor may request the Engineer to make final inspection of that unit. If the Engineer finds upon inspection that the unit has been satisfactorily completed in compliance with the contract, he may accept that unit as being completed and the Contractor may be relieved of further responsibility for that unit. Such partial acceptance shall in no way void or alter any of the terms of the contract.

(b) *Final Acceptance* — Upon due notice from the Contractor of presumptive completion of the entire project, the Engineer will make an inspection. If all construction provided for and contemplated by the contract is found to be completed to the Engineer's satisfaction, that inspection shall constitute the final inspection and the Engineer will notify the Contractor in writing of final acceptance as of the date of the final inspection.

If, however, the inspection discloses any work, in whole or in part, as being unsatisfactory, the Engineer will give the Contractor the necessary instructions for correction of same, and the Contractor shall immediately comply with and execute such instructions. Upon correction of the work, another inspection will be made that shall constitute the final inspection, provided the work has been satisfactorily completed. In such event, the Engineer will make the final acceptance and notify the Contractor in writing of this acceptance as of the date of final inspection.

105.17 Claims for Adjustment. If the Contractor deems that additional compensation is due him for work or material not clearly covered in the contract or not ordered by the Engineer as extra work. The Contractor shall notify the Engineer in writing of the intention to make claim for such additional compensation before beginning the work affected by such claim. If such notification is not given, and the Engineer is not afforded proper facilities by the Contractor for keeping strict account of actual cost as required, then the Contractor agrees to waive any claim for additional compensation. Such notice by the Contractor, and the fact that the Engineer has kept account of the cost as aforesaid, shall not in any way be construed as proving or substantiating the validity of the claim. If the claim, after consideration by the Engineer, is found to be just, it will be paid as extra work as provided herein for force account work. If the Engineer disapproves additional compensation for such work or material, the Contractor may appeal the matter to the Chief Administrative Officer of the Department for administrative determination by himself or his designees. Such appeal or determination does not in any way prejudice the Contractor's right to pursue legal remedies in a court of competent jurisdiction nor does it in any way hold the Engineer's decision or its effects in abeyance. Nothing in this subsection shall be construed as establishing any claim contrary to the terms of subsection 104.02.

105.18 Automatically Controlled Equipment. Whenever batching or mixing plant equipment is required to be operated automatically under the contract and a breakdown or malfunction of the automatic control occurs, the equipment may be operated manually for a period of .... hours (48 hours suggested) following the breakdown or malfunction, provided this method of operation will produce results otherwise meeting specifications.

SECTION 106—CONTROL OF MATERIAL

106.01 Source of Supply and Quality Requirements. The materials used on the work shall meet all quality requirements of the contract. In order to expedite the inspection and testing of materials, the Contractor shall notify the Engineer of his proposed sources of materials prior to delivery. At the option of the Engineer, materials may be approved at the source of supply before delivery is started. If it is found after trial that sources of supply for previously approved materials do not produce specified products, the Contractor shall furnish materials from other sources.

106.02 Local Material Sources. Possible sources of local materials may be designated on the plans and described in the special provisions. The quality of material in such deposits will be acceptable in general, but the Contractor shall determine the amount of equipment and work required to produce a material meeting the specifications. It is understood that it is not feasible to ascertain from samples the limits for an entire deposit, and that variations shall be considered as usual and are to be expected. The Engineer may order procurement of material from any portion of a deposit and may reject portions of the deposit as unacceptable.

The Department may acquire and make available to the Contractor the right to take materials from the sources designated on the plans and described under special provisions, together with the right to use such property as may be specified, for plant site, stockpiles, and hauling roads.

If the Contractor elects to use material from sources other than those designated, the Contractor shall acquire the necessary rights to take materials from the sources and shall pay all costs related thereto, including any which may result from an increase in length of haul. All costs of exploring and developing such other sources shall be borne by the Contractor. The use of material from other than designated sources will not be permitted until tests on preliminary samples indicate general acceptability of the material. Additional samples may be required of the Contractor for inspection and testing by the Engineer, prior to approval of and authorization to use the source.

Unless otherwise permitted, pits and quarries shall be excavated so that water will not collect and stand therein. Sites from which such material has been removed shall, upon completion of the work, be left in a neat and presentable condition. Where practicable, borrow pits, gravel pits, and quarry sites shall be located so they will not be visible from the highway.

106.03 Samples, Tests, Cited Specifications. All materials will be inspected, tested, and approved by the Engineer before incorporation in the work. Any work in which untested materials are used without approval or written permission of the Engineer shall be performed at the Contractor's risk. Materials found to be unacceptable and unauthorized will not be paid for and, if directed by the Engineer, shall be removed at the Contractor's expense. Unless otherwise designated, tests in accordance with the most recent cited standard methods of AASHTO or ASTM, in effect on the date of advertisement for bids, will be made by and at the expense of the Department. Samples will be taken by a qualified representative of the Department. All materials being used are subject to inspection, test, or rejection at any time prior to or during incorporation into the work. Copies of all tests will be furnished to the Contractor's representative at his request.

106.04 Certification of Compliance. The Engineer may permit use prior to sampling and testing of certain materials or assemblies accompanied by Certificates of Compliance, stating that such materials or assemblies fully comply with the requirements of the contract. The certificate shall be signed by the manufacturer. Each lot of such materials or assemblies delivered to the work must be accompanied by a Certificate of Compliance in which the lot is clearly identified.

Materials or assemblies used on the basis of Certificates of Compliance may be sampled and tested at any time and if found not to be in conformity with contract requirements will be subject to rejection whether in place or not.

The form and distribution of Certificates of Compliance shall be as approved by the Engineer.

106.05 Plant Inspection. The Engineer may undertake the inspection of materials at the source. Manufacturing plants may be inspected periodically for compliance with specified manufacturing methods and materials samples will be obtained for laboratory testing for compliance with materials quality requirements. This may be the basis for acceptance of manufactured lots as to quality.

In the event plant inspection is undertaken, the following conditions shall be met:

(a) The Engineer shall have the cooperation and assistance of the Contractor and the producer of the materials.
(b) The Engineer shall have full entry at all times to such parts of the plant as may concern the manufacture or production of the materials being furnished.
(c) If required by the Engineer, the Contractor shall arrange for an approved building for the use of the inspector; such building to be located conveniently near the plant, and conforming to the requirements of subsection 106.06.
(d) Adequate safety measures shall be provided and maintained.
(e) [**Optional**] Crushing or screening facilities shall be equipped with an automatic or semi-automatic mechanical sampling device.

It is understood that the Department reserves the right to retest all materials that have been tested and approved at the source of supply prior to incorporation into the work and to reject all materials which, when retested, do not meet the requirements of these specifications, or those established for the specific project.

106.06 Field Laboratory. The Contractor shall provide an inspector's shelter or field laboratory, consisting of a suitable building in which to house and use the equipment necessary to carry on the required tests. The building shall be provided and paid for in accordance with the contract requirements.

[Optional] 106.07 Foreign Materials. The Contractor shall, at no additional cost to the State, arrange for any required sampling and testing that the State is not equipped to perform. All testing shall generally be performed within the United States and be subject to witnessing by the Engineer. Certain materials or processes may necessitate the testing be performed or witnessed at the foreign source by State personnel. When the Engineer authorizes inspection at a foreign site, the Contractor shall reimburse the State for all expenses incurred outside the United States by the State's representative.

Each lot of foreign material shall be accompanied by a Certificate of Compliance prepared in accordance with requirements of subsection 106.04. Certified mill test reports shall be attached to the Certificate of Compliance for those materials for which mill test reports are required and shall clearly identify the lot to which they apply.

Structural materials requiring mill test reports will be accepted only from those foreign manufacturers who have previously established to the satisfaction of the Engineer the adequacy of their in-plant quality control to assure delivery of uniform material in conformance with contract requirements.

Adequacy of quality control shall be established at the option of the Engineer, by submission of detailed written proof of adequate control, or through an in-plant inspection by the Engineer or representative.

Structural materials will not be accepted unless they are properly identified with mill test reports and Certificates of Compliance.

106.08 Storage of Materials. Materials shall be so stored as to assure the preservation of their quality and fitness for the work. Stored materials, even though approved before storage, may again be inspected prior to their use in the work. Stored materials shall be located so as to facilitate their prompt inspection. Approved portions of the right-of-way may be used for storage purposes and for the placing of the Contractor's plant and equipment, but any additional space required therefor must be provided at the Contractor's expense. Private property shall not be used for storage purposes, without written permission of the owner or lessee, and if requested, copies of such written permission shall be furnished to the Engineer. All storage sites shall be restored to their original condition by and at the expense of the Contractor. This shall not apply to the stripping and storing of topsoil, or to other materials salvaged from the work.

106.09 Handling Materials. All materials shall be handled in such manner as to preserve their quality and fitness for the work. Aggregates shall be transported from the storage site to the work in tight vehicles so constructed as to prevent loss or segregation of materials after loading and measuring.

106.10 Unacceptable Materials. All materials not conforming to the requirements of the specifications shall be considered as unacceptable and will be rejected and removed immediately from the site of the work. Rejected material shall not be used until the defects have been corrected and approved by the Engineer.

106.11 Department-Furnished Material. The Contractor shall furnish all materials required to complete the work, except those specified to be furnished by the Department.

Material furnished by the Department will be delivered or made available to the Contractor at the points specified in the special provisions.

The cost of handling and placing all materials after they are delivered to the Contractor shall be included in the contract price for the item in connection with which they are used.

The Contractor will be held responsible for all material delivered and deductions will be made from any monies due for any shortages and deficiencies, from any cause whatsoever, and for any damage that may occur after delivery, and for any demurrage charges.

SECTION 107—LEGAL RELATIONS AND RESPONSIBILITY TO PUBLIC

107.01 Laws to be Observed. The Contractor shall keep fully informed of all Federal, State, and local laws, ordinances, and regulations and all orders and decrees of bodies or tribunals having any jurisdiction or authority, which in any manner affect those engaged or employed on the work, or which in any way affect the conduct of the work. He shall at all times observe and comply with all such laws, ordinances, regulations, orders, and decrees; and shall protect and indemnify the State and its representatives against any claim or liability arising from or based on the violation of any such law, ordinance, regulation, order, or decree, whether by the Contractor or the Contractor's employees.

107.02 Permits, Licenses, and Taxes. The Contractor shall procure all permits and licenses, pay all charges, fees, and taxes, and give all notices necessary and incidental to the due and lawful prosecution of the work.

107.03 Patented Devices, Materials, and Processes. If the Contractor employs any design, device, material, or process covered by letters of patent or copyright, the Contractor shall provide for such use by suitable legal agreement with the patentee or owner. The Contractor and the surety shall indemnify and save harmless the State, any affected third party, or political subdivision from any and all claims for infringement by reason of the use of any such patented design, device, material or process, or any trademark or copyright, and shall indemnify the State for any costs, expenses, and damages that it may be obliged to pay by reason of an infringement, at any time during the prosecution or after the completion of the work.

107.04 Restoration of Surfaces Opened by Permit. The right to construct or reconstruct any utility service in the highway or street or to grant permits for same, at any time, is hereby expressly reserved by the Department for the proper authorities of the municipality in which the work is done and the Contractor shall not be entitled to any damages either for the digging up of the street or for any delay occasioned thereby.

When an individual, firm, or corporation is authorized to work in the highway through a duly executed permit from the Department, the Contractor shall allow parties bearing such permits, and only those parties, to make openings in the highway. When ordered by the Engineer, the Contractor shall make all necessary repairs due to such openings and the work will be paid for as extra work, or as provided in these specifications, and will be subject to the same conditions as original work performed.

107.05 Federal Aid Participation. When the United States government participates in the cost of the work covered by the contract, the work shall be under the supervision of the State but subject to the inspection and approval of the proper officials of the United States government and in accordance with the applicable Federal statutes and rules and regulations. When any Federal laws, rules, or regulations are in conflict with any provisions of a federally assisted contract, the Federal requirements shall prevail, take precedence, and be in force over and against any such conflicting provisions.

Such inspection shall in no sense make the Federal government a party to this contract and will not interfere with the rights of either party hereunder.

107.06 Sanitary, Health, and Safety Provisions. The Contractor shall provide and maintain in a neat, sanitary condition such accommodations for the use of employees as necessary to comply with the requirements of the State and local Board of Health, or of other bodies or tribunals having jurisdiction.

Attention is directed to Federal, State, and local laws, rules, and regulations concerning construction safety and health standards. The Contractor shall not require any workers to work in surroundings or under conditions that are unsanitary, hazardous, or dangerous to health or safety.

107.07 Public Convenience and Safety. The Contractor shall at all times so conduct work as to assure the least possible obstruction to traffic. The safety and convenience of the general public and the residents along the highway and the protection of persons and property shall be provided for by the Contractor as specified under subsection 104.04.

107.08 Railway-Highway Provisions. If the plans require that materials be hauled across the tracks of any railway, the Department will arrange with the railway for any new crossings required or for the use of any existing crossings. If the Contractor elects to use crossings other than those shown on the plans, the Contractor shall make arrangements for the use of such crossings.

All work to be performed by the Contractor on the railroad right-of-way shall be performed at such times and in such manner as not to unnecessarily interfere with the movement of trains or traffic on the track of the railway company. The Contractor shall use all care and precaution to avoid accidents, damage, or unnecessary delay or interference with the railway company's trains or other property.

107.09 Construction Over or Adjacent to Navigable Waters. All work over, on, or adjacent to navigable waters shall be so conducted that free navigation of the waterways will not be interfered with and that the existing navigable depths will not be impaired except as allowed by permit issued by the U.S. Coast Guard and/or the U.S. Army Corps of Engineers, as applicable.

107.10 Barricades and Warning Signs. The Contractor shall provide, erect, and maintain all necessary barricades, suitable and sufficient lights, danger signals, signs, and other traffic control devices, and shall take all necessary precautions for the protection of the work and safety of the public. Highways closed to traffic shall be protected by effective barricades, and obstructions shall be illuminated during hours of darkness. Suitable warning signs shall be provided to properly control and direct traffic.

The Contractor shall erect warning signs in advance of any place on the project where operations may interfere with the use of the road by traffic, and at all intermediate points where the new work crosses or coincides with an existing road. Such warning signs shall be placed and maintained in accordance with the plans furnished. No signs, barricades, lights, or other protective devices shall be dismantled or removed without permission of the Engineer.

All barricades, warning signs, lights, temporary signals, and other protective devices shall conform with the *Manual on Uniform Traffic Control Devices for Streets and Highways* and Section 618—Traffic Control of these specifications.

107.11 Use of Explosives. When the use of explosives is necessary for the prosecution of the work, the Contractor shall exercise the utmost care not to endanger life or property, including new work. The Contractor shall be responsible for all damage resulting from the use of explosives.

The Contractor shall comply with all laws and ordinances, as well as with Title 29, Code of Federal Regulations, Part 1926, Safety and Health Regulations for Construction (OSHA), whichever is the most restrictive, with respect to the use, handling, loading, transportation, and storage of explosives and blasting agents.

The Contractor shall notify each property owner and public utility company having structures or facilities in proximity to the site of the work of any intention to use explosives. Such notice shall be given sufficiently in advance to enable them to take such steps as they may deem necessary to protect their property from injury.

107.12 Protection and Restoration of Property and Landscape. The Contractor shall be responsible for the preservation of all public and private property and shall protect carefully from disturbance or damage all land monuments and property marks until the Engineer has witnessed or otherwise referenced their location and shall not move them until directed.

When the Contractor's excavating operations encounter remains of prehistoric people's dwelling sites or artifacts of historical or archaeological significance, the operations shall be temporarily discontinued. The Engineer will contact archaeological authorities to determine the disposition thereof. When directed by the Engineer, the Contractor shall excavate the site in such a manner as to preserve the artifacts encountered and shall remove them for delivery to the custody of the proper state authorities. Such excavation will be considered and paid for as extra work.

The Contractor shall be responsible for all damage or injury to property of any character, during the prosecution of the work, resulting from any act, omission, neglect, or misconduct in this manner or method of executing the work, or at any time due to defective work or materials, and the Contractor's responsibility will not be released until the project has been completed and accepted.

When or where any direct or indirect damage or injury is done to public or private property by or on account of any act, omission, neglect, or misconduct in the execution of the work, or in consequence of the nonexecution thereof by the Contractor, the Contractor shall restore at the Contractor's own expense, such property to a condition similar or equal to that existing before such damage or injury was done, by repairing, rebuilding, or otherwise restoring as may be directed, or shall make good such damage or injury in an acceptable manner.

107.13 Forest Protection. In carrying out work within or adjacent to State or National Forests, the Contractor shall comply with all regulations of the State Fire Marshal, Conservation Commission, Forestry Department, or other authority having jurisdiction, governing the protection of forests and the carrying out of work within forests, and shall observe all sanitary laws and regulations with respect to the performance of work in forest areas. The Contractor shall keep the areas in an orderly condition, dispose of all refuse, obtain permits for the construction and maintenance of all construction camps, stores, warehouses, residences, latrines, cesspools, septic tanks, and other structures in accordance with the requirements of the Forest Supervisor.

The Contractor shall take all reasonable precaution to prevent and suppress forest fires and shall require employees and subcontractors, both independently and at the request of forest officials, to prevent and suppress and to assist in preventing and suppressing forest fires and to make every possible effort to notify a forest official at the earliest possible moment of the location and extent of any fire seen by them.

107.14 Responsibility for Damage Claims. The Contractor shall indemnify and save harmless the Department, its officers, and employees from all suits, actions, or claims of any character brought because of any injuries or damage received or sustained by any person, persons, or property on account of the operations of the said Contractor, or on account of or in consequence of any neglect in safeguarding the work; or through use of unacceptable materials in constructing the work; or because of any act or omission, neglect, or misconduct of said Contractor; or because of any claims or amounts recovered from any infringements of patent, trademark, or copyright; or from any claims or amounts arising or recovered under the Workmen's Compensation Act, or any other law, ordinance, order, or decree. So much of the money due the said Contractor under and by virtue of his contract as may be considered necessary by the Department for such purpose may be retained for the use of the State; or, in case no money is due, the surety may be held until such suit or suits, action or actions, claim or claims for injuries or damages as aforesaid shall have been settled and suitable evidence to that effect furnished to the Department; except that money due the Contractor will not be withheld when satisfactory evidence is produced that the Contractor is adequately protected by public liability and property damage insurance.

The Contractor shall procure and maintain at the Contractor's own expense, until final acceptance by the State of the work covered by the contract, insurance for liability for damages imposed by law, of the kinds and in the amounts herein provided, with insurance companies authorized to do business in the State, covering all operations under the contract, whether performed by the Contractor or by subcontractors. Before commencing the work, the Contractor shall furnish certificates of insurance in the form satisfactory to the Department, showing compliance with this subsection and which certify that the policies will not be changed or cancelled until 30-days written notice has been given to the Department.

The types and limits of insurance are as follows:

1. Workers' Compensation Insurance in accordance with prevailing laws.
2. Liability and property damage insurance in the following amounts:
   (a) Bodily injury liability:
   ($500,000 recommended), each person.
   ($1,000,000 recommended), each occurrence.
   (b) Property damage liability:
   ($500,000 recommended), each occurrence.
   ($1,000,000 recommended), aggregate.

107.15 Third Party Beneficiary Clause. It is specifically agreed between the parties executing this contract that it is not intended by any of the provisions of any part of the contract to create the public or any member thereof a third party beneficiary hereunder, or to authorize anyone not a party to the contract to maintain a suit for personal injuries or property damage pursuant to the terms or provisions of the contract.

107.16 Opening Sections of Project to Traffic. Opening of sections of the work to traffic prior to completion of the entire contract may be desirable from a traffic service standpoint, or may be necessary due to conditions inherent in the work, or by changes in the Contractor's work schedule, or necessary due to conditions or events unforeseen at the time of the contract. Such openings due to any of the foregoing conditions shall be made when ordered by the Engineer. Under no condition shall such openings constitute acceptance of the work or a part thereof, or a waiver of any provisions of the contract.

Special provisions shall state, insofar as possible, which sections shall be opened prior to completion of the contract. On any section opened by order of the Engineer, whether covered in the special provisions or not, the Contractor shall not be required to assume any expense entailed in maintaining the road for traffic. Such expense shall be borne by the Department, or compensated for in a manner provided in subsection 109.04. On such portions of the project that are ordered by the Engineer to be opened for traffic in the case of unforeseen necessity that is not the fault of the Contractor, compensation for additional expense, if any, to the Contractor and allowance of additional time, if any, for completion of any other items of work on the portions of the project ordered by the Engineer to be opened shall be as set forth in a change order mutually agreed on by the Engineer and the Contractor.

If the Contractor is dilatory in completing shoulders, drainage structures, or other features of the work, the Engineer may so notify the Contractor in writing and establish therein a reasonable period of time in which the work should be completed. If the Contractor is dilatory, or fails to make a reasonable effort toward completion in this period of time, the Engineer may then order all or a portion of the project opened to traffic. On such sections which are so ordered to be opened, the Contractor shall conduct the remainder of the construction operations so as to cause the least obstruction to traffic and shall not receive any added compensation due to the added cost of the work by reason of opening such section to traffic.

On any section opened to traffic under any of the above conditions, whether stated in the special provisions or opened by necessity of Contractor's operations, or unforeseen necessity, any damage to the highway not attributable to traffic which might occur on such section (except slides) shall be repaired by and at the expense of the Contractor. The removal of slides shall be done by the Contractor on a basis determined by the Engineer prior to the removal of such slides.

107.17 Contractor's Responsibility for Work. Until final written acceptance of the project by the Engineer, the Contractor shall have the charge and care thereof and shall take every precaution against injury or damage to any part thereof by the action of the elements or from any other cause, whether arising from the execution or from the nonexecution of the work. The Contractor shall rebuild, repair, restore, and make good all injuries or damages to any portion of the work occasioned by any of the above causes before final acceptance and shall bear the expense thereof, except damage to the work due to unforeseeable causes beyond the control of and without the fault or negligence of the Contractor, including but not restricted to acts of God, such as earthquake, tidal wave, tornado, hurricane, or other cataclysmic phenomenon of nature, or acts of the public enemy or of governmental authorities.

In case of suspension of work from any cause whatever, the Contractor shall be responsible for the project and shall take such precautions as may be necessary to prevent damage to the project, provide for normal drainage, and shall erect any necessary temporary structures, signs, or other facilities. During such period of suspension of work, the Contractor shall properly and continuously maintain in an acceptable growing condition all living material in newly established plantings, seedings, and soddings furnished under the contract, and shall take adequate precautions to protect new tree growth and other important vegetative growth against injury. When work is suspended for the reasons delineated in the last sentence of subsection 104.04(b), all of the foregoing costs during the period of suspension shall be borne by the Contractor.

107.18 Contractor's Responsibility for Utility Property and Services. Prior to commencing work, the Contractor shall make arrangements to protect the properties of railway, telegraph, telephone, and power companies, or other property from damage that could result in considerable expense, loss, or inconvenience.

The Contractor shall cooperate with the utility owners in the removal and rearrangement of any underground or overhead utility lines or facilities to minimize interruption to service and duplication of work by the utility owners.

In the event utility services are interrupted as a result of accidental breakage, the Contractor shall promptly notify the proper authorities and cooperate with them until service has been restored. Work undertaken around fire hydrants shall not commence until provisions for continued service has been made and approved by the local fire authority.

107.19 Furnishing Right-of-Way. The Department will be responsible for the securing of all necessary rights-of-way in advance of construction. Any exceptions will be indicated in the contract.

107.20 Personal Liability of Public Officials. The Commissioner, Engineer, or their authorized representatives are acting solely as agents and representatives of the State when carrying out the provisions of the specifications or exercising the power or authority granted to them under the contract and there shall not be any liability on them either personally or as officials of the State.

107.21 No Waiver of Legal Rights. Upon completion of the work, the Department will expeditiously make final inspection and notify the Contractor of acceptance. Such final acceptance, however, shall not preclude or estop the Department from correcting any measurement, estimate, or certificate made before or after completion of the work, nor from recovering from the Contractor or surety or both, overpayments sustained by failure on the part of the Contractor to fulfill the obligations under the contract. A waiver on the part of the Department of any breach of any part of the contract shall not be held to be a waiver of any other or subsequent breach.

The Contractor, without prejudice to the terms of the contract, shall be liable to the Department for latent defects, fraud, or such gross mistakes as may amount to fraud, or as regards the Department's rights under any warranty or guaranty.

107.22 Environmental Protection. The Contractor shall comply with all Federal, State, and local laws and regulations controlling pollution of the environment. Necessary precautions shall be taken to prevent pollution of streams, lakes, ponds, and reservoirs with fuels, oils, bitumens, chemicals, or other harmful materials and to prevent pollution of the atmosphere from particulate and gaseous matter.

Fording of live streams shall not be permitted unless the Contractor's plan for such operation meets the approval of the Engineer and results in minimum siltation to the stream. Unless approved by the Engineer, mechanized equipment shall not be operated in live streams except as required to construct channel changes and temporary or permanent structures that are part of the contract.

When work areas or pits are located in or adjacent to live streams, such areas shall be separated from the main stream by a dike or barrier to keep sediment from entering a flowing stream. Care shall be taken during the construction and removal of such barriers to minimize siltation of the stream.

Water from aggregate washing or other operations containing sediment shall be treated by filtration, settling basins, or other means sufficient to reduce the sediment content to not more than that of the stream or lake into which it is discharged.

Other requirements relating to temporary and permanent erosion and water pollution controls are set forth in Section 208-Water Pollution Control of these specifications.

SECTION 108—PROSECUTION AND PROGRESS

108.01 Subletting of Contract. The Contractor shall not sublet, sell, transfer, assign, or otherwise dispose of the contract or contracts or any portion thereof, or of the right, title, or interest therein, without written consent of the Engineer. If such consent is given, the Contractor will be permitted to sublet a portion of the work, but shall perform with his own organization work amounting to not less than 30 percent of the total contract cost, unless a higher percentage is specified in the contract. Any items designated in the contract as "specialty items" may be performed by subcontract and the cost of any such specialty items performed by subcontract may be deducted from the total cost before computing the amount of work required to be performed by the Contractor's own organization. No subcontracts, or transfer of contract, shall relieve the Contractor of liability under the contract and bonds.

108.02 Notice to Proceed. The "Notice to Proceed" will stipulate the date on which it is expected the Contractor will begin the construction and from which date contract time will be charged. Commencement of work by the Contractor constitutes a waiver of this notice.

108.03 Prosecution and Progress. The Contractor, when required, shall furnish the Engineer with a "Progress Schedule" for approval. The progress schedule may be used to establish major construction operations and to check on the progress of the work. The Contractor shall provide sufficient materials, equipment, and labor to guarantee the completion of the project in accordance with the plans and specifications within the time set forth in the proposal.

If the Contractor falls significantly behind the submitted schedule, the Contractor shall submit a revised schedule for completion of the work within the contract time and modify the operations to provide such additional materials, equipment, and labor necessary to meet the revised schedule. Should the prosecution of the work be discontinued for any reason, the Contractor shall notify the Engineer at least 24 hours in advance of resuming operations.

108.04 Limitation of Operations. The Contractor shall conduct the work at all times in such a manner and in such sequence as will assure the least interference with traffic. He shall have due regard to the location of detours and to the provisions for handling traffic. The Engineer may require the Contractor to finish a section on which work is in progress before work is started on any additional sections if the opening of such section is essential to public convenience.

108.05 Character of Workmen. The Contractor shall at all times employ sufficient labor and equipment for prosecuting the several classes of work to full completion in the manner and time required by these specifications.

All workmen shall have sufficient skill and experience to perform properly the work assigned to them. Workmen engaged in special work or skilled work shall have sufficient experience in such work and in the operation of the equipment required to perform the work satisfactorily.

Any person employed by the Contractor or by any subcontractor who, in the opinion of the Engineer, does not perform his work in a proper and skillful manner or is intemperate or disorderly shall, at the written request of the Engineer, be removed forthwith by the Contractor or subcontractor employing such person, and shall not be again employed in any portion of the work without the approval of the Engineer.

Should the Contractor fail to remove such person or persons as required above, or fail to furnish suitable and sufficient personnel for the proper prosecution of the work, the Engineer may suspend the work by written notice until compliance with such orders.

108.06 Methods and Equipment. All equipment which is proposed to be used on the work shall be of sufficient size and in such mechanical condition as to meet requirements of the work and to produce a satisfactory quality of work. Equipment used on any portion of the project shall be such that no injury to the roadway, adjacent property, or other highways will result from its use.

When the methods and equipment to be used by the Contractor in accomplishing the construction are not prescribed in the contract, the Contractor is free to use any methods or equipment that will accomplish the contract work in conformity with the requirements of the contract.

When the contract specifies the use of certain methods and equipment, such methods and equipment shall be used unless others are authorized by the Engineer. If the Contractor desires to use a method or type of equipment other than specified in the contract, he may request authority from the Engineer to do so. The request shall be in writing and shall include a full description of the methods and equipment proposed and of the reasons for desiring to make the change. If approval is given, it will be on the condition that the Contractor will be fully responsible for producing work in conformity with contract requirements. If, after trial use of the substituted methods or equipment, the Engineer determines that the work produced does not meet contract requirements, the Contractor shall discontinue the use of the substitute method or equipment and shall complete the remaining work with the specified methods and equipment. The Contractor shall remove the deficient work and replace it with work of specified quality, or take such other corrective action as the Engineer may direct. No change will be made in basis of payment for the construction items involved nor in contract time as a result of authorizing a change in methods or equipment under these provisions.

108.07 Determination and Extension of Contract Time. The number of days allowed for completion of the work included in the contract will be stated in the proposal and contract, and will be known as the "Contract Time."

When the contract time is on a working day basis, the Engineer will furnish the Contractor a weekly statement showing the number of days charged to the contract for the preceding week and the number of days specified for completion of the contract. The Contractor will be allowed 1 week in which to file a written protest, setting forth in what respect said weekly statement is incorrect, otherwise the statement shall be deemed to have been accepted by the Contractor as correct.

When the contract time is on a calendar day basis, it shall consist of the number of calendar days stated in the contract, counting from the effective date of the Engineer's order to commence work, including all Sundays, holidays, and non-work days. All calendar days elapsing between the effective dates of any orders of the Engineer to suspend work and to resume work for suspensions not the fault of the Contractor shall be excluded.

When the contract completion time is a fixed calendar date, it shall be the date on which all work on the project shall be substantially completed.

The number of days for performance allowed in the contract as awarded is based on the original quantities as defined in subsection 102.04. If satisfactory fulfillment of the contract requires performance of work in greater quantities than those set forth in the proposal, the contract time allowed for performance shall be increased on a basis commensurate with the amount and difficulty of the added work.

If the Contractor finds it impossible for reasons beyond his control to complete the work within the contract time as specified or as extended in accordance with the provisions of this subsection, the Contractor may, at any time prior to the expiriation of the contract time as extended, make a written request to the Engineer for an extension of time, setting forth the reasons he believes will justify the granting of the request. The Contractor's plea that insufficient time was specified is not a valid reason for extension of time. If the Engineer finds that the work was delayed because of conditions beyond the control and without the fault of the Contractor, the time for completion may be extended in such amount as the conditions justify. The extended time for completion shall then be in full force and effect the same as though it were the original time for completion.

When final acceptance has been duly made by the Engineer as prescribed in subsection 105.16, the daily time charge will cease.

108.08 Failure to Complete on Time. For each calendar day or work day, as specified, that any work shall remain uncompleted after the contract time prescribed for the completion of the work, the sum specified in the contract will be deducted from any money due the Contractor, not as a penalty, but as liquidated damages. However, any adjustment of the contract time for completion of the work granted under the provisions of subsection 108.07 will be considered in the assessment of liquidated damages.

Permitting the Contractor to continue and finish the work or any part of it after the time fixed for its completion or after the date to which the time for completion may have been extended, will in no way operate as a waiver on the part of the Department of any of its rights under the contract.

The Department may waive such portions of the liquidated damages as may accrue after the work is in condition for safe and convenient use by the traveling public.

(Explanatory note—not a part of the specification.)

Time is an essential element of the contract and it is important that the work be pressed vigorously to completion. The cost to the Department of the administration of the contract, including engineering, inspection, and supervision, will be increased as the time occupied in the work is lengthened. Loss will accrue to the public due to delayed completion of the contemplated facility.

The following schedule of liquidated damages shall be considered as a guide.

The Department may, at its discretion, use a higher schedule of liquidated damages if extreme public need, such as restoration of facilities after disaster, or an accrual of extremely high traffic benefits indicate high desirability of early completion of the facilities covered in the contract. The Department may also, at its discretion, use a lower schedule of liquidated damages, but in any case, such lower schedule shall be sufficient to recover excess engineering charges.

Original Contract Amount		**Daily Charge**	
From More Than	**To and Including**	**Calendar Day or Fixed Date**	**Work Day**
$ 0	$ 25,000	$ 45	$ 63
25,000	50,000	75	105
50,000	100,000	110	154
100,000	500,000	150	210
500,000	1,000,000	225	315
1,000,000	2,000,000	300	420
2,000,000	5,000,000	450	630
5,000,000	10,000,000	600	840
10,000,000	—	700	980

Rates for liquidated damages will be set forth in the contract. When the contract time is on either the calendar day or fixed calendar date basis, the schedule for calendar days shall be used. When the contract time is on a work day basis, the schedule for work days will be used.

108.09 Default of Contract. If the Contractor:

(a) Fails to begin the work under the contract within the time specified in the Notice to Proceed, or

(b) Fails to perform the work with sufficient workmen and equipment or with sufficient materials to assure the prompt completion of said work, or
(c) Performs the work not in accordance with the contract requirements and/or refuses to remove and replace rejected materials or unacceptable work, or
(d) Discontinues the prosecution of the work, or
(e) Fails to resume work that has been discontinued within a reasonable time after notice to do so, or
(f) Becomes insolvent or is declared bankrupt, or commits any act of bankruptcy or insolvency, or
(g) Allows any final judgment to remain unsatisfied for a period of 10 days, or
(h) Makes an assignment for the benefit of creditors, or
(i) Fails to comply with contract requirements regarding minimum wage payments or EEO requirements, or
(j) For any other cause whatsoever, fails to carry on the work in an acceptable manner, the Engineer will give notice in writing to the Contractor and the surety of such delay, neglect, or default.

If the Contractor or surety, within a period of 10 days after such notice, does not proceed in accordance therewith, then the Department will, upon written notification from the Engineer of the fact of such delay, neglect, or default, and the Contractor's failure to comply with such notice, have full power and authority without violating the contract, to take the prosecution of the work out of the hands of the Contractor. The Department may appropriate or use any or all materials and equipment on the ground as may be suitable and acceptable and may enter into an agreement for the completion of said contract according to the terms and provisions thereof, or use such other methods as in the opinion of the Engineer will be required for the completion of said contract in an acceptable manner.

All costs and charges incurred by the Department, together with the cost of completing the work under contract, will be deducted from any monies due or which may become due said Contractor. If such expense exceeds the sum which would have been payable under the contract, then the Contractor and the surety shall be liable and shall pay to the Department the amount of such excess.

108.10 Termination of Contract. The Department may, by written order, terminate the contract or any portion thereof after determining that for reasons beyond the control of either the Contractor or the Department, the

Contractor is prevented from proceeding with or completing the work and that termination would be in the public interest. Such reasons for termination may include, but need not be necessarily limited to, executive orders of the president relating to prosecution of war or national defense, national emergency which creates a serious shortage of materials, orders from duly constituted authorities relating to energy conservation, and restraining orders or injunctions obtained by third-party citizen action resulting from national or local environmental protection laws or where the issuance of such order or injunction is primarily caused by acts or omissions of persons or agencies other than the Contractor.

When the Department orders termination of a contract effective on a certain date, all completed items of work as of that date will be paid for at the contract bid price. Payment for partially completed work will be made either at agreed prices or by force account methods described elsewhere in these specifications. Items that are eliminated in their entirety by such termination shall be paid for as provided in subsection 109.05 of these specifications.

Acceptable materials obtained by the Contractor for the work but which have not been incorporated therein, may, at the option of the Department be purchased from the Contractor at actual cost delivered to a prescribed location, or otherwise disposed of as mutually agreed.

After receipt of Notice of Termination from the Department, the Contractor shall submit, within 60 days of the effective termination date, a claim for additional damages or costs not covered above or elsewhere in these specifications. Such claim may include such cost items as reasonable idle equipment time, mobilization efforts, bidding and project investigative costs, overhead expenses attributable to the project terminated, legal and accounting charges involved in claim preparation, subcontractor costs not otherwise paid for, actual idle labor cost if work is stopped in advance of termination date, guaranteed payments for private land usage as part of original contract, and any other cost or damage item for which the Contractor feels reimbursement should be made. The intent of negotiating this claim would be that an equitable settlement figure be reached with the Contractor. In no event, however, will loss of anticipated profits be considered as part of any settlement.

The Contractor agrees to make cost records available to the extent necessary to determine the validity and amount of each item claimed.

Termination of a contract or portion thereof shall not relieve the Contractor of contractual responsibilities for the work completed, nor shall it relieve the surety of its obligation for and concerning any just claim arising out of the work performed.

SECTION 109—MEASUREMENT AND PAYMENT

109.01 Measurement of Quantities. All work completed under the contract will be measured by the Engineer according to U.S. standard measure or by the metric system when the contract so provides.

A station when used as a definition or term of measurement will be 100 linear feet.

The method of measurement and computations to be used in determination of quantities of material furnished and of work performed under the contract will be those methods generally recognized as conforming to good engineering practice.

Unless otherwise specified, longitudinal measurements for area computations will be made horizontally, and no deductions will be made for individual fixtures having an area of 9 square feet or less. Unless otherwise specified, transverse measurements for area computations will be the neat dimensions shown on the plans or ordered in writing by the Engineer.

Structures will be measured according to neat lines shown on the plans or as altered to fit field conditions.

All items which are measured by the linear foot, such as pipe culverts, guardrails, underdrains, etc., will be measured parallel to the base or foundation upon which such structures are placed, unless otherwise shown on the plans.

In computing volumes of excavation, the average end area method or other acceptable methods will be used.

The thickness of plates and galvanized sheet used in the manufacture of corrugated metal pile, metal plate pipe culverts and arches, and metal cribbing will be specified and measured in decimal fractions of inches.

The term "ton" will mean the short ton consisting of 2,000 pounds avoirdupois. All materials that are measured or proportioned by weight shall be weighed on accurate, approved scales by competent, qualified personnel at locations designated by the Engineer. If material is shipped by rail, the car weight may be accepted, provided that only the actual weight of material is paid for. However, car weights will not be acceptable for material to be passed through mixing plants. Trucks used to haul material being paid for by weight shall be weighed empty daily at such times as the Engineer directs, and each truck shall bear a plainly legible identification mark.

Materials to be measured by volume shall be hauled in approved vehicles and measured therein at the point of delivery. Vehicles for this purpose may be of any size or type acceptable to the Engineer, provided that the body is of such shape that the actual contents may be readily and accurately determined. All vehicles shall be loaded to at least their water level capacity and all loads shall be leveled when the vehicles arrive at the point of delivery.

When requested by the Contractor and approved by the Engineer in writing, material specified to be measured by the cubic yard may be weighed and converted to cubic yards for payment purposes. Factors for conversion from weight measurement to volume measurement will be determined by the Engineer and agreed to by the Contractor before such method of measurement of pay quantities is used.

Bituminous materials will be measured by the gallon or ton.

Volumes will be measured at 60°F or will be corrected to the volume at 6°F using ASTM D 1250 for asphalts or ASTM D 633 for tars.

Net certified scale weights or weights based on certified volumes in the case of rail shipments will be used as a basis of measurement, subject to correction when bituminous material has been lost from the car or the distributor, wasted, or otherwise not incorporated in the work.

When bituminous materials are shipped by truck or transport, net certified weights or volume subject to correction for loss or foaming, may be used for computing quantities.

Cement will be measured by the ton or hundredweight.

Timber will be measured by the thousand feet board measure (M.F.B.M.) actually incorporated in the structure. Measurement will be based on nominal widths and thicknesses and the extreme length of each piece.

The term "lump sum" when used as an item of payment will mean complete payment for the work described in the contract.

When a complete structure or structural unit (in effect, "lump sum" work) is specified as the unit of measurement, the unit will be construed to include all necessary fittings and accessories.

Rental of equipment will be measured in hours of actual working time and necessary traveling time of the equipment within the limits of the project. If special equipment has been ordered by the Engineer in connection with force account work, travel time and transportation to the project will be measured. If equipment has been ordered held on the job on a standby basis by the Engineer, half-time rates for the equipment will be paid.

When standard manufactured items are specified such as fence, wire, plates, rolled shapes, pipe conduit, etc., and these items are identified by gage, unit weight, section dimensions, etc., such identification will be considered to be nominal weights or dimensions. Unless more stringently controlled by tolerances in cited specifications, manufacturing tolerances established by the industries involved will be accepted.

Scales for the weighing of highway and bridge construction materials that are required to be proportioned or measured and paid for by weight, shall be furnished, erected, and maintained by the Contractor, or be certified permanently installed commercial scales.

Scales shall be accurate within 1/2 percent of the correct weight throughout the range of use. The Contractor shall have the scales checked under the observation of the inspector before beginning work and at such other times as requested. The intervals shall be uniform in spacing throughout the graduated or marked length of the beam or dial and shall not exceed 1/10 of 1 percent of the nominal rated capacity of the scale; but not less than 1 pound. The use of spring balances will not be permitted.

Beams, dials, platforms, and other scale equipment shall be so arranged that the operator and inspector can safely and conveniently view them.

Scale installations shall have available ten standard 50-pound weights for testing the weighing equipment or suitable weights and devices for other approved equipment.

Scales must be tested for accuracy and serviced before use at a new site. Platform scales shall be installed and maintained with the platform level and rigid bulkheads at each end.

Scales Overweighing — (indicating more than true weight) will not be permitted to operate and all materials received subsequent to the last previous correct weighing acccuracy test will be reduced by the percentage of error in excess of 1/2 of 1 percent.

In the event inspection reveals the scales have been underweighing, they shall be adjusted and no additional payment to the Contractor will be allowed for materials previously weighed and recorded.

All costs in connection with furnishing, installing, certifying or testing, and maintaining scales; for furnishing check weights, and scale house, and for all other items specified in this section for the weighing of highway and bridge construction materials for proportioning or payment shall be included in the unit contract prices for the various pay items of the project.

When the estimated quantities for a specific portion of the work are designated as pay quantities in the contract, they shall be the final quantities for which payment for such specific portion of the work will be made, unless the dimensions of said portions of the work shown on the plans are revised by the Engineer. If revised dimensions result in an increase or decrease in the quantities of such work, the final quantities for payment will be revised in the amount represented by the authorized changes in the dimensions.

109.02 Scope of Payment. The Contractor shall receive and accept compensation provided for in the contract as full payment for furnishing all materials and for performing all work under the contract in a complete and acceptable manner and for all risk, loss, damage, or expense of whatever character arising out of the nature of the work or the prosecution thereof, subject to the provisions of subsection 107.21.

If the "Basis of Payment" clause in the specifications relating to any unit price in the bid schedule requires that the said unit price cover and be considered compensation for certain work or material essential to the item, this same work or material will not also be measured or paid for under any other pay item that may appear elsewhere in the specifications.

109.03 Compensation for Altered Quantities. When the accepted quantities of work vary from the quantities in the bid schedule, the Contractor shall accept payment at the original contract unit prices for the accepted quantities of work done. No allowance except as provided in subsections 104.02 and 108.10 will be made for any increased expenses, loss of expected reimbursement, or loss of anticipated profits suffered or claimed by the Contractor resulting either directly from such alterations or indirectly from unbalanced allocation among the contract items of overhead expense and subsequent loss of expected reimbursements or from any other cause.

109.04 Extra and Force Account Work. Extra work performed in accordance with the requirements and provisions of subsection 104.03 will be paid for at the unit prices or agreed prices stipulated in the order authorizing the work, or the Department may require the Contractor to do such work on a force account basis to be compensated in the following manner:

(a) *Labor* — For all labor and foremen in direct charge of the specific operations, the Contractor shall receive the rate of wage (or scale) agreed upon in writing before beginning work for each and every hour that said labor and foremen are actually engaged in such work.

The Contractor shall receive the actual costs paid to, or in behalf of, workmen by reason of subsistence and travel allowances, health and welfare benefits, pension fund benefits, or other benefits, when such amounts are required by collective bargaining agreement or other employment contract generally applicable to the classes of labor employed on the work.

An amount equal to ..... percent (35 percent suggested) of the sum of the above items will also be paid the Contractor.

(b) *Bond, Insurance, and Tax* — For property damage, liability, and workmen's compensation insurance premiums, unemployment insurance contributions, and social security taxes on the force account work, the Contractor shall receive the actual cost, to which cost ..... percent (10 percent suggested) will be added. The Contractor shall furnish satisfactory evidence of the rate or rates paid for such bond, insurance, and tax.

(c) *Materials* — For materials accepted by the Engineer and used, the Contractor shall receive the actual cost of such materials delivered on the work, including transportation charges paid (exclusive of machinery rentals as hereinafter set forth), to which cost ..... percent (15 percent suggested) will be added.

(d) *Equipment* — For any machinery or special equipment (other than small tools), including fuel and lubricants, plus transportation costs, the use of which has been authorized by the Engineer, the Contractor shall receive the rental rates agreed upon in writing before such work is begun for the actual time that such equipment is in operation on the work.

(e) *Miscellaneous* — No additional allowance will be made for general superintendence, the use of small tools, or other costs for which no specific allowance is herein provided.

(f) *Subcontracting* — For administration costs in connection with approved subcontract work, the Contractor shall receive an amount equal to ..... percent (5 percent suggested) of the total cost of such work computed as set forth above.

(g) *Compensation* — The Contractor's representative and the Engineer shall compare records of the cost of work done as ordered on a force account basis.

(h) *Statements* — No payment will be made for work performed on a force account basis until the Contractor has furnished the Engineer with duplicate itemized statements of the cost of such force account work detailed as follows:

1. Name, classification, date, daily hours, total hours, rate, and extension for each laborer and foreman.
2. Designation, dates, daily hours, total hours, rental rate, and extension for each unit of machinery and equipment.
3. Quantities of materials, prices, and extensions.
4. Transportation of materials.
5. Cost of property damage, liability, and workmen's compensation insurance premiums, unemployment insurance contributions, and social security tax.

Statements shall be accompanied and supported by receipted invoice for all materials used and transportation charges. However, if materials used on the force account work are not specifically purchased for such work but are taken from the Contractor's stock, then in lieu of the invoices, the Contractor shall furnish an affidavit certifying that such materials were taken from stock, that the quantity claimed was actually used, and that the price and transportation claimed represent the actual cost to the Contractor.

The additional payment, based on the percentage stated above, shall constitute full compensation for all items of expense not specifically designated. The total payment made as provided above shall constitute full compensation for such work.

109.05 Eliminated Items. Should any items contained in the proposal be found unnecessary for the proper completion of the work, the Engineer may, upon written order to the Contractor, eliminate such items from the contract, and such action shall in no way invalidate the contract. When a Contractor is notified of the elimination of items, the Contractor will be reimbursed for actual work done and all costs incurred, including mobilization of materials prior to said notification.

109.06 Progress Payments. Progress payments will be made at least once each month as the work progresses. More frequent payments may be made during any period when, in the judgment of the Department, the value of work performed during such period is of sufficient amount to warrant same. Said payments will be based on estimates prepared by the Engineer of the value of the work performed and materials complete in place in accordance with the contract and for materials delivered in accordance with subsection 109.07.

No progress payment will be made when the total value of the work done since the last estimate amounts to less than $..... ($500.00 suggested).

From the total of the amounts ascertained as payable an amount equivalent to ..... percent (10 percent suggested) of the whole will be deducted and retained by the Department until after completion of the entire contract in an acceptable manner. The balance, or an amount equivalent to ..... percent (90 percent suggested) of the whole, less all previous payments, shall be certified for payment. It is provided, however, that after ..... percent (50 percent suggested) of the work has been completed, the Department may make any of the remaining progress payments in full.

When not less than 95 percent of the work has been completed, the Engineer may, at his discretion and with the consent of the surety, prepare an estimate from which will be retained an amount not less than twice the contract value or estimated cost, whichever is greater, of the work remaining to be done, and the remainder less all previous payments and deductions will then be certified for payment to the Contractor. After acceptance of the contract, the amount, if any, payable for a contract item of work in excess of any maximum value for progress payment purposes set forth in the contract for said item, will be paid.

109.07 Payment for Material on Hand. Partial payments may be made for materials to be incorporated in the work, provided the materials meet the requirements of the plans and specifications and are delivered on, or in the vicinity of, the project site or stored in acceptable storage places. Partial payments shall not exceed .... percent (85 percent suggested) of the contract bid price for the item or the amount supported by copies of invoices, freight bills, or other supporting documents required by the Engineer, including a duly executed Certification of Title in lieu of paid invoices. The quantity paid shall not exceed the corresponding quantity estimated in the contract.

No partial payment will be made on living or perishable plant materials until planted as specified in the contract.

Approval of partial payment for stockpiled materials will not constitute final acceptance of such materials for use in completing items of work.

109.08 Payment of Withheld Funds. Attention is directed to subsection 109.06—Progress Payments, and in particular to the retention provisions of said subsection.

Upon the Contractor's request, the Department will make payment of funds withheld from progress payments if the Contractor deposits in escrow securities eligible for the investment of State funds or bank certificates of deposit, upon the following conditions:

(a) The Contractor shall bear the expenses of the Department and the State Treasurer in connection with the escrow deposit made.

(b) Securities or certificates of deposit to be placed in escrow shall be subject to approval of the Department and unless otherwise permitted by the escrow agreement, shall be of a value of at least 100 percent of the amounts of retention to be paid to the Contractor pursuant to this section.

(c) The Contractor shall enter into an escrow agreement satisfactory to the Department.

(d) The Contractor shall obtain the written consent of the surety to such agreement.

109.09 Acceptance and Final Payment. When the project has been accepted as provided in subsection 105.16, the Engineer will prepare the final estimate of work performed. If the Contractor approves the final estimate or files no claim or objection to the quantities therein within 30 days of receiving the final estimate, the State will process the estimate for final payment. With approval of such final estimate by the Contractor, payment will be made for the entire sum found to be due after deducting all previous payments and all amounts to be retained or deducted under the provisions of the contract.

If the Contractor files a claim in accordance with contract requirements, it shall be submitted in writing in sufficient detail to enable the Engineer to ascertain the basis and amount of such claim. In such cases the final sum determined by the Engineer to be due will be paid pending study of the claim. Upon final adjudication of the claim, any additional payment determined to be due the Contractor will be placed on a supplemental estimate and processed for payment.

All prior partial estimates and payments shall be subject to correction in the final estimate and payment.

Chapter 5

The Bidding Documents

5.1 Introduction

After the Plans, Technical Sections, and General Conditions are substantially complete, attention turns to the Bidding Documents. If construction is ready to go forward, qualified responsible bidders must be attracted. This can be accomplished by publicizing the project. For contractors interested in submitting a proposal, the Bidding Documents present information and instructions governing the preparation and submission of proposals, and awarding of the Contract. Bidding Documents serve to present a uniform procedure for all bidders to follow. Since the bids will be prepared on a common basis, it will enable the Owner to better evaluate proposals before awarding the Contract.

As similarly mentioned previously in Article 4.1 Introduction of the General Conditions, the arrangement and titles of Subsections and majority of Articles presented in this Chapter follow the pattern established by the AASHTO Guide Specifications. Sections 102 and 103 of the AASHTO General Provisions deal with the Bidding Documents. Articles of these Sections are referred to and discussed herein. Sections 102 and 103 have been reproduced and inserted at the end of this Chapter for direct reference. The Articles not directly related to the AASHTO Guide Specifications merit consideration because of their frequent occurrence in construction contract documents.

Further discussion and guidelines to assist the specification writer in presenting Articles of the Bidding Documents can be found in Chapter 12, Presenting the Bidding Documents.

5.2 Bidding Requirements and Conditions

A. Notice to Contractors.

This Notice is also titled Advertisement for Bids or Invitation for Bids. An Invitation for Bids would apply to privately funded projects, in which bids would be solicited from selected contractors. The purpose of the Notice is to present factual information to help a potential bidder make the correct decision on whether to bid it or not. When an unusual or large project is being designed, it is most likely that news concerning it will be spread by word of mouth, and interested contractors may know most of the story in advance of a formal advertisement. As for other projects, few contractors would learn about the job were it not publicized. Projects funded by public monies are generally required by law to be made available for public competition, to ensure that the funds will be spent fairly. In private work, the Owner does not have to publicly advertise for bids. He is free to negotiate directly with whomever he chooses.

For publicly funded projects, the Notice to Contractors is normally inserted in local newspapers, engineering and construction publications, and other media available to contractors. It may be repeated several days in succession or repeated weekly for a month. This Notice will generally include information on:

1. Project and Contract identification.
2. Name and address of Owner.
3. Description of the Work and its location.
4. Major items of work and estimated quantities.
5. Availability of the right-of-way and site utilities.
6. Estimated value of the Work.
7. Location for examination and procurement of contract documents. A specified fee is normally charged for each set of documents. In most cases the fee is refunded upon return of the documents in good condition.
8. Date, time, and place for receipt of bids.
9. Name and address of person authorized to receive bids.
10. Proposal guaranty (see Article 5.2J, Proposal Guaranty).
11. Contract bonds (see Article 5.3E, Requirements of Contract Bonds).
12. Prequalification (see Article 5.2B, Prequalification of Bidders).
13. Limitation on the amount of Work that may be subcontracted.
14. Notice of pre-bid conference (see Article 5.2G, Pre-Bid Conference).
15. Owner's right to reject bids.
16. Governing laws and regulations, such as Affirmative Action Requirements and Equal Employment Opportunity.

After digesting the information in the Notice, a contractor is in a better position to decide whether his organization, equipment and experience are adequate to handle the Work, and whether he wants the job.

Illustration of a Notice to Contractors is presented in Appendix A, as Exhibit A. It has been reproduced from Contract Specifications, February 3, 1978, for construction of the Lexington Market Station Structure, part of the Baltimore Region Rapid Transit System, State of Maryland, Department of Transportation.

B. Prequalification of Bidders.

Refer to Article 102.01, same title, of the AASHTO Sample at the end of this Chapter. Owners can be exposed to financial risk when they award a construction contract. Getting the "wrong" contractor on the job can bring problems and increased costs.

Contracts for public works projects are awarded to the lowest responsible bidder. In the open competition for public works contracts, a contractor who furnishes the required surety; who has a record free from defaults; and who has not been proven dishonest; can establish himself as a responsible bidder. Thus, a contractor inexperienced in the proposed work, or who may have insufficient plant and equipment to do the work expeditiously, or who may be overextended beyond his capacity, can still be in competition with responsible, qualified contractors. The employment of such a contractor can result in slow progress and an unsatisfactory quality of work. These problems would generally not be encountered in private construction, because competent bidders can be selected without restriction. To eliminate or minimize potential problems associated with the awarding of contracts for public works, most owners will require bidders to prequalify themselves. In addition to the required information specified in referenced AASHTO Article 102.01, the Owner may want to know a contractor's current work load. This procedure will enable the Owner to determine, before issuing the Bidding Documents, if a contractor is competent to satisfactorily perform the proposed Work. Some owners may require bidders to submit this information along with their proposal.

Many Public Agencies will maintain current lists of contractors qualified to bid on their projects. On these lists, contractors are classified as to the types of projects they are prequalified for, and the maximum size of contract they are capable of handling. The contractors are required to update their prequalification record at least once each year.

Illustration of a qualification form (Contractor's Questionnaire Pre-Award Evaluation Data) to be completed and submitted along with the proposal, is presented in Appendix A, as Exhibit B. It has been reproduced from the same Contract Specifications for construction of the Lexington Market Station Structure that was referred to in Article 5.2A, Notice to Contractors.

C. Contents of Proposal Forms.

Refer to Article 102.02, same title, of the AASHTO Sample at the end of this Chapter. Included among the proposal forms is the Bid Schedule or Unit Price Schedule, which lists the payment items covering the Work. On this form, the bidder indicates his bid prices for doing the Work. Discussion and illustration of a Bid Schedule is presented in Article 5.2F, Interpretation of Quantities in Bid Schedule.

Some of the additional forms that may be included with the proposal forms are:

1. Contractor's Questionnaire Pre-Award Evaluation Data Form (see Article 5.2B, Prequalification of Bidders).

2. Bid Bond Form (see Article 5.2J, Proposal Guaranty).
3. Free Competitive Bidding Affidavit Form. On this form, the bidder certifies that he has not entered into any agreement, participated in any collusion, or taken any action in restraint of free competitive bidding in connection with the Contract.
4. Schedule for Participation By Minority Contractors Form. On this form the bidder lists the minority contractors who will be involved in the Contract, if awarded. The information includes name, address, type of work to be performed, and agreed-upon price, for each listed minority contractor.

D. Issuance of Proposal Form.

Refer to Article 102.03, same title, of the AASHTO Sample at the end of this Chapter. Various reasons for refusing to issue Bidding Documents to a contractor may be presented by an owner, as illustrated in the referenced AASHTO Article.

E. Examination of Plans, Specifications, Special Provisions, and Site of Work.

Refer to Article 102.05, same title, of the AASHTO Sample at the end of this Chapter. After having digested the information in the Notice to Contractors, a contractor interested in bidding the job will proceed to purchase a set of the Contract Documents. The referenced AASHTO Article advises the bidder to examine them carefully before preparing and submitting his proposal. He is also advised to visit the site of the Work and familiarize himself with existing conditions. By so doing, the bidder acquaints himself with the local weather, transportation conditions, and soil conditions. For those projects located in remote areas or those having unusual site conditions, the Owner will usually schedule a guided tour of the site for interested bidders. Attention is directed to the provision in referenced AASHTO Article 102.05 which states that when a bidder submits his bid he acknowledges that he has inspected the site and has satisfied himself with the site conditions and the requirements of the Contract. He therefor cannot claim ignorance of visible site conditions that existed when he submitted his bid, nor of any misunderstanding of the requirements of the Contract Documents. It is, of course, in the Owner's interest to have bidders fully inform themselves of all aspects of the proposed Work. An informed bidder submits a more realistic bid.

1. Addenda. In the process of examining the Documents, a bidder may encounter an ambiguity, inconsistency, discrepancy, omission, or error. If the bidder desires an interpretation or correction, he is generally required to submit a written request to the Owner. If the request is valid, an interpretation or

correction will be issued in the form of an addendum which is mailed or delivered to all parties who purchased a set of the Documents.

A word about the subject of addenda. It is not uncommon for changes to be made to Contract Documents after they have been issued to bidders. These changes may result from changes in the design; from the discovery of errors or discrepancies in the Contract Documents; from responding to contractors' inquiries submitted during the bidding period; from responding to questions raised at the pre-bid conference (see Article 5.2G, Pre-Bid Conference); or if the Owner is postponing the bid opening date. Modifications to the Contract Documents to incorporate these changes are issued as an Addendum to all holders of the Contract Documents. The addendum is distributed early enough before the bid opening date to give the bidders sufficient time to study the changes and modify their bids if necessary. Cutoff time for issuing addenda to bidders is generally seven to ten days before bid opening. Once issued, an addendum becomes a part of the Contract Documents. On larger contracts which have a bidding period of six to twelve weeks, as many as three or four addenda may be issued. Under the procedure of issuing addenda, all bidders are furnished identical information. A standard instruction concerning addenda would read:

> The Owner may modify the Contract Documents not later than _____days prior to the date fixed for receipt of bids, by issuing an addendum to all purchasers of the Contract Documents. Bidders must acknowledge receipt of all addenda on the Bid Form.

An addendum performs the same function as a change order (see Article 4.3B, Changes). The timing of the change establishes how it will be processed. If the change is made during the bidding period, it is handled by an addendum. If the change is made after the Contract has been executed, it is handled by a change order.

2. Subsurface Data. Before designing work that is to be performed below the ground surface, a preliminary subsurface investigation is usually made by the Owner or for him. One standard procedure is taking boring samples of the soil at various depths below the surface. Materials in the samples are then classified and described. This information is then tabulated in the form of boring logs drawn to a vertical scale, describing the different materials and indicating their depths below the ground surface. Some contracts include Boring Logs in the Bidding Documents by exhibiting them on the Plans. Otherwise they are made available to bidders for examination along with the boring samples. Most bidders desire to visually examine the material in the boring samples.

F. Interpretation of Quantities in Bid Schedule.

Refer to Article 102.04, same title, of the AASHTO Sample at the end of this Chapter. Each element of work is either designated as a payment item, or it is included for payment in another item. The Bid Schedule on which bidders indicate their prices for performing the Work lists all of the unit price and lump sum payment items in the Contract.

The Unit Price Schedule shown in Figure 5.1 presents a simple, abbreviated example illustrating the basic parts of a bid schedule. The items of work presented in this particular example relate to roadway construction.

In the first column of the Unit Price Schedule, a number is assigned to each payment item. Identifying each item with an item number makes it easier to refer to and locate in a Bid Schedule, particularly when the Schedule may contain as many as 50 or more pay items.

The second column identifies the Technical Section in which the reader will find measurement and payment clauses for the specific item, as well as material and workmanship requirements.

The third column lists the descriptive title of each item. This title will immediately give the reader a fairly good idea of the type of work it involves.

The fourth column lists the estimated quantity of units for each pay item. These quantities have been determined by the designer and are based on his knowledge of the proposed work. Since these quantities are estimated, the figures are rounded off to facilitate the preparation of a bid. The quantities also serve to give the bidder a general idea of the magnitude of the major items of work. The actual final quantities may be greater or less than the estimated quantities. Excessive variations from the estimated quantities, and how they are handled, are explained in Article 4.3D, Variations in Estimated Quantities. Having all bidders base their bid prices on the same quantities of work will enable the Owner to make a fair comparison of the bids. Note that the estimated quantity for a lump sum payment item is indicated as one.

The fifth column lists the unit of measurement for each individual item. The unit of measurement is compatible with the work of the item. Excavation for example, would be measured by the cubic yard and not by the square yard.

The sixth column tells the story. It is here that the bidder indicates his prices for doing the Work. Again, note that a unit price is not applicable for a lump sum payment item.

The seventh and last column is a reflection of the Unit Price and Estimated Quantity columns. It presents the total bid price for each payment item and is the product of the unit price multiplied by the estimated quantity for that item. No multiplication is required for a lump sum item. The sum of all the figures in this column represents the bidder's total bid price for the Contract. This figure becomes the basis for comparison of all the bids submitted.

The author wishes to clarify a common misconception in identifying the quantities shown in a Bid Schedule as being "approximate." The word "approximate" is defined in the dictionary as being nearly exact. On the other hand, "estimated" quantities as presented in the Unit Price Schedule, are just what their name implies; an evaluation or judgement made from incomplete data. In Article 12.2F,

CONTRACT TITLE AND CONTRACT NO.
UNIT PRICE SCHEDULE

Item No.	Sect. No.	Description	Est. Qty.	Unit	Unit Price	Total
1	701	Mobilization	1	Lump Sum		$_____
2	201	Clearing and Grubbing	20	Acre	$_____	$_____
3	202	Roadway Excavation	24,500	Cubic Yard	$_____	$_____
4	702	15″ RC Pipe—Class IV	850	Linear Foot	$_____	$_____
5	703	Catch Basin —Type A	8		$_____	$_____
6	601	Reinforcing Steel	2,500	Pound	$_____	$_____
7	602	Concrete — Class 3000	150	Cubic Yard	$_____	$_____
8	501	Cement Concrete Pavement	22,400	Square Yard	$_____	$_____
9	801	Fertilizing and Seeding	15,800	Square Yard	$_____	$_____
			TOTAL BID PRICE		$__________	

Figure 5.1. Unit Price Schedule Form.

Interpretation of Quantities in Bid Schedule; Paragraph 3. "Approximate" and "Estimated" Quantities presents a more detailed explanation of this difference.

A Unit Price Schedule may contain sophisticated elements such as fixed prices, fixed quantities, alternatives, and options. These are discussed in Article 12.2F, Interpretation of Quantities in Bid Schedule.

G. Pre-Bid Conference.

It is essential that all bidders have the same understanding of the Contract for which they are preparing proposals. Toward this end, an owner will arrange a pre-bid conference for bidders to attend, at which bidders will have the opportunity to ask questions for interpretation or clarification of the Contract Documents. This conference also provides the Owner an opportunity to present the objectives for his proposed project and to explain the various forms included in the Bidding Documents, such as affirmative action requirements, equal employment opportunity, and insurance requirements. In addition, it enables the Owner to point out the unusual or particularly critical items in the Contract.

The pre-bid conference is generally held two to three weeks before the bid opening date. At this conference, representatives of the Designer, including the specification writer, are present. Answers to relevant questions presented at this conference are issued in writing in the form of an Addendum to all purchasers of the Contract Documents.

H. Preparation of Proposal.

Refer to Article 102.06, same title, of the AASHTO Sample at the end of this Chapter. Instructions for preparing a Proposal are quite important to bidders. The material presented in the referenced AASHTO Article is self-explanatory and covers the subject adequately, with one possible exception; bidders should be reminded to acknowledge in the space provided for in the Proposal Form, the receipt of Addenda issued by the Owner. A deviation from these instructions can result in the Proposal being judged irregular and not eligible for consideration. Examples of irregular proposals are presented in following Article 5.2I.

I. Irregular Proposals.

Refer to Article 102.07, same title, of the AASHTO Sample at the end of this Chapter. Concerning unit bid prices, when a bid is a balanced bid, each bid item is priced to reflect its share of the cost of the Work and its share of the Contractor's profit. In an unbalanced bid, a contractor will increase the prices on certain items and make corresponding price reductions in other items, so that the total amount of the bid remains unchanged.

Following are two reasons illustrating why a contractor would attempt to unbalance his bid:

1. The Contractor can build up his working capital early in the Contract by increasing the bid prices of early work items such as clearing and grubbing, building demolition, and roadway excavation. Corresponding reductions would be made in the prices for work done late in the Contract, such as paving and landscaping.

2. The Contractor believes the Engineer's estimated quantities for certain items are too low. By unbalancing his bid in favor of these items, he can realize an increased profit in payment for the larger actual quantities of these items.

In the rush to meet a submittal deadline, the possibility of omissions or irregularities occurring in a proposal will increase and may, unfortunately, result in the bid being rejected. In addition to the reasons presented in referenced AASHTO Article 102.07, a bid may also be disqualified for:

1. Failure to sign the Proposal.
2. Failure to fill in all the blank spaces in the bid schedule.
3. The Proposal not being accompanied by the required bid security.
4. Failure to acknowledge receipt of addenda.
5. The Proposal being submitted by a bidder who did not prequalify.
6. The Proposal mailed, but not received until after the time of bid opening.
7. The required list of available plant and equipment not being furnished.
8. Failure to provide designation of subcontractors, when required.

J. Proposal Guaranty.

Refer to Article 102.08, same title, of the AASHTO Sample at the end of this Chapter. This guaranty is also referred to as Bid Security or Bid Bond. The Proposal Guaranty is to indemnify the Owner against losses caused by failure of the low bidder to execute the Contract (see Article 5.3G, Failure to Execute Contract). It provides protection against an irresponsible bidder who submits a low bid and then fails to follow through. The guaranty is generally furnished in the form of a bid bond. Other forms may be a cashier's check or a certified check. The required amount of the guaranty may vary from five to twenty percent of the total amount bid, depending on the size of the Contract.

Illustration of a Proposal Guaranty form (Bidder's Bond) is presented in Appendix A, as Exhibit C. It has been reproduced from the Contract Specifications Book dated April 1983 for construction of the Mt. Lebanon Tunnel of the Stage 1 Light Rail Transit System for the Port Authority of Allegheny County, Pennsylvania.

K. Delivery of Proposals.

Refer to Article 102.09, same title, of the AASHTO Sample at the end of this Chapter. The reason for sealed bids to be submitted in special envelopes with contents clearly indicated is to prevent them from being opened by mistake before the official date and time of the bid opening. It is important to the bidder that he get his bid into the proper hands and on time. A bid received late will be disregarded in most cases, regardless of the circumstances.

Included in information to be the furnished on the outside of the sealed envelope, would be the Contract name and number, date and time of bid opening, and name of bidder.

L. Withdrawal or Revision of Proposals.

Refer to Article 102.10, same title, of the AASHTO Sample at the end of this Chapter. After having submitted his bid, the bidder may discover an error or he may have second thoughts about some of the prices he proposed. Then again, because of some unexpected development, a bidder may be unable to take on additional work. This Article permits the bidder to withdraw or revise his proposal. Withdrawal of a bid generally does not prevent the bidder from submitting a new bid. Of course, no action can be taken if the request is received after the time specified.

M. Combination or Conditional Proposals.

Refer to Article 102.11, same title, of the AASHTO Sample at the end of this Chapter. Combination proposals are applicable to multicontract projects; particularly where the work of individual contracts is similar in nature. An example would be construction of two adjacent sections of highway. Bidders would be given the option of bidding on each contract separately, or on a combination of the two contracts, or on all three options. There could be three separate proposals in the Bidding Documents: a proposal for Contract A; another for Contract B; and a third for combined Contracts A and B. The Bid Schedule in a proposal for combined Contracts A and B would list the payment items of both contracts. The estimated quantity for a payment item common to both contracts would be equal to the sum of the quantity in each contract.

A conditional proposal can come about when an owner is not in a position to guarantee award of a contract within the normal time of 30 to 60 days. One reason could be uncertainty on the availability of funds. Bidders would be requested to make their proposals conditional on award being made within the number of days after bid opening, specified in their proposal.

N. Public Opening of Proposals.

Refer to Article 102.12, same title, of the AASHTO Sample at the end of this Chapter. Publicly funded projects require that all bids received before the time specified for opening bids, be opened and read aloud at a public bid opening. This enables bidders and other interested parties to witness the proceedings. Being present at the bid opening enables them to tabulate the figures if they so desire. Those proposals received after the time set for the public opening are returned unopened to their senders.

Arrangements governing the opening of proposals for privately funded projects are established by the individual owner. Some owners may prefer to open the bids in private.

O. Disqualification of Bidders.

Refer to Article 102.13, same title, of the AASHTO Sample at the end of this Chapter. In addition to the reasons presented in the referenced AASHTO Article, the following may also be considered by some owners as sufficient reason for the disqualification of a bidder: 1) Lack of competency, adequate plant, and other equipment, as revealed by the qualification statement required of bidders; or 2) If the bidder has defaulted under previous contracts with the Owner.

P. Material Guaranty.

Refer to Article 102.14, same title, of the AASHTO Sample at the end of this Chapter. The requirements specified in the referenced AASHTO Article place the successful bidder on notice that he may have to furnish complete statements of the origin, composition, and manufacture, of all materials to be used in the Work. These requirements might for example, be presented by an Owner who, before awarding the Contract to the successful bidder, desires to be assured that certain or all materials to be incorporated in the Work will be produced or manufactured in the United States.

Additional information on samples is presented in Article 3.2.2B, Samples, and in Article 4.5D, Samples, Tests, Cited Specifications.

Q. Non-Collusive Bidding Certification.

Refer to Article 102.15, same title, of the AASHTO Sample at the end of this Chapter. When public money is involved in a construction contract, a noncollusion affidavit is generally required in the interest of free competiton.

R. Escalation Clauses.

Escalation clauses are clauses that allow reimbursement to the Contractor to ease the adverse impact of future but unknown steep increases in the costs of labor or material during inflationary periods. These clauses are most often included in contracts for work that will take a long period of time to complete; generally a duration of more than 18 months.

Traditionally, contractors have been willing to submit bids that absorb the effects of inflation on costs of labor and materials. A contractor can only counter the risk of increased wages and prices by including contingency allowances in his bid prices. For long term contracts during unstable economic conditions, the risk may be great enough to influence some contractors against bidding. Contractors who submit bids are sometimes compelled to include such exorbitant contingencies that their bids often far exceed the Owner's budgeted cost. The Owner can alleviate this situation by assuming an equal or greater share of the risk through escalation clauses. This may be considered as an alternative to Article 1.3B, Cost Reimbursable Contract. An explanation of how this arrangement may be provided in the Contract is presented in Part II, Article 12.2R, Escalation Clauses.

5.3 Award and Execution of Contract

A. Consideration of Proposals.

Refer to Article 103.01, same title, of the AASHTO Sample at the end of this Chapter. After being opened and read aloud, the proposals are checked for mathematical accuracy. In addition, they are reviewed to determine that there are no irregularities, as outlined in previous Articles 5.2I, Irregular Proposals, and 5.2.O, Disqualification of Bidders.

In the event of a discrepancy between prices written in words and those written in figures, the written words will usually govern.

After bids have been opened, there are occasions when a low bidder will claim that he made an honest error. If he can support such a claim with satisfactory evidence, he is usually permitted to withdraw his bid and have his bid guaranty returned to him.

Some mistakes are obvious. For example, if a low bid is wildly lower than the next low bid, it is obvious that a mistake has been made. The wrong placement of a decimal point is another mistake that is obvious. Concerning the notification of a mistake, many Specifications will require that a bidder must notify the Owner within 48 hours after bids have been opened.

Examples of technicalities that may be waived by an owner when considering proposals are: 1) Failure of the bidder to sign the bid bond; 2) Date on Proposal does not agree with the bid due date; or 3) Failure to initial a correction in the bid.

B. Award of Contract.

Refer to Article 103.02, same title, of the AASHTO Sample at the end of this Chapter. A bidder is bound by his offer after the bids have been opened. If his bid is successful, he is obligated to accept the Contract when offered. What happens if a contract is not awarded within the time period specified? The low bidder usually has two options. He can withdraw his bid without penalty, or the time of award can be extended by mutual agreement with the Owner.

C. Cancellation of Award.

Refer to Article 103.03, same title, of the AASHTO Sample at the end of this Chapter. The contents of the referenced AASHTO Article, which seem to be a standard provision, relieves the Owner of any responsibility for cancelling the award at any time before execution of the Contract, when he deems it to be in his best interests.

D. Return of Proposal Guaranty.

Refer to Article 103.04, same title, of the AASHTO Sample at the end of this Chapter. After bids have been opened and the two or three lowest bidders determined, there is no further need for the Owner to retain proposal guaranties of the other bidders.

E. Requirements of Contract Bonds.

Refer to Article 103.05, same title, of the AASHTO Sample at the end of this Chapter. If every contractor was sincere, capable, and financially responsible, and if every set of Contract Documents was flawless, there would be no reason to assure the satisfactory completion of a contract. However, since this combination of conditions is virtually unattainable, the Owner has to consider the possibility of an uncompleted contract. This situation can subject the Owner to excessive costs plus delayed completion of his project. Also, if the defaulting Contractor has not paid all his bills, the Owner may have to face liens filed against his project. It therefore becomes necessary for the Owner to receive assurance that his project will be completed and that he will be protected against mechanic's liens.

This protection is provided in the Contract by requiring the Contractor to furnish surety bonds. In a surety bond, one party guarantees the performance by another party of an obligation. To explain this relationship, the Surety (bonding company) promises to the Owner (obligee) that if the Contractor (principal or obligor) does not perform as required in the Contract, the Surety will step in and guarantee its completion. Surety bonds include the types listed in the following paragraphs:

1. Performance bond. To ensure completion of the Contract, the Contractor is required to furnish a performance bond in a sum equal to the full amount of the Contract. This bond is a guaranty to the Owner that the Contractor will perform in accordance with the terms and conditions of the Contract. Should the Contractor default on his Contract, the Surety guarantees that the Work will be completed within the amount of money contracted for. Illustration of a Per-

formance Bond Form is presented in Appendix A, as Exhibit D. It has been reproduced from the Contract Specifications for construction of the Mount Lebanon Tunnel, previously mentioned in Article 5.2J, Proposal Guaranty.

2. Payment bond. Protection against claims and liens resulting from failure of the Contractor to pay all his bills and obligations is provided by requiring the Contractor to furnish a payment bond (sometimes called a Labor and Materialman's Bond). The specified sum of a payment bond may vary from 50 percent to 100 percent of the amount of the Contract, depending on the Owner's requirements. This bond guarantees that the Contractor will pay all bills for work done and materials supplied, in connection with the Contract. It covers all people working for the Contractor, including his employees, subcontractors, and suppliers. Illustration of a Payment Bond Form is presented in Appendix A, as Exhibit E. It also has been reproduced from the Contract Specifications for construction of the Mount Lebanon Tunnel.

Construction contract bonds also serve to provide a guaranty of payment to subcontractors and those who furnish labor, materials, or equipment. They give assurance that once started, the Contract will be completed. Generally, a Surety will only write a bond if it believes that the Contractor has sufficient financial resources and the competence to perform the Work.

F. Execution and Approval of Contract.

Refer to Article 103.06, same title, of the AASHTO Sample at the end of this Chapter. Some owners may require that, in addition to returning the signed Contract and Contract bonds, the successful bidder also include the necessary liability insurance policies (see Article 4.6N, Responsibility for Damage Claims). Other owners may specify that insurance policies have to be furnished before the Contractor occupies the site.

Some owners, unable to execute the Contract within the time period specified, may give the successful bidder an option of either withdrawing his bid without penalty or of accepting an extended period for the Owner to execute the Contract.

G. Failure to Execute Contract.

Refer to Article 103.07, same title, of the AASHTO Sample at the end of this Chapter. The referenced AASHTO Article is pretty much a standard provision in publicly funded, competitively bid contracts. Its contents are self-explanatory.

Failure of the low bidder to execute the Contract makes it necessary for the Owner to begin the award process all over again or to readvertise the Work, if he wants his project to go forward. Either way, this involves additional effort, expense, and delay for the Owner. Attaching the low bidder's proposal guaranty as liquidated damages is therefore understandable.

Pages 157 through 163 have been reproduced from the AASHTO Guide Specifications for Highway Construction 1985, Sections 102 and 103. The Articles of these Sections are concerned with the Bidding Documents and are referred to herein.

SECTION 102—BIDDING REQUIREMENTS AND CONDITIONS

102.01 Prequalification of Bidders. At least .... days (10 days suggested) prior to submitting a bid, the bidder shall be required to file an experience questionnaire and a confidential financial statement, certified by a public accountant (certified suggested). The statement shall include a complete report of the bidder's financial resources and liabilities, equipment, past record, and personnel.

Bidders intending to consistently submit proposals shall prequalify at least once a year. Prequalification may be changed during that period upon the submission of additional favorable reports or upon unsatisfactory performance. Prequalification may authorize a Contractor to bid on individual projects of a given size or for a particular kind of work.

102.02 Contents of Proposal Forms. Upon request, the Department will furnish the prospective bidder with a proposal form. This form will state the location and description of the contemplated construction, show the appropriate estimate of the various quantities and kinds of work to be performed or materials to be furnished, and will have a schedule of items for which unit bid prices are invited. The proposal form will state the time in which the work must be completed, the amount of the proposal guaranty, and the date, time, and place of the opening of proposals. The form will also include or designate any special provisions or requirements which vary from or are not contained in the standard specifications.

All papers bound with or attached to the proposal form are considered a part thereof and must not be detached or altered when the proposal is submitted.

The plans, specifications, and other documents designated in the proposal form will be considered a part of the proposal whether attached or not.

The prospective bidder will be required to pay the Department the sum stated in the Notice to Contractors for each copy of the proposal form and each set of plans.

102.03 Issuance of Proposal Form. The Department reserves the right to disqualify or refuse to issue a proposal form if a bidder is in default for any of the following reasons:

(a) Lack of competency and adequate machinery, plant, and other equipment, as revealed by the financial statement and experience questionnaire required under subsection 102.01.
(b) Uncompleted work which, in the judgment of the Department, might hinder or prevent the prompt completion of additional work if awarded.
(c) Failure to pay, or satisfactorily settle, all bills due for labor and material on former contracts in force at the time of issuance of proposals.
(d) Failure to comply with any prequalification regulations of the Department.
(e) Default under previous contracts.
(f) Unsatisfactory performance on previous work.

102.04 Interpretation of Quantities in Bid Schedule. The quantities appearing in the bid schedule are approximate and are prepared for the comparison of bids. Payment to the Contractor will be made for the actual quantities of work performed and accepted or materials furnished in accordance with the contract. The scheduled quantities of work to be done and materials to be furnished may be increased, decreased, or omitted as hereinafter provided.

102.05 Examination of Plans, Specifications, Special Provisions, and Site of Work. The bidder is expected to examine carefully the site of the proposed work, the proposal, plans, specifications, supplemental specifications, special provisions, and contract forms before submitting a proposal. The submission of a bid shall be considered prima facie evidence that the bidder has made such examination and is satisfied as to the conditions to be encountered in performing the work and as to the requirements of the plans, specifications, supplemental specifications, special provisions, and contract.

Boring logs and other records of subsurface investigations are available for inspection by bidders. It is understood that such information was obtained and is intended for State design and estimating purposes only. It is made available to bidders that they may have access to identical subsurface information available to the State, and is not intended as a substitute for personal investigation, interpretations, and judgment of the bidders.

102.06 Preparation of Proposal. The bidder shall submit the proposal on the forms furnished by the Department. The bidder shall specify a unit price in words or figures, or both if required, for each pay item for which a quantity is given and shall show the products of the respective unit prices and quantities written in figures in the column provided, and the total amount of the proposal obtained by adding the amounts of the several items. All the words and figures shall be in ink or typed. In case of a discrepancy between the prices written in words and those written in figures, the prices written in words shall govern.

When an item in the proposal contains a choice to be made by the bidder, the bidder shall indicate the choice in accordance with the specifications for that particular item, and no further choice will be permitted.

The bidder's proposal must be signed with ink by the individual, by one or more members of the partnership, by one or more members or officers of each firm representing a joint venture, by one or more officers of a corporation, or by an agent of the Contractor legally qualified and acceptable to the State. If the proposal is made by an individual, the name and post office address must be shown; by a partnership, the name and post office address of each partnership member must be shown; as a joint venture, the name and post office address of each member or officer of the firms represented by the joint venture must be shown; by a corporation, the name of the corporation and the business address of its corporate officials must be shown.

102.07 Irregular Proposals. Proposals will be considered irregular and may be rejected for the following reasons:

(a) If the proposal is on a form other than that furnished by the Department, or if the form is altered or any part thereof is detached.
(b) If there are unauthorized additions, conditional or alternate bids, or irregularities of any kind which may tend to make the proposal incomplete, indefinite, or ambiguous as to its meaning.
(c) If the bidder adds any provisions reserving the right to accept or reject an award, or to enter into a contract pursuant to an award.
 This does not exclude a bid limiting the maximum gross amount of awards acceptable to any one bidder at any one bid letting, provided that any selection of awards will be made by the Department.
(d) If the proposal does not contain a unit price for each pay item listed except in the case of authorized alternate pay items.
(e) If the Department determines that any of the unit bid prices are significantly unbalanced to the potential detriment of the Department.

102.08 Proposal Guaranty. No proposal will be considered unless accompanied by a guaranty of the character and in an amount not less than the amount indicated in the proposal form.

102.09 Delivery of Proposals. Each proposal shall be submitted in a special envelope furnished by the Department. The blank spaces on the envelope shall be filled in correctly to clearly indicate its content. When an envelope other than the one furnished by the Department is used, it shall be of the same general size and shape and be marked to clearly indicate its contents. When sent by mail, the sealed proposal shall be addressed to the Department in care of the official in whose office the bids are to be received. All proposals shall be filed prior to the time and at the place specified in the Notice to Contractors. Proposals received after the specified time will be returned to the bidder unopened.

102.10 Withdrawal or Revision of Proposals. A bidder may withdraw or revise a proposal after it has been deposited with the Department, provided the request for withdrawal or revision is received by the Department, in writing or by telegram, before the time set for receipt of bids.

102.11 Combination or Conditional Proposals. If the Department so elects, proposals may be issued for projects in combination and/or separately, so that bids may be submitted either on the combination or on separate units of the combination. The Department reserves the right to make awards on combination bids or separate bids to the advantage of the Department. No combination of bids, other than those specified in the proposals by the Department, will be considered. Separate contracts will be written for each individual project included in the combination.

Conditional proposals will be considered only when stated in the special provisions.

102.12 Public Opening of Proposals. Proposals will be opened and read publicly at the time and place indicated in the Notice to Contractors. Bidders, their authorized agents, and other interested parties are invited to be present.

102.13 Disqualification of Bidders. Either of the following reasons may be considered as being sufficient for the disqualification of a bidder and the rejection of the bidder's proposal or proposals:

(a) More than one proposal for the same work from an individual, firm, or corporation under the same or different name.
(b) Evidence of collusion among bidders. Participants in collusion will receive no recognition as bidders for any future work of the Department until they shall have been reinstated as a qualified bidder.

102.14 Material Guaranty. The successful bidder may be required to furnish a complete statement of the origin, composition, and manufacture of any or all materials to be used in the construction of the work, together with samples, which may be subjected to the tests provided for in these specifications to determine their quality and fitness for the work.

102.15 Non-Collusive Bidding Certification. Every bid submitted to the State Highway Department shall contain the following statement subscribed and affirmed by the bidder as true under the penalties of perjury. This statement will appear in the bid proposal in the form of a certification and must be signed by the bidders and submitted with the bid documents.

Non-Collusive Bidding Certification

(a) By submission of this bid, each bidder and each person signing on behalf of any bidder, certifies as to its own organization, under penalty of perjury, that to the best of their knowledge and belief:
  (1) The prices in this bid have been arrived at independently without collusion, consultation, communication, or agreement with any other bidder or with any competitor for the purpose of restricting competition.
  (2) Unless required by law, the prices which have been quoted in this bid have not been knowingly disclosed and will not knowingly be disclosed by the bidder, directly or indirectly, to any other bidder or competitor prior to opening of bids.
  (3) No attempt has been made or will be made by the bidder to induce any other person, partnership, or corporation to submit or not to submit a bid for the purpose of restricting competition.

(b) A bid shall not be considered for award nor shall any award be made where there has not been compliance with (a)(1)(2)(3) above. If the bidder cannot make the foregoing certification, the bidder shall so state and shall furnish with the bid a signed statement which sets forth in detail the reasons why the certification cannot be made. Where (a)(1)(2) and (3) above have not been complied with, the bid shall not be considered for award nor shall any award be made unless the head of the State Highway Department, or designee, determines that such disclosure was not made for the purpose of restricting competition.

The fact that a bidder (a) has published price lists, rates, or tariffs covering items being procured, (b) has informed prospective customers of proposed or pending publication of new or revised price lists for such items, or (c) has sold the same items to other customers at the same prices being bid, does not constitute a disclosure within the meaning of subparagraph one(a).

The bid made to the State Highway or Transportation Department shall be deemed to have been authorized by the board of directors of the bidder. Such authorization shall be deemed to include the signing and submission of the bid and the inclusion therein of the certificate as to non-collusion on the part of the corporation.

The signers of this proposal hereby tender to the State Highway or Transportation Department this sworn statement that the named Contractor has not, either directly or indirectly, entered into any agreement, participated in any collusion, or otherwise taken any action to restrain free competitive bidding in connection with this proposal.

SECTION 103—AWARD AND EXECUTION OF CONTRACT

103.01 Consideration of Proposals. After the proposals are opened and read, they will be compared on the basis of the summation of the products of the quantities shown in the bid schedule by the unit bid prices. The results of such comparisons will be immediately available to the public. In the event of a discrepancy between unit bid prices and extensions, the unit bid price shall govern.

The Department reserves the right to reject any or all proposals, to waive technicalities, or to advertise for new proposals.

103.02 Award of Contract. The award of contract will be made within ..... (30 days suggested) calendar days after the opening of proposals to the lowest responsible and qualified bidder whose proposal complies with all the requirements prescribed. The successful bidder will be notified, by letter mailed to the address shown on the proposal, of the acceptance of the bid and the award of the contract.

103.03 Cancellation of Award. The Department reserves the right to cancel the award of any contract before the execution of said contract without any liability against the Department.

103.04 Return of Proposal Guaranty. All proposal guaranties, except those of the two lowest bidders, will be returned immediately following the opening and checking of the proposals. The retained proposal guaranties of the two lowest bidders will be returned after a satisfactory bond has been furnished and the contract has been executed.

103.05 Requirement of Contract Bond. At the time of the execution of the contract, the successful bidder shall furnish a surety bond or bonds in a sum equal to the contract amount. The form of the bonds and the security shall be acceptable to the Department.

103.06 Execution and Approval of Contract. The contract shall be signed by the successful bidder and returned, together with the contract bond, within 15 days after the contract has been mailed to the bidder. If the contract is not executed by the Department within .... (15 days suggested) following receipt of the signed contracts and bonds, the bidder shall have the right to withdraw the bid without penalty. No contract shall be considered as effective until it has been fully executed by all the parties thereto.

103.07 Failure to Execute Contract. Failure to execute the contract and file acceptable bonds within 15 days after the contract has been mailed to the bidder shall be just cause for the cancellation of the award and the forfeiture of the proposal guaranty to the Department, not as a penalty, but in liquidation of damages sustained. Award may then be made to the next lowest responsible bidder, or the work may be readvertised.

Chapter 6

Classifications and Types of Specifications

6.1 Classification of Specifications

A. Introduction.

Most Specifications for construction contracts are based on Specifications that have been preprinted by owners, and which are maintained and updated periodically. There are two commonly known classifications for this type of preprinted document, Standard Specifications and Master (Guide) Specifications. In preparing a set of Specifications for a specific Contract, preprinted Standard Specifications are modified to present only requirements that are applicable to the specific Contract. One of the advantages of working with preprinted Specifications is that they serve as a check against the accidental omission of a coverage.

In the process of developing a set of Specifications for a specific Contract, two additional classifications come into use, Outline Specifications and Contract Specifications.

B. Standard Specifications, Supplemental Specifications, and Special Provisions.

Owners such as Federal, State, and other public agencies who regularly award construction contracts establish a set of Standard Specifications applicable to their construction projects. These Specifications contain all of the conceivable requirements governing the particular Owner's standard items of construction. These standard requirements establish a uniformity of administrative procedures and quality in materials and workmanship for constructing their facilities. In addition to Technical Sections, the Standard Specifications include General Conditions and Bidding Documents.

Standard Specifications are generally published in a bound volume. Updated editions are printed at intervals of three to ten years, depending on the needs of the particular Owner. In order to maintain a current edition up-to-date between printings, interim changes to the Standard Specifications will be issued by the Owner in the form of printed Supplements. When the book of Standard Specifications is next printed, the changes issued in the Supplemental Specifications are then incorporated in the new edition. Samples of Standard Specifications are readily available in books of Standard Specifications published by the various State Departments of Transportation.

Multicontract construction projects of five to ten or more years duration, such as rapid transit systems or toll highways, are also suitable for developing project Standard Specifications. In these instances, there would probably be only one edition of the document.

Standard Specifications and their Supplements are unable to provide all of the requirements needed for a specific Contract. To fill this gap, another set of Specifications called the Special Provisions are prepared by the specification writer for the specific Contract. The Special Provisions present requirements applicable only to the specific Contract. They amend the Standard and Supplemental Specifications by adding to them, deleting from them, or modifying them. Arrangement of Sections and subsections in the Special Provisions follow the same order as in the Standard Specifications. Additional information on the use of Standard Specifications and the preparation of Special Provisions is presented in Article 8.4, Using Standard Specifications, and in Article 13.4, Standard Specifications and Special Provisions.

Specifications for a public works contract will in most cases, consist of two or three separate documents, namely the (1) Standard Specifications, (2) Supplemental Specifications when applicable, and (3) Special Provisions.

C. Master (Guide) Specifications.

A Master Specification, or Guide Specification as it is sometimes called, is all inclusive in its coverage. It will incorporate requirements to cover all conceivable conditions and situations that can be anticipated for a particular item of work. It may also include instructional notes for the specification writer. A Standard Specification differs somewhat from a Master Specification in that a Standard Specification only incorporates requirements applicable to the type of work normally required by the particular Owner.

Since Master Specifications are not published in a bound volume, they are suited for word processing. Consequently, when they are being updated, changes can be incorporated directly into the document. The procedure for utilizing Master Specifications in the preparation of Contract Specifications is to have the word processor print out the desired Master Specification. The specification writer then edits this printout by deleting inapplicable portions of the text and inserting new requirements. The edited Master Specification is then returned to the word processor operator for finalizing. After the changes have been made, the final printout will represent a Contract Specification. Use of a Master Specification thus eliminates the procedure of incorporating Standard Specifications with their

Supplemental Specifications and then preparing Special Provisions to cover specific requirements of the Contract.

Several Federal Agencies have Master Specifications, among them being the Navy Department, the Corps of Engineers, and the Federal Aviation Administration. National Master Specifications are also sponsored by the Construction Specifications Institute.

D. Outline Specifications.

In the course of preparing the Technical Sections for a Contract, their preparation will advance in stages of completion until they reach the final (100 percent) stage. Some Owners will desire to receive advance information on proposed materials, standards of workmanship, finishes, and equipment. This advance information would be presented in an early submittal of the Specifications, commonly referred to as Outline or Preliminary Specifications. Each proposed Technical Section is presented in an outline form. The presentation would include a tentative Table of Contents. Outline Specifications fall within the 30 to 35 percent completion stage of a set of Contract Specifications, as described in Article 14.6, Scheduling.

The following illustration of an Outline Specification deals with membrane waterproofing for an underground station structure:

SECTION 07100
MEMBRANE WATERPROOFING

1. DESCRIPTION
   A. Applied to exterior surface of walls and roof slab.
2. MATERIALS
   A. Butyl rubber sheet membrane, 1/8 inch thick.
   B. Protection Courses
      (1) Horizontal Surfaces: Concrete Class 2500 with welded wire fabric reinforcement.
      (2) Vertical Surfaces: Asphalt impregnated fiber board, 1/2 inch thick.
3. CONSTRUCTION REQUIREMENTS
   A. Surface preparation, application of membrane waterproofing, and temporary protection, to be in accordance with manufacturer's instructions.
   B. Permanent Protection Courses
      (1) Horizontal Surfaces: Cast-in- place concrete two inches thick with wire fabric reinforcement.
      (2) Vertical Surfaces: Fiber board, adhered to surface.
4. METHOD OF MEASUREMENT
   A. Membrane Waterproofing: S.Y.
   B. Concrete Protection Course: S.Y.
   C. Fiber Board Protection Course: S.Y.

E. Contract Specifications.

Contract Specifications are Specifications that have been prepared for a specific construction contract. When Standard Specifications are used in preparing a set of Contract Specifications, the normal procedure is to incorporate the applicable Standard Specifications and their Supplements, by direct reference. Necessary modifications to these Standard Documents are accomplished through the Special Provisions, as indicated in Article 6.1B, Standard Specifications, Supplemental Specifications, and Special Privisions.

6.2 Types of Specifications

A. Introduction.

Technical requirements are not all specified in the same manner. They are presented in different ways, to best suit the Owner's needs. The types of Specifications reflecting these variations are referred to as Proprietary, Descriptive, and Performance Specifications.

B. Proprietary Specifications.

A designer will, after many years in business, acquire first hand knowledge on the merits or poor performance of various products and items of equipment. He utilizes this knowledge to guide him in specifying subsequent projects. He will designate a product or item of equipment by its brand name, manufacturer, and model number, knowing that it will perform satisfactorily. Presenting requirements in this manner is simple, concise, and allows little chance of misinterpretation. Proprietary Specifications are thus sometimes called Standard Brand Specifications. A contractor who has complied with the Specification and has furnished what was specified, normally cannot be held responsible if the product or equipment should fail to perform as intended.

Specifying a particular product having features peculiar to one manufacturer or producer is generally not acceptable in public works contracts. The objective in public works contracts is to broaden competition in procurement and avoid any charges of favoritism brought by other manufacturers. There can be exceptions. It is permissible in a contract to specify materials or equipment by brand name when work connecting to existing systems and equipment is required to be of the same manufacturer as the existing construction. Also, there may be times when it is impractical or impossible to make an exhaustive study within the allowable time limit or budget and come up with a set of performance criteria. In such a situation, it has been permissible for the designer to list two or more acceptable manufacturers. In order not to restrict competition to listed manufacturers, the words "or approved equal" are added (see Article 8.7, Trade Names and "Or Approved Equal"). An illustration of specifying a product by listing multiple manufacturers would read: Sealing compound shall be "Sealing and Dielectric Compound" as manufactured by AMP Special Industries, Inc., Valley Forge, Pennsylvania; "Tac-Tape" as manufactured by Royston Laboratories Inc., Pittsburgh, Pennsylvania; "Cold Seal Butyl Tape" as manufactured by Utility Products Company, King of Prussia, Pennsylvania; or an approved equal.

C. Descriptive Specifications.

Descriptive Specifications are most common in civil works construction contracts. In addition to providing a detailed description of required materials and their properties without specifying brand names, they provide construction details and the desired quality of workmanship. Descriptive Specifications are sometimes called Prescriptive Specifications, or Materials and Methods Specifications.

When the Contractor is given detailed instructions on the method to be followed in performing an operation, he generally cannot be held responsible for the end result so long as he followed the Specifications. This subject is discussed in greater detail in Article 8.16, Methods and Results.

D. Performance Specifications.

Performance Specifications are sometimes called End Product Specifications or End Result Specifications. They specify Standards and criteria, and the end results to be obtained. Performance Specifications do not normally prescribe dimensions, materials, finishes, or methods of manufacture. Instead, the desired quality, function, and other characteristics of the product or item of equipment, are specified. By specifying the performance requirements, it becomes the responsibility of the Contractor to select the materials and methods to achieve the end result.

This method of specifying is often used when items of equipment such as pumps, compressors, and motors, are to be provided. The Specification will prescribe the purpose for which the item is to be used, the conditions of operation, and (in the case of heating and air conditioning systems) prescribe certain requirements for its installation.

E. Descriptive Versus Performance Specifications.

These two types of Specifications will each have its advantages and its limitations.

1. Descriptive Type Specifications. These are generally used by owners such as State Departments of Transportation, who are in the business of regularly awarding construction contracts. From experience gained over the years in administering their contracts, they have accumulated much knowledge on the behavior and performance of materials used and on construction methods. They

can, therefore, designate a specific material or method of construction with assurance that the desired result will be obtained.

There are also some disadvantages to using this type of Specification. In requiring a specific procedure to be followed, it will be difficult to hold the Contractor accountable for the results if he has faithfully complied with the requirements. Again, specifying the method to be followed can prevent the contractor from using his ingenuity and coming up with an improved method. Giving the Contractor free rein to select his own method for producing the required results can sometimes be an economic benefit to the Owner.

2. Performance Type Specifications. This type of Specification is helpful in public works contracts, because the use of a brand name for specifying a manufactured product or item of equipment is not encouraged. Using this type of Specification for an item of equipment involves the presentation of a detailed description of its required characteristics and performance standards.

This type of Specification is also helpful in eliminating a lot of detailed instructions for construction procedures, and at the same time gives the Contractor the opportunity to use his ingenuity. To illustrate: in the compaction of roadway embankment, instead of specifying the type and size of compacting equipment the Contractor must use, and the number of passes he is to make with this equipment on each lift of embankment material deposited, it would simply be stated that the Contractor is to compact each lift to not less than 95 percent of its maximum density as determined by ASTM D 698, Standard Test Methods for Moisture-Density Relations of Soils and Soil Aggregate Mixtures Using 5.5 Lb. Rammer and 12-In. Drop.

3. Combined Descriptive and Performance Type Specifications. The type of Specification most commonly used incorporates both descriptive and performance type requirements. Under this arrangement, the Owner and Designer have the opportunity to draw on their background of experience, and at the same time allow the Contractor freedom of choice to use his expertise in other areas. Specifying minimum requirements on materials and procedures based on experience acquired in previous projects can provide reasonable assurances to the Owner that the desired results can be obtained.

By specifying gradation limits for roadway embankment material, and the allowable maximum loose thickness of each layer spread, the Designer is assured that the embankment can be compacted to produce the specified minimum density. Although the Specification limits the Contractor to material that will be acceptable, and to the maximum thickness of each layer of uncompacted material, it does allow him the freedom to select his compaction equipment and to establish his spreading and compaction procedures.

Chapter 7

National Reference Standards

7.1 Introduction

The purpose of a construction reference Standard is to establish a minimum level of quality for materials, fabrication practices, or construction procedures. Reference Standards are prepared and published by professional societies, national industry associations, and government agencies. Standards of workmanship for certain trade groups will also be found among these publications. The publications perform a great service to the industry by standardizing test methods and the more common construction procedures. Being able to refer to a national Standard relieves a specification writer of the task of presenting the detailed requirements and instructions that would otherwise have to be written into the Contract Specifications.

7.2 Standards Making Organizations

Three organizations that sponsor the more commonly used reference Standards are:

A. American Society for Testing and Materials (ASTM).

ASTM is located at 1916 Race Street, Philadelphia, Pennsylvania 19103. This is probably the largest source of volunteer consensus Standards on characteristics and performance of materials, products, systems, and services. The Society operates through more than 140 main technical committees with more than 2,000 subcommittees. The committees function under regulations that ensure balanced representation among producers, users, consumers, and general interest participants.

The 1987 Annual Book of Standards consists of 66 volumes which are updated and reprinted each year. In addition to specifying the acceptable chemical and physical properties of materials, the ASTM Specifications set up standard procedures for acceptance testing of the materials. The test methods cover sampling and selection of specimens and describe the testing procedures for determining properties, composition, and performance of the materials, products, or systems specified.

B. American Association of State Highway and Transportation Officials (AASHTO).

AASHTO is located at 444 North Capitol Street, N.W., Suite 225, Washington D.C. 20001. The Subcommittee on Materials has members representing each of the 50 states, the Commonwealths of Puerto Rico and the Northern Mariana Islands, the District of Columbia, the United States Department of Transportation, the New Jersey Turnpike Authority, The Massachusetts Metropolitan District Commission, The Port Authority of New York and New Jersey, six Canadian Provinces and two Territories. The current edition of Standards for Transportation Materials is published in two parts; Part I dealing with Specifications for Materials, and Part II dealing with Methods of Sampling and Testing. Interim Specifications are published each year, and a revised edition of the Standards is published every four years. Many of the Standards are technically identical with or similar to those of ASTM. In these cases, the ASTM designation number is also shown in the heading of the Standard. AASHTO also publishes Standard Specifications for Design and Construction of Highway Bridges.

C. American Concrete Institute (ACI).

ACI can be contacted through P.O. Box 19150, Detroit, Michigan 48219-0150. The Institute is a nonprofit organization of engineers, architects, scientists, constructors, and individuals, associated in their technical interest with the field of concrete. One of the purposes of the Institute is the development of Standards for the design and construction of concrete structures. Standards are prepared by technical committees. One of the publications of the Institute is the ACI Manual of Concrete Practice, a five volume compilation of current ACI Standards and Committee Reports. It is revised annually.

ACI Standards are indicated as Codes, Specifications, and Standard Practices. Standards designated as Specifications can be referred to and made a part of the Contract Specifications. Standards designated as Standard Practices present recommended and acceptable materials and methods for use in design, inspection, or the preparation of Contract Specifications.

D. Other Organizations.

Following is a listing of some other organizations that sponsor Standards for public works projects.

1. American National Standards Institute (ANSI), Inc. 1430 Broadway, New York, New York 10018. ANSI is a coordinator of America's voluntary Standards system. It serves as a clearinghouse and information center for American National Standards and international Standards. ANSI publishes an annual catalog of all approved American National Standards.

2. American Water Works Association (AWWA), 6666 West Quincy Avenue, Denver, Colorado 80235. AWWA publishes Standards for materials, equipment, and construction procedures involving potable water systems.

3. Steel Structures Painting Council (SSPC), 4400 Fifth Avenue, Pittsburgh, Pennsylvania 15213-2683. SSPC publishes Standards covering methods, equipment, and systems for the surface preparation and painting of steel structures.

4. American Railway Engineering Association (AREA), 50 F Street, N.W., Room 7702E, Washington, D.C. 20001. AREA publishes recommended practices governing the construction of railways.

5. American Welding Society, (AWS), Inc., 2501 N.W. 7th Street, Miami, Florida 33125. AWS publishes Standards for structural welding, including welder qualifications.

6. Additional Organizations. Additional organizations are listed in Appendix B, Sources of Information for the Specification Writer.

7.3 Presenting Reference Standards in the Specifications

The following examples illustrate the use and presentation of reference Standards in the Specifications.

A. Material Standard.

Because of the wide use of concrete in construction projects, concrete aggregates are probably one of the most frequently specified materials. A requirement for coarse aggregate would read: "Coarse aggregate shall conform to the requirements of ASTM C 33, Standard Specification for Concrete Aggregate."

This Standard presents the requirements for coarse aggregate, including general characteristics, gradation, allowable limits for deleterious substances, physical properties, limitations on alkali reaction with Portland cement, method of sampling the aggregate, and standard methods for performing the various acceptance tests.

B. Test Method Standard.

In the performance of various construction operations, the Contractor is required to produce certain specified end results. One of the more commonly required end results refers to compacted roadway embankment. In this operation, embankment material is spread in layers, with each layer being compacted before the next layer of material is spread over it. The acceptability of a compacted layer is determined by obtaining its "in-place" density. A requirement for acceptance of embankment compaction would read: "Each layer of embankment shall be compacted to a density of not less than 95 percent of maximum density.

Maximum density will be determined in accordance with AASHTO Designation T99, Standard Methods of Test for the Moisture-Density Relations of Soils Using a 5.5 Lb. Rammer and a 12-In. Drop. The density of the compacted soil in place will be determined in accordance with AASHTO Designation T191, Standard Method of Test for Density of Soil In-Place by the Sand-Cone Method."

To explain this briefly; before the placement of embankment gets underway, a sample of the soil to be used for embankment is taken to a testing laboratory. There, its maximum density is established in accordance with AASHTO T99. After a layer of embankment material has been spread and compacted, its degree of compaction is determined by taking a sample from the compacted layer and obtaining its density, all in accordance with the requirements of AASHTO T191. If the "in-place" density is found to be 95 percent or more of its maximum density, the compacted layer of embankment is acceptable and the Contractor is permitted to spread the next layer of material. If the "in-place" density is found to be less than specified, the compacted layer is not acceptable and requires additional compaction.

C. Construction Procedure Standard.

Portland cement concrete placed during cold weather is subject to damage if the proper precautions are not taken. The American Concrete Institute has developed a Standard to be followed. A Specification requirement governing concreting procedures in cold weather would read: "When the air temperature is expected to drop to 40 degrees Fahrenheit or lower during the placing and curing of concrete, the protective measures to be taken shall conform to the applicable requirements of ACI 306R, Cold Weather Concreting." Protective measures presented in this ACI Standard range from preparation of the area that is to receive the concrete, to heating the materials prior to their being placed in the mixer, and to protection of the concrete during and after its placement.

7.4 Building Codes

A brief word about building codes. A building code is a legal document adopted by governmental bodies. It consists of a series of Standards and Specifications designed to protect the lives of people (both inside and outside buildings) from fire and other hazards and to protect the health and safety of the general public. A building code provides a set of minimum requirements that apply to the design, construction, repair, demolition, and occupancy of buildings. It is enforced by municipal and county building departments. There are three major building codes in the United States, each of which is dominant in a different part of the country. They are:

A. Standard Building Code.

Written by the Southern Building Code Congress International, Inc. (SBCC), 900 Montclair Road, Birmingham, Alabama 35213. Dominant in the southern part of the country.

B. Basic Building Code.

Written by The Building Officials and Code Administrators, International (BOCA), 4051 West Flossmoor Road, Country Club Hills, Illinois 60477. Dominant in the eastern and midwestern parts of the country.

C. Uniform Building Code.

Written by The International Conference of Building Officials (ICBO), 5360 South Workman Mill Road, Whittier, California 90601. Dominant on the west coast and in some midwestern areas of the country.

Part II

Preparing and Presenting Engineering Construction Specifications

Chapter 8

Procedures and Practices in Specification Writing

8.1 Introduction

Completion of a construction contract to the satisfaction of all parties concerned is no simple task. This is borne out by the proliferation of construction claims. In a court of law, the written word will generally take precedence over the drawings. Preparation of the Specifications for a construction contract thus constitutes one of the critical steps in the design process. Some individuals maintain that specifications writing is nothing more than assembling material that has been previously written. Those who have been involved in court litigation and have been interrogated on the witness stand know differently. Each set of Specifications must be tailored to fit the requirements of a specific construction contract.

A questionnaire survey of the construction industry was conducted by the American Society of Civil Engineers Committee on Specifications. Responses from 223 contractors around the country were published September 1978 in the ASCE Journal of the Construction Division as Paper 14001, Summary Report of Questionnaire on Specifications (Contractor Returns). One question asked the Contractors to indicate, from their past experience, requirements in project Specifications that they found to be unfair, annoying, and meaningless, as far as the Work was concerned. Most common complaints were:

a. Specifications that attempt to cover up omissions by using the clause "as directed by the Engineer."
b. Specifications that are obsolete, unclear, poorly written, or irrelevant.
c. The use of Standard Specifications that do not apply.
d. Specifying methods instead of end results and then requiring the Contractor to be responsible for producing the desired results.
e. Specifying materials that are unavailable and methods that are obsolete.
f. Making the Contractor responsible for errors of the Designer.
g. Unreasonable retainage of monies or unreasonable delay in the release of retainage.
h. Tolerances which are unreasonable.

The Engineer's liability is no longer limited to the contract between him and his client; it may now extend to third parties. The courts do not require or expect perfection from the Engineer. However, as the member of a profession, the Engineer has a duty to exercise the reasonable technical skill, ability, and competence that would be expected of an Engineer in a similar situation. It should be kept in mind that in the courts, the written Specifications are generally interpreted literally.

Among the many things that a Specification must do if it is to accomplish its desired purpose, the most important are that it present in a clear and concise manner what the Engineer requires in order to fulfill his design, to protect the Owner against poor workmanship and inferior materials, and at the same time allow the Contractor maximum freedom to do the Work in his own way, provided he produces a satisfactory end product. The ultimate goal is to prepare a set of Specifications that will get the job done to the satisfaction of all parties concerned, without litigation or excessive costs.

8.2 Study the Plans

Before beginning to write Specifications for a contract, the Plans should first be studied. It is essential for the writer to become thoroughly familiar with details of the proposed Work that he will be specifying. This will also familiarize him with the items of construction, which will in turn help him establish the Technical Sections required. In studying the Plans, he should look for unusual site conditions or unusual items of construction to be called to the attention of bidders.

Repeating information that is already shown on the Plans should be avoided. When information on the Plans is repeated in the Specifications, it can sometimes be stated incorrectly and result in discrepancies and ambiguities. The General Conditions of the Contract will usually include a statement to the effect that any requirement shown on the Plans and not described in the Specifications, or described in the Specifications and not shown on the Plans, shall have the effect of having been provided for in both.

There should be no discrepancies between the Plans and the Specifications. If the Specifications should contain the clause "as shown on the Plans," it must be verified by the specification writer. Examples of a discrepancy between the Specifications and the Plans were noted by the author in the final review of a set of Contract Documents. They portray a specification writer's limited familiarity with the Plans. To illustrate:

A. In the Technical Section for Portland Cement Concrete Pavement, it was specified that the Contractor was to maintain 1,500 linear feet of forms in place, in advance of his paving operations. In reviewing the Plans, it was noted that the individual length of the many portions of pavement to be constructed did not exceed 1,200 feet.

B. The Technical Section for Site Preparation required removal of trees, stumps, brush and roots, from within the limits of grading including ditches and channels. A study of the Plans showed the site of the Work to be located in paved areas within the City.

8.3 Work Closely With the Designer

It is essential that the specification writer work closely with the designer. It should be kept in mind that the designer has "lived" with his design for months. He is the one most familiar with it. If the specification writer should lack any details or information, he must not make any assumptions. He should request the designer to supply him with the missing details or information. Knowing the intent or reason for a particular detail on the Plans will help the specification writer prepare a more realistic requirement for that item of work. On the other hand, when specification input is supplied by the designer, it should not automatically be incorporated verbatim. The specification writer should ensure that the requirements are proper and the presentation acceptable.

8.4 Using Standard Specifications

Standardization is all around us and Construction Specifications cannot be excluded (see Article 6.1B, Standard Specifications, Supplemental Specifications, and Special Provisions). The need for Standard Specifications is understandable; they produce uniformity in the finished product and at the same time reduce production costs. On the other hand, the use of Standard Specifications may tend to create complacency on the part of the specification writer and eliminate independent thinking.

A Standard Specification will usually contain some material that does not apply to the particular Contract. When a Standard Specification is incorporated into and made a part of the Contract Specifications, every requirement in that Standard Specification will automatically apply to the Contract, unless it is modified or deleted by the Special Provisions. Some specification writers may incorporate a Standard Specification into the Contract Specifications without making the appropriate review for its conformance to the requirements of the Contract. If not corrected, confusion develops. Therefore, when Standard Specifications are being incorporated into a Contract: 1) The specification writer should ensure that the latest published edition of the Standard Specifications and its complete set of Supplements, are available; and 2) The Standard Specifications and Supplements being incorporated should be carefully reviewed. Requirements that do not apply to the Contract should be deleted or modified. This procedure should be consistent throughout the Documents. Modifying the Standard Specifications in one place and failing to modify in another, can defeat the intended purpose. This precautionary review should be particularly applied to Public Works Standard Specifications, which have a tendency to keep getting more voluminous. This results from a practice of adding requirements because of problems encountered on previous jobs, and making few or no deletions.

Additional information on the procedure of incorporating a Standard Specification into a set of Contract Specifications is presented in Article 13.4, Standard Specifications and Special provisions.

8.5 Utilizing Specifications of Previous Contracts

Few Specifications are now prepared "from scratch." If a Standard Specification is not being incorporated, the specification writer will most likely select a Specification from a previous contract involving similar construction, and modify it to suit the needs of the Contract being specified. Some critics may call this a "cut and paste" operation. Nevertheless, this method offers an excellent source of Specification material, provided it fits the situation.

Finding a Specification for normal items of construction such as roadway excavation, embankment, cast-in-place concrete, and roadway pavement, generally presents no problem. The problem begins when a Specification has to be prepared for an item of construction that is infrequently specified, such as cofferdams, caissons, or tremie concrete. It is, therefore, good policy to compile and maintain a file listing Specifications prepared for unusual items of construction. Having this list saves precious time by eliminating the need to search through the completed contracts files. Additional information on office files is presented in Article 14.12B, Completed Contract Specifications.

When looking for a previously prepared Specification, one should be selected that is commensurate with the magnitude of the work being contemplated. For example; a concrete Specification for dam construction would not be appropriate to use for construction of a concrete box culvert.

When utilizing a Specification prepared for another contract, the precautions to be taken are similar to those presented in Article 8.4, Using Standard Specifications.

8.6 Using Reference Standards

Materials should be defined by their reference to national Standards, whenever it is possible. In referring to a published Standard, its date of issue should not be specified because then the Specification will require continual updating. Instead, the Bidding Documents should include the statement: "When reference is made to a published Standard it shall be understood to refer to the edition current on the date of issuance of the Bidding Documents."

When making reference to a Standard, the number and title of the Standard should be given. If there is occasion to refer to the Standard again in the same

Section, it will not be necessary to repeat the title; the number of the Standard will suffice. Availability of the Standard should be investigated to ensure that it has not been discontinued. When specifying a Standard, it should be reviewed to eliminate any possible conflicts with the job requirements.

8.7 Trade Names and "Or Approved Equal"

Specifying a product by a specific brand name is frowned upon by public agencies, as previously explained in Article 6.2B, Proprietary Specifications. When there is no established Specification Standard or when a Standard does not specify all of the desired characteristics, two or more trade names are generally permitted to be used. Three trade names, if available, are usually preferred by a public agency. Each trade name should be identified with the name of the manufacturer, model name and number, type, size, and any other classifications that are applicable. If the manufacturer is not nationally known, his address or the name and address of a local distributor should be included. Most of this information is usually available from the designer.

When trade names are listed, public agencies generally require the words "or approved equal" to be added after the listing. This gives other manufacturers and suppliers an opportunity to prove that their product will also meet the requirements. When using the "or approved equal," it is necessary to spell out the functional physical and chemical characteristics that are essential to the product's intended end use. Possession of these characteristics will be the basis for acceptability of a proposed substitution. Otherwise, a contractor may be tempted to offer a cheaper and inferior substitute in the hope of getting it approved. Approval of an inferior substitute can expose the Engineer to liability if the approved substitute does not perform.

8.8 Specifying New Products

The specification writer should be cautious when specifying new products or old products in new applications. He should seek out the latest information including test literature on the material or product to be specified. Manufacturers usually publish their product's specifications. Although these specifications should not be accepted blindly, they can be a starting point in evaluating the product. If the product is referenced to a national Standard such as ASTM or ANSI, its physical and chemical properties can be evaluated. The specification writer should obtain from the manufacturer a list of comparable projects on which the product has been previously used, including names of the owner and designer. The writer should verify its use record. If untested or untried products lack an adequate experience record, they should be checked out very carefully.

8.9 Know Your Subject

It is rare for a specification writer to be knowledgeable in all of the items of construction that he may be required to specify. Nor is he expected to be familiar with every construction practice. He should, however, exercise greater care when having to specify work dealing with an operation that he is not knowledgeable in. If he is a member of a Specification Department within his Firm, he should consult with other members of the Department for guidance. Each member of the Department will have an intimate working knowledge in some specific area. Collectively, the Department should be knowledgeable in most areas. Other sources of information are the Design Departments; previous Contract Specifications of comparable construction; the Construction Department if the Firm provides technical inspection; industry literature; and the technical publications. In any event, the specification writer should not attempt to "bull his way through"; it can be disastrous.

8.10 Disclosure of Known Information

Information that can affect the Contractor's operations should not be withheld. This includes information known or information that the Designer should have known. If this information is not disclosed in the Contract Documents, it can result in additional costs to the Contractor and be cause for a future claim against the Owner.

A. Information normally possessed by the Designer and made available to the bidder, would include:

1. Information obtained from preliminary subsurface investigations made for purposes of design (see Article 12.2E, Examination of Plans, Specifications, Special Provisions, and Site of Work; Paragraph 4, Subsurface Data).
2. Information on existing utilities and available construction drawings of structures designated to be demolished or protected by the Contractor (see Article 12.2E, Examination of Plans, Specifications, Special Provisions, and Site of Work; Paragraph 2, Reference Drawings and Information).
3. Schedule of train movements and other requirements of a Railroad Company that may limit the Contractor's operations (see Article 11.6I, Railway-Highway provisions).
4. Requirements for maintaining and protecting marine traffic, and other limitations governing the Contractor's operations in a navigable waterway (see Article 11.6J, Construction Over, In, or Adjacent to Navigable Waters).
5. Other requirements that may affect and limit the Contractor's operations (see Article 11.7F, Limitation of Operations).

B. "Information that the Designer should have known" means knowledge that would have become known to him had he investigated sources of information available to him. This would include records in his own or other departments, or any other sources of information to which he might have access.

8.11 Tolerances

Variability in construction materials and workmanship is always present and it cannot be ignored in the Specifications (see Article 3.2.3D, Allowable Tolerances). Specifying rigid limits with no tolerances to cover the variations in natural and manufactured products can only lead to arguments, claims, and lawsuits. It would also be economically unfair to the Owner to specify Standards and limits that are unnecessary. When there is doubt about specific tolerance limits, the designer should be consulted.

One example of rigid limits is the use of the often-repeated phrase "true to line and grade." In a contract for construction of concrete spillways, the Specifications stated that concrete forms had to be erected true to line and grade. The inspector enforced this requirement literally and the Contractor had to erect his forms with extreme accuracy, incurring extra costs. The Contractor sued, maintaining that the wording of formwork requirements should not have been read literally but should have been interpreted in accordance with trade practice. He also pointed out that the concrete Specification provided tolerances in the line and grade of the finished spillways. The Contractor argued that he should have been allowed the same tolerances for the formwork as that permitted for the finished structure. The court ruled that the Contractor was entitled to recover his extra costs.

8.12 Quality Control and Quality Assurance

Although quality control and quality assurance are both concerned with the quality of materials and workmanship, they represent two separate and distinct areas of responsibility. Control of quality is the Contractor's responsibility. He establishes the procedures to control and guide his operations so that they will produce the desired results. Quality assurance is a function of the Owner's site representative. It is his responsibility to monitor the work of the Contractor for conformance to the requirements of the Contract. This subject is also discussed in Article 4.4A, Introduction. In line with this distinction, and in order that the Contractor may know what to expect, the Specifications should indicate in the appropriate Sections, the tests to be performed by the Engineer and the methods to be used, for checking compliance with the requirements.

Quality assurance should not be confused with quality control by assigning responsibility for the quality assurance of some of the Work to the Contractor.

This separation of responsibility was not very well handled in one set of Contract Specifications reviewed by the author. In the Specifications, one of the Technical Sections was titled Testing Laboratory Services. In this Section the Contractor was required to provide the services of a testing agency to perform specified services and testing. Some of these duties included continuous batch plant inspection; tests for in-place density of compacted embankment; and casting, curing, and breaking concrete test cylinders. To assume that a profit-oriented contractor given these quality assurance responsibilities will faithfully police his work for conformance, and voluntarily replace or repair any work that does not conform, would be somewhat naive. The first loyalty of the testing agency performing these inspections and tests for quality assurance would be to the Contractor who is paying for its services.

8.13 Identifying and Controlling Risks

A. General.

Risk is associated with the possibility of loss or injury. Construction involves many risks, known and unknown. Both the Owner and Contractor are exposed to risks, with the greater exposure focusing on the Contractor. With Contract Documents being prepared by the Owner, there can be a tendency to slant Contract language in favor of the Owner. Conceivable risks may be assigned to the Contractor by the use of disclaimer or exculpatory clauses. These clauses attempt to hold the Owner and Engineer blameless, and make the Contractor responsible for their actions. Each evasive clause introduced in Contract Documents is a risk imposed on the Contractor which he must translate into an increase in his bid price.

B. Identifying the Risks.

Risks should be distributed fairly between Owner and Contractor; particularly those risks related to the unexpected. There are the normal risks or uncertainties for the Contractor that obviously cannot be put in the category of the unexpected. Some of these normal risks are by nature acts of God, such as hurricanes, earthquakes, and extreme weather. Other normal risks include the unavailability of materials, labor problems, transportation costs, and accidents.

The critical risks for the Contractor are those that are difficult to anticipate, such as conditions encountered below the ground surface that differ from the information presented or difficult to anticipate; excessive variation between the actual quantities of work performed and the estimated quantities presented in the Unit price Schedule on which the Contract bid was based; and delays caused by the Owner.

A more productive policy for generating more bids at lower prices would be one that recognizes risks that the Owner is better able to assume and which he can better control; leaves to the bidder those risks over which the bidder has control; and shares those risks which neither can control.

C. Controlling the Risks.

The Contractor should not be held responsible for the adequacy of the Plans and Specifications; nor should the Owner disclaim responsibility for subsurface exploratory data obtained and presented to bidders; nor should the Contractor be required to notify the Engineer before performing any work, when the Plans or Specifications are at variance with any laws, ordinances, rules, or regulations; nor should the Contractor be responsible to satisfy himself of the accuracy of dimensions given on the Plans. Responsibilities that are clearly those of the Contractor include project scheduling, procurement of material and equipment, construction methods, selection of subcontractors, job safety, and timely completion of the Contract.

Risks and responsibilities that should be borne by the Owner include full disclosure of known information (see Article 8.10, Disclosure of Known Information); site access; prompt checking of the Contractor's drawings; and timely progress payments. Additional risks that should be assumed by the Owner include:

1. Differing Subsurface Conditions—when the Contractor encounters a subsurface condition that was neither anticipated nor planned for by the Contractor or Owner, the Contractor should be given consideration for the resulting additional costs and time. A procedure for handling this situation is explained in Article 4.3C, Differing Subsurface Conditions. A bidder should not be responsible for obtaining his own subsurface data. The reasoning is presented in Article 12.2E, Examination of Plans, Specifications, Special Provisions, and Site of Work; Paragraph 4, Subsurface Data.
2. Escalation Clauses—Contracts of long duration can experience steep increases in the cost of labor and material. These unknown additional costs can be shared by both the Owner and Contractor by including an escalation clause in the Specifications. This arrangement is explained in Articles 5.2R and 12.2R, Escalation Clauses.
3. Variations in Estimated Quantities—It is often difficult, if not impossible, for the Designer to estimate (with any degree of accuracy) quantities for the bidding period. When final pay quantities of major unit price items differ appreciably from the estimated quantities, improper compensation can result. This situation can be controlled by providing an arrangement for renegotiating the Contract unit price of a major item when its final pay quantity differs from the estimated quantity by more than a specified percentage. This risk and a method

of control are more fully explained in Article 4.3D, Variations in Estimated Quantities.

After it has been determined what risks the Owner is prepared to accept in the interest of obtaining the lowest practical bids, the Specifications should clearly indicate what risks the Owner is assuming; what risks are to be shared; and what risks the Contractor must assume. Specifications that are not clear on an equitable distribution of risks will most likely result in costly litigation.

8.14 Payments and Payment Items

A. Payments.

Prompt payments will help reduce a Contractor's financing needs. In addition to normal progress payments for work performed, payment should be provided for the following situations, when applicable:

1. Mobilization of plant and equipment. When plant and equipment that must be initially assembled on a project represent a substantial investment by the Contractor, a Mobilization payment item should be established. Article 10.4, Mobilization, presents the reasoning for establishing this item and the arrangement of payment.

The need for a Mobilization item was demonstrated to the author on the Reservoir Project described earlier in Article 4.3C, Differing Subsurface Conditions. The Unit Price Schedule in the Dam Contract did not include an item for Mobilization. Before the Contractor could begin his concrete operation he had to first assemble and erect on the site an automatic concrete batching and mixing plant, a vacuum cooling plant for cooling the concrete aggregates, and a 1200-foot span cableway with moveable head tower. The Contractor stated that this preparation involved an investment of over a million dollars. At the time (1959), this represented approximately 25 percent of the value of the Contract. According to the terms of the Contract, the Contractor would not get any return on this outlay until concrete placement would begin for the foundation of the dam. This operation did not get underway until 5-1/2 months after the Contractor had come on the site. Although the Contractor had no legal recourse, the author believes that the absence of a Mobilization item for early reimbursement had a dampening effect on relations between the staffs of the Contractor and the Engineer.

2. Materials and equipment stored on site. Payment for early purchase may be applicable when certain materials are in short supply or when critical items of equipment are involved. Failure to provide a provision in the Contract for early payment presented the author with an additional problem on the same Reservoir Project. The Contractor for construction of the Chemical Building

purchased the processing equipment early and had it delivered to the site. Cost was in the hundreds of thousands of dollars. The Specifications provided no payment for the equipment until it was permanently installed. Permanent installation of equipment is normally not made until the building is closed in and made weathertight. Being in short supply of funds, the Contractor decided not to wait for the structure to be closed in. He had the equipment uncrated and installed in place. He then enclosed it with plastic sheeting. This provided no protection against corrosion of finished surfaces and motors. Consequently, after final inspection, the Contractor was required to replace damaged motors and perform other corrective work. Additional discussion and the conditions for advance payment, are presented in Article 4.8I, Payment for Material on Hand.

B. Payment Items.

As mentioned earlier, the designer usually establishes the initial list of Contract payment items. This list should be reviewed for coverage of the Work. If changes are necessary, the designer should be consulted, because the preliminary cost estimate is based on his list of payment items. The criteria for determining whether an item of work is to be paid for on a unit price or lump sum basis has been presented in Article 1.3A, Fixed Price Contract.

The costs of all related operations of an item of construction should not be included in one Contract unit price for that item solely for the purpose of simplifying measurement and payment requirements. An example of this is the practice of some designers to include the cost of temporary trench sheeting into the unit price of the related pipe item. This arrangement provides a tempting incentive for the Contractor to omit installing trench sheeting when he feels it is not necessary, because it means "money in his pocket." Omitting temporary sheeting in trenches is a dangerous practice. Trench sheeting should be paid for separately under its own payment item classification. Additional discussion on the consequences of this practice, and a procedure for controlling it, are presented in Article 10.7F. Measurement and Payment; Paragraph 4a, Temporary Sheeting.

8.15 Limit the Engineer's Involvement in the Work

The Engineer should not be authorized to direct any of the Contractor's operations; nor should he be required to approve the Contractor's construction equipment, methods, temporary construction, or safety procedures. He should not be involved in areas that are the sole responsibility of the Contractor. This can only serve to expose the Engineer to litigation. The following two examples, noted by the author in a final review of Contract Specifications, illustrate improper involvement of the Engineer.

In presenting requirements for Temporary First Aid Facilities, the Specification stated: "Stretchers and blankets shall be located on the jobsite where required by the Engineer, and shall be maintained, protected and readily accessible for use at all times." It was recommended that the designation "Engineer" be changed to read "Contractor's Safety Engineer."

The second example involved the Engineer in directing the Contractor's operation. A Section entitled Underdrains concerned installation of perforated plastic pipe underdrain for roadway construction. In the subsection of Execution, the Specification stated: "Pipe for underdrain shall not be delivered on any part of the Project until directed by the Engineer. Pipe underdrain shall be constructed where indicated on the Contract Drawings, and as directed by the Engineer."

8.16 Methods and Results

When possible, it is good practice to avoid specifying the method to be followed by the Contractor in his performance of an operation. If the method is specified, and the final result is unsatisfactory, the Contractor cannot be held responsible so long as he can prove that he followed the Specifications. In this respect, if a method has to be specified, a required end result should not be specified. A common practice of specifying both method and result concerns placement of earth embankment. The Contractor is instructed to spread the material in loose layers of a specified maximum thickness, told the type and capacity of compaction equipment to be used, and the number of passes to make on each lift. He is then told that the compacted embankment must meet a specified minimum in-place density.

Where possible, the Specifications should relate to performance and end results, rather than how to achieve the end result. The more detailed the requirements are in controlling the Contractor's methods of performing the Work, the more responsible the Owner becomes for the outcome of the Work. Responsibility for the end result will rest with the party outlining the method that is to be followed. Additional information on Methods and Results is presented in Article 3.2.3F, End Results.

8.17 Presenting Instructions and Requirements

A. Specifications provide the vehicle for transmitting instructions to the Contractor. The instructions are not to be considered suggestions. Although the specification writer may have definite reasons for the requirements he specifies, he should not explain these reasons; they can become a cause of controversy. Justification for a requirement should have no place in the Specifications. Additional information on the purpose of Specifications is presented in Article 2.2, Function of the Specifications.

B. Once a requirement is stated it should not be repeated in detail. Although it may sometimes be desired to repeat a requirement for emphasis, repetitions can be a source of error in the Specifications. When corrections become necessary, they may be made in one place and overlooked in another where the same requirement has been repeated. Repetitions increase the content of the Specifications and there is always the possibility of mismatching the repetitions through omission or poor proofreading.

C. Special or unusual terms should be defined (see Article 4.2C, Definitions). Some examples of terms that require clarification are:

1. Fixed (Plan) Quantity
2. Specialty Item
3. Substantial Completion
4. Working Day

D. Cross references should be kept to a minimum and, when made, they should be specific. When reference is made to a particular Article, the Article number, title, and the title of Section containing it should be stated. Reference made to a specific Article or Paragraph implies reference to all paragraphs in that Article or to all subparagraphs in the Paragraph cited, unless otherwise noted.

8.18 Multi-Contract Projects

When two or more prime contractors are to occupy the same area, additional instructions are necessary to define the limits of their respective responsibilities and coordinate their various activities. One prime contractor has no control over the activities of another prime contractor. The Owner alone has this authority and it is his responsibility to coordinate the work of prime contractors and clearly define their responsibilities. Areas that require consideration include:

a. Cooperation between the Contractors, including preparation of a joint schedule of operations.
b. Designation of the Owner's representative (Engineer or Construction Manager), who will have authority to schedule and coordinate the work of prime contractors on the site.
c. Defining the limits of work and sequence of construction for each prime contract.
d. Availability of work areas.
e. Designation of temporary facilities, such as water, electrical power, sanitary facilities, and first aid facilities, for joint use including the responsibility for providing and maintaining them.

Additional discussion on multicontract projects is presented in Articles 4.4J and 11.4J, Cooperation Between Contractors.

Chapter 9

Specification Language

9.1 Introduction

An ideal situation in construction contracts would be one in which the Contract Documents clearly define the Work, the Contractor interprets the Specifications exactly as they were intended, he prepares his bid on this basis, and the Contract is completed to everyone's satisfaction. Unfortunately, it does not happen this way.

When there is a disagreement over the meaning of Specification language, the courts will most likely follow the fundamental rule that the language is interpreted against the interests of the party who wrote it. The determining factor is the meaning of the words as written, not what the specification writer thought he was saying.

It is not difficult to prepare a written statement which can be interpreted differently from what was intended. The writer knows what he means and he believes he presents it clearly. Someone else, however, may get a different meaning from it. As Specification language becomes more complex, the possibilities for ambiguity increase and the intent of the Contract becomes lost.

Good writing involves two basic elements; the writer must first have something to say and, secondly, it must be presented clearly, simply, and forcefully. If readers are to get the message, the Specifications must be presented in understandable language.

9.2 Guidelines

A. Since Contract Specifications are part of a legal document, fewer words will generally mean less risk of legal problems. Simple, short sentences will be appreciated better than long, complex sentences. An excess of words will frequently result in confusion, and may discourage careful reading. At the other extreme, incomplete sentences may often lead to misinterpretation. A good rule to follow is to mention everything that is necessary to minimize ambiguity but to say it in the fewest words consistent with that purpose. The following three examples, noted by the author in a final review of Contract Specifications, illustrate excessive and meaningless wording.

1. "Unsuitable materials shall be materials which do not meet the requirements of Articles 2.1 . . . ." This definition of unsuitable materials offered nothing new, and could have been deleted.
2. The following two paragraphs concerning the handling of electrical equipment should have been combined into one sentence:

   "F. Loading shall be done in a manner to prevent misalignment of parts.
   G. Unloading shall be done in a manner to prevent misalignment of parts."

3. "Cleanouts shall be that such as manufactured by Zurn, Josam, Wade, or approved equal." The words "that such" could have been deleted without detracting from the intent.

B. Introducing more than one thought into a sentence should be avoided. A sentence containing more than one instruction can lead to confusion. Wherever possible, important ideas or words should be placed near the beginning or near the end of a sentence. These are the locations that will get the reader's attention. The middle of a sentence usually gets less of the reader's attention. Each sentence should be so framed that a reasonable person could not read into it a meaning other than that intended by the writer.

C. There are only two parties to the Contract; the Owner, as represented by the Engineer, and the Contractor. Generally, references in the Specifications should be made only to the Owner, the Engineer, and the Contractor. Reference should not be made to Subcontractors, Engineer's Representative, or Inspector. Reference to the Owner, Engineer, and Contractor, should be by singular number and masculine gender. Reference may be made to Others, where referring to work to be performed by contractors or agencies not a party to the Contract. Where possible, the Others referred to should be identified. Designation of Bidder can be used when referring to activities before award of the Contract.

D. In addition: 1) The elementary rules of grammar should be followed; 2) Specified requirements should be definitive and in the positive, not in the negative. A good illustration of this guideline is contained in the following statement noted by the author during a final review of Contract Specifications. It concerned instrument monitoring which involved the reading of instruments by the Contractor for detecting ground movement and the movement of structures that could develop as a result of tunnel excavation operations: "Instrument monitoring is the reading of the installed instruments not read by the Engineer at defined time intervals." Suggested wording was: Instrument monitoring is the reading by the Contractor of the installed instruments, at defined time intervals; 3) A word should be repeated rather than using a synonym, because a synonym may have a slightly different meaning; 4) Names may be repeated when necessary. Using pronouns can sometimes cause confusion; 5) Nomenclature should be the same as that used on the Plans; and 6) When reference is made to work that is to be performed by contractors not a party to the Contract, it should be referred to as "work by others."

9.3 Proper Use of Terms

Words or terms improperly used can cause confusion and problems. They also may be an indication of the writer's unfamiliarity with the subject. Several examples are given in the following articles.

A. "Will" versus "Shall."

When an obligation on the part of the Contractor is indicated, it is to be associated with "shall." The word "must" should not be used. When there is an expression of intent on the part of the Owner or Engineer, it is to be associated with the word "will."

B. "Inspection" versus "Supervision."

The Engineer "inspects" the Work to ensure that it conforms to and is in accordance with the requirements of the Contract. He does not "observe" it; he "inspects" it. The Contractor "supervises," because he controls and directs the various construction operations.

C. "Conformance to" versus "In Accordance With."

Materials "conform to," and workmanship is "in accordance with," the Contract requirements.

D. "Option" versus "Alternative."

When it is the intention to grant the Contractor a choice, he is given the "option" to select. "Alternative" is used where the Owner or the Engineer retains control of the decision. When Bidders are required to submit "alternative" prices or bids, the Owner makes the final selection.

E. "Consists of" versus "Includes."

One term restricts and the other does not. "Consists of" is used where a fixed quantity or a specific list of items is required. For example: The data to be submitted shall "consist of. . . ." In this instance, the Contractor cannot be required to submit more than is listed. When minimum requirements are specified or major items are listed, and it is clear that additional items or requirements may be necessary, then "includes" is used. For example: The work "includes. . . ," or the submittal shall "include at least the following. . . ." With the proper use of these terms, there is no need to use the term "but not necessarily limited to."

F. "At the Contractor's Expense."

This term should not be used, as it may imply that the other work specified is not at the Contractor's expense. If there is a possibility of the Contractor misunderstanding who is to pay for a specific operation or material, the term "at no additional cost to the Owner" is preferable.

G. "Grout" versus "Mortar."

Interchanging these two terms is a common fault of specification writers, yet they have separate meanings. Terms like "grout mortar" and "dry packed grout," are incorrect descriptions. Although grout and mortar have the same basic ingredients of portland cement, sand, and water, their major difference is their consistency. One is flowable, the other is not. As defined in the American Concrete Institute's Publication, *Cement and Concrete Terminology:* "Grout is a mixture of cementitious material and aggregate to which sufficient water is added to produce pouring consistency without segregation of the constituents."

H. "Temporary."

This term should be used when referring to work that is to be provided by the Contractor for use during construction, and which is to be removed before completion of the Contract. This term should not be used when referring to work that is to be provided by the Contractor for use during construction and which is to be left in place for use by a subsequent contractor. Such work should be referred to as "interim" Work (See Article 4.2C, Definitions).

I. "Compacted Layers."

This term is incorrectly stated, as the following sentence will illustrate. The example concerned the placement of bituminous concrete for a roadway base course. It read: "The base course shall be placed in compacted layers not exceeding a maximum depth of five inches." The material cannot be placed in a compacted state. Compaction is obtained *after* the material is in place. The sentence should have read: "The base course shall be placed in layers with each layer compacted to a maximum thickness of five inches."

J. "Furnish and Place" versus "Construct."

In a final review of Contract Specifications, the author noted the following description of work specified in a Section entitled Cement Concrete Sidewalk: "The work specified in this Section consists of furnishing and placing Cement Concrete Sidewalks." The term "furnish and place" is associated with a prefabricated product or material that involves only a placing operation. A cement concrete sidewalk does not fit this classification as many construction operations are involved. The Specification should have used the term "constructing" instead of "furnishing and placing."

9.4 Ambiguous Wording

Ambiguous wording in the Specifications can be confusing because it can be interpreted differently by different readers. If the Contractor reasonably interprets an ambiguous requirement, it will usually be his interpretation that will prevail in the courts, and not that of the Engineer. This points up the necessity for clear and definite wording. When there is a doubt in the Contractor's mind as to what is intended, it provides an opening for him to propose substitutions and alternatives in his proposal and during construction.

Many words and terms of indefinite meaning place the decision of acceptability on the Engineer in the field. The Contractor, in preparing his bid, has no way of determining what the Engineer in the field may actually require. To offset this uncertainty, the Contractor will in most cases add a contingency sum for his protection, thus increasing the construction cost.

Some examples of indefinite words which should not be used are: 1) "reasonable"; 2) "workmanlike"; 3) "etc."—this implies unlimited additions. It is vague and indefinite; 4) "and/or"; 5) "similar"—this can mean identical or exactly alike. It can also mean a general likeness or resemblance.

The following are examples of expressions subject to different interpretation: 1) "workmanlike manner"; 2) "to the extent necessary"; 3) "reasonable time"; and 4) "first class workmanship."

Several examples of expressions that contribute nothing to the Specifications are: 1) "to the satisfaction of the Engineer"; 2) "satisfactory to the Engineer"; 3) "in an approved manner"; and 4) "as approved by the Engineer." These expressions require approval decisions by the Engineer, which can sometimes cause delays. Since all work on the project is subject to the approval of the Engineer, expressions like these can be omitted.

9.5 Vocabulary

Care should be exercised in the selection of words. Unfamiliar words, and words having more than one meaning, should be avoided. Complicated, high sounding words are not advised. This can be illustrated by presenting examples from Specifications that had been reviewed by the author. The words in question are in italic type.

A. In a General Requirements Section entitled Shop Drawings, Product Data and Samples, the Article titled Product Data contained the following requirement: "Certify compatibility of product with all other products with which it is to perform or with which it is to be placed *in juxtaposition*." Substituting the words "side by side" or "in contact" would have been more helpful.

B. This example concerned the laying of pipe in a trench. The Specification read: "Place buried pipe in *unmade* ground, with not less than four feet cover." The word "undisturbed" was substituted for "unmade."

C. This example concerned the compaction of backfill, and read: "Backfill shall be compacted to a density at least equal to that of the *contiguous* fill." A suitable substitute would have been "adjacent fill."

9.6 Spelling

Use of "thru" for "through" and other short-cut spelling used by part of the trades is not recommended. Standard spellings should be followed. Incorrect spelling and typographical errors in the Specifications can have serious consequences. Mistakes of this type can sometimes completely change the meaning of a sentence and cause disputes and extra costs.

9.7 Abbreviations

Standard abbreviations having only one meaning and which are to be used throughout the Specifications, should be listed in the General Conditions (see Article 4.2B, Abbreviations). An abbreviation that will be used in only one Section may be identified either at the beginning of the Section or the first place in the Section where it is to be used. If an abbreviation is to be used only once or twice, or at widely spaced intervals, it may be better to spell out the words.

The use of abbreviations that may have more than one meaning, such as "pf," which can mean power factor or point of frog, should be avoided.

Abbreviations that are used only on the Plans, such as "NIC" (not in Contract), should not be used in the Specifications.

Some abbreviations, like "in." for "inch" and "ft." for "feet," provide little saving in typing or space. This type of abbreviation should be avoided.

9.8 Capitalization

Standard capitalization, such as the first word of a sentence, proper nouns, names of agencies, titles, names of persons, and names of days, months, and holidays, should be adhered to. When used in the Specifications, the following words and terms should be capitalized:

a. Addendum (No.)
b. Bidding Documents
c. Bid Item
d. Change Order
e. Construction Manager
f. Contract (when it applies to the Contract of which the Specifications are a part)
g. Contract Documents
h. Contractor (a party to the Contract)
i. Engineer
j. General Conditions
k. Notice to Contractors
l. Notice to Proceed
m. Payment Bond
n. Performance Bond
o. Plans (when it applies to the Plans of the Contract)
p. Special Provisions
q. Specifications
r. Standard Specifications
s. Titles of Sections and Articles in the Specifications.

When there is a doubt, it is best not to capitalize. Also, one should be consistent in capitalizations throughout the Specifications.

9.9 Punctuation

Compared with most other types of writing, the mandatory and legal character of Specifications requires precision in wording and punctuation. An error in punctuation or a misplaced comma can sometimes change the entire meaning of a sentence and provide the basis for a claim.

Standard English punctuation should be used to convey the message clearly and to ensure proper interpretation by those reading the Specifications. Commas should be used sparingly. Sentences should be planned so as to require a minimum of punctuation. Punctuation that does not clarify the text should be omitted.

9.10 Numbers

The following procedures are recommended:

a. Numbers one through nine should be expressed in words. Numbers 10 and above should be expressed in figures.

b. When numbers are in a sequence they should all be presented in the same form. For example, "eight, nine, and ten" or "9, 11, 13 and 14."
c. A number should not be expressed in both words and figures. For example, "three hundred (300)."
d. A zero should be placed in front of the decimal point for figures less than one. For example, "0.14 inch."
e. Typewriter characters should not be used for fractions, such as "¼." They can be illegible on a bad copy. Also, there is no way to make them consistent with fractions that are not on special keys, such as "3/8." It is best to write out all fractions with full numbers and a slash mark. For example, "three and one-half inches" should be presented as "3-1/2 inches."

9:11 Symbols

The use of symbols can lead to typing errors which may not be caught during proofreading. They may also be difficult to read on a badly reproduced copy. Words such as "feet," "inches," "and," "percent," "degrees," "plus," and "pounds" should be used instead of using ′, ″, &, %, °, +, and #.

The symbol "$," for dollars, may be used in the Bid Schedule.

Chapter 10

Specifying the Technical Sections

10.1 Introduction

Requirements specified in the Technical Sections should be technically adequate and economically sound. To be technically adequate, the requirements should present complete details of the materials to be furnished. This should include a description of the methods for selecting and testing samples of the materials, specifying the tests and other measures that the Engineer will take to assure compliance with Contract requirements, and clearly stating the results desired. The Specifications must be coordinated with the Plans if conflicts and ambiguities in the requirements are to be avoided. To be economically sound, material and construction requirements should be adequate for the use intended, and economical. The Specifications should not ask for more than is required.

It would be impractical if not impossible, in this Chapter, to attempt a detailed discussion of all items involved in engineered construction. Accordingly, coverage will be limited to those items of construction most commonly encountered. The author will endeavor to introduce the specification writer to a working relationship with the items of construction that are to be discussed here. Chapter 3, which describes the Technical Section and explains the purpose and makeup of each of the five subsections, should be consulted with use of this Chapter.

The Standard Specifications issued by the State Departments of Transportation represent a good source of information for specifying items of engineered construction.

10.2 General Guidelines

A. Disposition of Removed Structures and Materials (see Article 3.2.3B, Disposition of Removed Material)

Whenever structures, equipment, or materials are specified for removal by the Contractor, instructions on their disposition must also be furnished. Three procedures for disposal are normally available, as discussed in the following paragraphs.

1. Salvage: The Specifications indicate the specific items of removed equipment or materials that are to be turned over to the Owner. This can include the hauling of excavated material and stockpiling it at a designated location off the site.

2. Reuse in the Work: The most common example of this type of disposal involves soil material removed in an excavation and found suitable for reuse in embankment or backfill.

3. Disposal Off the Site: Materials and equipment not disposed of under paragraphs 1 and 2 above become the property of the Contractor and he is required to dispose of them off the site. Requirements and additional information governing the use of private property by the Contractor, for disposal of material, are presented in Articles 4.3L and 11.3L, Disposal of Material Outside the Work Site.

B. Samples.

Where possible, samples of proposed materials should be required from the Contractor. The initial approval of samples, before proposed materials are purchased by the Contractor, can minimize delays and costs. Additional information on Samples is presented in:

1. Article 3.2.2B, Samples
2. Article 4.5D, Samples, Tests, Cited Specifications
3. Article 11.5D, Samples, Tests, Cited Specifications

C. Tests and Testing.

The Contractor should not be assigned the responsibility for such quality assurance functions as batch plant inspection; casting, curing, and breaking concrete test cylinders; confirming in-place density of compacted embankment; or taking cores to verify thickness of pavement courses. Testing for quality assurance is, and should remain, a responsibility of the Owner's site representative. The Contractor however, should not be relieved of his responsibility for quality control testing. Test results should be specified to accompany material samples.

Those tests to be performed by the Engineer, and the frequency of taking them, should be specified. Having this advance information, the Contractor can be prepared for any resulting delays and disruptions to his operations. Additional information on Tests and Testing is presented in:

1. Article 3.2.2E, Tests (Contractor)
2. Article 3.2.3H, Tests (Engineer)
3. Article 4.5D, Samples, Tests, Cited Specifications
4. Article 8.12, Quality Control and Quality Assurance
5. Article 11.5D, Samples, Tests, Cited Specifications

D. Tolerances.

If Specifications are to be realistic, allowable tolerances must be specified. Any effort to reduce natural variations in materials will only increase costs. Additional information on Tolerances is presented in:

1. Article 3.2.2C, Tolerances
2. Article 3.2.3D, Allowable Tolerances
3. Article 4.4E, Conformity With Plans and Specifications
4. Article 8.11, Tolerances
5. Article 11.4E, Conformity With Plans and Specifications

E. Knowledge of Construction Practice.

It is always helpful to have a working knowledge of the item of construction being specified. Working with Specifications that are unrealistic or impossible to comply with makes the Engineer's job of enforcing the Specifications that much more difficult. Additional information on Knowledge of Construction Practices is presented in Articles 8.9, Know Your Subject, and 15.3, Field Experience.

F. Consistency in Presentation.

Information and instructions should be presented in a style that is consistent throughout the Technical Sections. When Special Provisions to a set of Standard Specifications are being prepared, the style of presentation should follow that of the Standard Specifications. Inconsistencies in presentation (and their recommended corrections) are illustrated in subsequent Articles 10.2H through 10.2L, which discuss the five subsections of a Technical Section.

G. Specification Preparation Without Benefit of Client's Standard Specifications.

When the Client does not maintain a set of Standard Specifications, it can sometimes be difficult to obtain previously prepared Specifications for a comparable project (see Article 8.5, Utilizing Specifications of Previous Contracts). In this situation, an available source of information is the published volume of Standard Specifications for Road and Bridge Construction, prepared by the local State Department of Transportation (DOT).

The State DOT's Standard Specifications are especially helpful in providing information on locally available materials. Also, requirements in these Specifications are familiar to contractors who have done previous work in the State.

Technical Sections of these Standard Specifications can be incorporated into and made a part of the Contract Specifications by direct reference. This can be accomplished in a manner similar to that used when utilizing a client's Standard Specification. Information on this procedure is presented in Article 13.4, Standard Specifications and Special provisions.

On the other hand, if there is no desire to incorporate a Technical Section of the DOT Specifications, the specification writer can utilize what information he desires, reword it as necessary, and incorporate it.

H. Description (See Article 3.2.1, Same Title).

Since the Description subsection is the first portion of a Technical Section to be read, it should provide a concise but complete description of the work involved in the Section. It may be necessary for purposes of clarification to identify related work that is not included in the Section.

When preparing a set of Special Provisions, the opening statement in each Special Provision Section must make reference to the specific Standard Specification that it is incorporating. An example of citing a specific Standard Specification is illustrated in Article 13.4A, Presentation.

When incorporating a Technical Section of the local State DOT Standard Specifications, a sample wording for the opening statement would be: "All work under this Section and the payment therefor, shall be in accordance with the requirements of Section 205, Roadway Excavation of the ______ State Department of Transportation, Standard Specifications for Road and Bridge Construction, except as modified herein."

Repetitive use of the following wording or similar wording in the opening statement should be avoided: "The Contractor shall provide all materials and perform all labor in connection with. . . ." This type of statement contributes nothing and constitutes superfluous wording. A statement to this effect is generally included in the General Conditions of the Contract and, as such, applies to all the Work. An example of wording for an opening statement is presented in Article 3.2.1, DESCRIPTION. Note the use of "includes" rather than "consists of," for reasons presented in Article 9.3, Proper Use of Terms.

Consistency in wording for similar presentations should be maintained throughout the various Sections. When different wording is used it can cause confusion and misinterpretation. The author examined a published copy of Contract Specifications and compared the wording of the opening statement for each Technical Section. What he found provides a good illustration of what not to do. The following twelve different wordings were used for beginning the opening statement in the various Sections:

1. "The work described in this Section consists of. . . ."
2. "The work in this Section consists of. . . ."

3. "The work of this Section consists of. . . ."
4. "This work consists of. . . ."
5. "The work specified in this Section consists of. . . ."
6. "The work of this Section shall consist of. . . ."
7. "This work shall consist of. . . ."
8. "The work shall consist of. . . ."
9. "This Section shall consist of. . . ."
10. "This specification covers. . . ."
11. "The document covers. . . ."
12. "The Pilot Wire Cable consists of. . . ." (The Section title was Pilot Wire Cable.)

In describing the work of a Section, the Contractor should be given a general idea of the work involved.

An example of an inadequate description was noted in the final review of a set of Contract Specifications. The Section was titled Reinforced Cement Concrete Pipe and the opening statement read: "This work shall consist of furnishing and laying reinforced cement concrete pipe of the size and kind herein specified, including excavation, backfill, materials necessary for the placement of pipe, the removal of excess or unacceptable material and the satisfactory completion of the work." In addition to containing many unnecessary words, this opening statement gave no indication of where the pipe was to be laid, nor for what purpose. Investigation indicated that the pipe was to be used in constructing storm sewers. A simpler and more informative opening statement would have been: "This work includes furnishing and laying reinforced cement concrete pipe for storm sewers, including excavation, backfill and all else necessary to complete the work as specified."

I. Materials (See Article 3.2.2, Same Title).

1. National reference Standards should be used where possible (see Article 8.6, Using Reference Standards).

2. Material classifications that can be misinterpreted, or not completely understood, should be defined. For example, classifications like unclassified excavation, rock excavation, common excavation, select backfill, impervious backfill, and granular backfill, should be defined. An explanation of the various classifications of earthwork is presented in Article 10.7A, Classifications and Definitions.

3. When the same material is specified in more than one Section, detailed requirements are rarely repeated; they are normally referred to the basic Section. When referring requirements to another Section, care should be exercised in the selection of wording. If the referral is not worded properly, the Contractor may be presented with an opportunity to request payment under both Sections for the

same material. Double payment for the same material is discussed in Article 4.8D, Scope of Payment.

4. Additional information on Materials is presented in:

a. Article 4.5B, Source of Supply and Quality Requirements.
b. Article 4.5K, Unacceptable Materials.
c. Article 5.2P, Material Guaranty.
d. Article 11.5B, Source of Supply and Quality Requirements.
e. Article 11.5K, Unacceptable Materials.
f. Article 12.2P, Material Guaranty.

J. Construction Requirements (See Article 3.2.3, Same Title).

1. Arranging requirements in the order of sequence of construction will help reduce the possibility of overlooking a requirement that should be specified.
2. Limitations or restrictions that can affect performance of the work specified in the Section should be clearly spelled out.
3. Protective measures to be taken by the Contractor should be specified when the Work may be affected by freezing weather.
4. When manufactured products are involved and when applicable, it should be specified that their installation shall be in accordance with the manufacturer's written instructions.

K. Method of Measurement (See Article 3.2.4, Same Title).

It should be stated that measurements for payment will be made only for work that has been acceptably performed within the limits shown on the Plans, or ordered by the Engineer. This is to discourage contractors from claiming payment for work that may be unacceptable or for work that may have been performed outside the designated limits. Each presentation of Method of Measurement should clearly indicate how the measurements will be taken; whether from Plan dimensions or of the work as actually constructed. The limits of these measurements should also be defined.

Consistency in wording is important, as stated earlier in this Chapter. The following example noted in the final review of a set of Contract Specifications illustrates inconsistency in wording that can occur among the various Sections. There were seven different wordings for beginning the opening statement under the Method of Measurement, as follows:

1. "This work shall be measured for payment. . . ."
2. "Measurement for this item. . . ."
3. "The measured quantity under this item. . . ."

4. "The quantity to be measured for payment under this item. . . ."
5. "The amount of material to be paid for under the Contract. . . ."
6. "The unit of quantity to be paid for under this item. . . ."
7. "The quantity to be paid for under this work. . . ."

The suggested wording for beginning the opening statement for unit price items (as previously illustrated in Article 3.2.4, METHOD OF MEASUREMENT) is: "The quantity of (name of item) to be paid for will be the number of (unit). . . ." This presenation can be used throughout those Sections where unit price items are involved. When lump sum items have to be dealt with there are no measurements for payment to be made. In this situation the wording can read: "Mobilization will not be measured separately for payment." or "Mobilization will be paid for on a lump sum basis wherein no measurements will be made."

It should be noted that, in addition to the suggested wording for unit price items, each presentation must also outline how the measurements will be taken and how the quantities for payment will be computed. It should not be simply stated that the item will be measured by the linear foot or by the square foot, and let it go at that. This can spell trouble. This is illustrated in the following case history: The work involved temporary steel sheet piling for the support of excavation. Sheet piling was to be paid for on a unit price basis. Under Method of Measurement it was simply stated that the support of excavation would be measured by the square foot; no other instructions were given. Since this was to be a temporary installation it was the Contractor's responsibility to design the sheet pile wall. He designed a wall with a much deeper embedment of sheet piling than was considered necessary. Temporary sheet piling is normally paid for on the basis of the area of the exposed face of the sheeting, with measurements being taken in a plane parallel to the face of the wall. The Specifications, however, failed to specify the vertical limits of measurements to be taken and the Contractor seized this opportunity to claim payment for an area based on the total length of sheeting installed, including embedment.

Additional information on the subject of Method of Measurement is presented in Article 4.8B, Measurement of Quantities.

L. Basis of Payment (See Article 3.2.5, Same Title).

Every item of work acceptably performed by the Contractor has to be accounted for as regards payment for it. If it is not paid for separately, it has to be included in payment for another item. When defining payment for an item of work, related work whose costs are also included in the Contract price for that item should be specified. Examples illustrating this are presented in Article 3.2.5. The specification writer should also identify those items of related work whose costs are included for payment under other payment items.

In presenting payment items, their names should be identical to the names listed in the Unit Price Schedule.

10.3 Organizing the Sections

Establishing a tentative list of required Technical Sections should be one of the first steps in preparing Contract Specifications. This can be accomplished by studying available Contract Plans and listing the various items of construction involved. The Plans will probably only be about 50 percent complete at this stage, but they should present enough advance information to enable the specification writer to get started. Notes or other information accumulated by the specification writer from correspondence or preliminary briefings will also help. Having this information on items of construction, it should not be too difficult to put together a tentative list of Sections.

Most owners of civil works construction projects maintain their own Standard Specifications (see Article 6.1B, Standard Specifications, Supplemental Specifications, and Special Provisions). State Departments of Transportation generally have patterned Section arrangement and Section titles after the standards established by AASHTO. Recent projects involved with the construction of rapid transit systems have patterned their Standard Specifications after the 16 division format sponsored by the Construction Specifications Institute.

Many counties, municipalities, public authorities, and private owners do not have their own Standard Specifications. In these situations, the specification writer has to determine the procedure he will follow in establishing Section titles and numbers, and arrangement of the Sections. Some guidelines are offered in the following paragraphs.

A. Consult the current published Standard Specifications of the local State DOT (see Article 10.2G, Specification Preparation Without Benefit of Client's Standard Specifications).

B. If new Section titles have to be established, select titles that are brief and descriptive of the work involved. For example: 1) ROADWAY EXCAVATION AND EMBANKMENT; 2) STORM DRAINAGE; or 3) CAST-IN-PLACE CONCRETE.

C. In establishing Section coverage, the intent should be to include in the Section all work that is related to the particular construction item. Thus all instructions relating to the item would be presented in that Section. If this line of thinking is not followed, the reader may have to thumb through pages of other Sections in order to get complete information on a particular item. The following example, noted in a final review of Contract Specifications for a rapid transit project, illustrates a deviation from this line of thinking: The Specifications contained three Sections with these Titles:

Section 2610-1052, UNDERDRAINS
Section 2615-0252, UNDERDRAIN OUTLETS
Section 2615-1252, UNDERDRAIN CLEANOUTS

Since the work of these three Sections concerned underdrains, the requirements should have been consolidated into one Section titled UNDERDRAINS. In preparing multiple Sections dealing with the same item of construction, the specification writer can mistakenly duplicate requirements or duplicate payment for the same work. Multiple Sections can also cause misinterpretations by the Contractor. At the least, this arrangement requires reading the contents of three Sections in order to determine the requirements.

D. Avoid establishing a separate Section for specifying the testing requirements. Tests and testing should be included in the same Section with the requirements specified for the same item of work.

Additional information on Section arrangement is presented in Articles 3.3, Section Arrangement, and 13.3A, Arrangement of Sections.

10.4 Mobilization

When a payment item for Mobilization is included in a Contract, it is there for the purpose of reimbursing the Contractor for his start-up expenses at a time when he needs it most: in the early stage of his Contract. This item covers costs for the establishment of offices and shops, temporary facilities, storage areas, and the movement of personnel, equipment, and supplies to the site.

Payment for Mobilization, which is on a lump sum basis, begins with the first progress payment estimate. If this item was to be presented as a typical bid item, some bidders might be tempted to unbalance their bid by increasing the bid price for this item in order to receive an early payment much larger than their actual start-up costs. This can be controlled however by utilizing one of the arrangements described below.

A. A Fixed Price is established for this item, thus preventing the submittal of an unbalanced price for Mobilization. The Fixed Price is inserted in the Unit Price Schedule, requiring all bidders to use this Fixed Price in arriving at their total Contract bid figure.

B. The item for Mobilization is presented as a bid item. Bidders are instructed in the Bidding Documents that the price bid for this item cannot exceed a specified percentage (between five and ten percent, depending on the size of the Contract) of the total Contract bid price.

Various procedures for progress payments on this item have been used. The author prefers the following Basis of Payment for its simplicity and timely payment: Mobilization will be paid for at the Contract Lump Sum price as listed in the Unit Price Schedule. Fifty percent of the Lump Sum price will be paid on the first monthly payment estimate following the Contractor's moving in of all the necessary facilities indicated in the Description that will enable him to begin work on Contract items. The remaining fifty percent will be paid at the rate of ten percent in each of the next five monthly payment estimates. The Lump Sum price also includes the costs of demobilization upon completion of the Work.

Additional information on Mobilization is presented in Article 8.14A, Payments; Paragraph 1, Mobilization of Plant and Equipment.

10.5 Demolition

Demolition of existing structures is usually necessary when the construction right-of-way passes through a populated area. The requirements governing demolition in urban areas are naturally more numerous and exacting than those for rural areas. The public must be protected, as well as adjacent properties. The following checklist may be useful when preparing requirements for building demolition:

a. Indicate when the structures to be demolished are to be made available to the Contractor.
b. Determine the local regulations governing demolition.
c. Submittals required from the Contractor, before beginning demolition of buildings, may include: 1) Evidence that all utilities and services have been discontinued; 2) Details for capping and plugging the disconnected utilities; and 3) Construction details of temporary sidewalk canopies required for protection of the public.
d. Contractor to obtain razing permits, if required.
e. Indicate whether the Contractor will be allowed use during the Contract of any buildings situated outside the limits of construction.
f. Extermination of rodents prior to demolition, if buildings are located in populated areas.
g. Indicate whether use of explosives is permitted or prohibited. This would be applicable when concrete structures are involved.
h. Contractor is prohibited from holding public sales on the site of salvaged materials or equipment.
i. All glass is to be removed from buildings before beginning demolition.
j. Control of dust is critical in populated areas.
k. Indicate whehter Contractor will be permitted to move any buildings off the site, or whether they all must be demolished in place.
l. Payment is normally made on a lump sum basis per each building or structure.

Additional information on Demolition is presented in Articles 3.2.3A, Preliminary Preparations, and 3.2.3C, Procedures.

10.6 Clearing and Grubbing

This work is normally limited to areas of construction. Tree stumps can remain in place under certain conditions. This would be permitted in areas of embankment where the fill exceeds a specified minimum thickness; also when clearing reservoir areas that will be under water. On certain projects, the Contractor may be permitted to bury his tree stumps on the site. Usable timber from wooded areas is generally salvaged.

Clearing and Grubbing is generally paid for on a lump sum basis. When the limits of work cannot be clearly defined, payment is established on the basis of a price per acre.

10.7 Earthwork

A. Classifications and Definitions.

The classification of materials to be encountered in the Work or furnished by the Contractor should be clearly defined. A misinterpretation of a classification can affect work performance and cause problems in determining proper payment under the many Earthwork items. Many State DOT Standard Specifications present definitions of their standard classifications.

1. Earth Excavation and Rock Excavation.

Earth excavation and rock excavation should be defined when they are each being paid for separately. When boring information indicates little or no rock and the excavation is designated as unclassified, it should be defined as the removal of "all materials of whatever nature encountered." One set of Specifications defined earth excavation as the removal of "all kinds of materials." Because of this unclear definition, one contractor was able to collect his additional costs of removing rock in excavating for pipe trenches. The court ruled that the phrase "all kinds of materials" was ambiguous, especially when considered in connection with the meaning of earth excavation as commonly accepted in construction contracts.

2. Embankment or Fill Versus Backfill.

Embankment or fill should not be confused with backfill. When spaces from excavation are being refilled, the operation should be referred to as backfilling and the material used as backfill material. When soil material is being placed upon the surface of existing ground and the resulting construction will be higher than the adjacent ground surface, this should be referred to as embankment or fill and the material used, as embankment or fill material.

3. Borrow Material.

The designation of "borrow material" is a reference to the source of supply and not to the characteristics of the material. Borrow material is material obtained from sources off the site. It generally has to meet specific gradation requirements, such as those specified for an impervious or a granular material.

B. Control of Soil Erosion.

There are public laws and regulations controlling pollution of the environment. Temporary erosion control measures have to be specified to control the run-off from "raw" slopes of cuts and embankments. A description of the various measures for controlling soil erosion can usually be found in the local State DOT Standard Specifications.

Additional information on controlling pollution of the environment is presented in Articles 4.6V and 11.6V, Environmental Protection.

C. Excavation.

1. Notification to Utilities.

The Contractor should be advised in the Description subsection of his responsibility to give timely notice to the Utilities of his proposed excavation operation. Requirements for digging test pits to confirm location of the underground facilities should include sheeting and bracing, dewatering, backfilling, and surface restoration. If pits are to be excavated in paved areas, surface restoration under the item of Test Pits should consist of temporary pavement. Additional information on Notification to Utilities is presented in Articles 4.6R and 11.6R, Contractor's Responsibility for Utility Property and Services.

2. Disposition of Materials.

Materials unsuitable for reuse in the Work must of course be disposed of off the site. Suitable material may be reused in the Work for embankment, fill, backfill, or select material, as applicable. Suitable material that is surplus can be disposed of in various ways. It may be used in the Work to flatten slopes of embankments; disposed of off the site at a location designated by the Owner; or it may become the property of the Contractor for his disposal off the site. Additional information on Disposition of Materials is presented in: a) Article 3.2.3B, Disposition of Removed Material; b) Article 4.3L, Disposal of Material Outside the Work Site; c) Article 10.2A, Disposition of Removed Structures and Materials; and d) Article 11.3L, Disposal of Material Outside the Work Site.

3. Roadway Excavation.

Site conditions or other limitations that may affect the Contractor's excavation operations should be mentioned. If the Contractor is required to follow a specific procedure, this should be spelled out. Otherwise the Contractor should be permitted to pursue his own methods to produce the desired results.

Existing topsoil should be removed before beginning the excavation operation. The Specifications should indicate the thickness of topsoil to be removed, usually four to six inches. The stripped topsoil should be stockpiled on the site if it is suitable for reuse in the Work as determined by the Engineer. Unsuitable topsoil should be disposed of off the site.

When excavating in rock, the Contractor must comply with State and Local regulations governing the use of explosives. Regulations cover storage of explosives, transportation from magazine to jobsite, and safety in handling. Specific requirements include the use of flagmen to warn and keep persons and traffic from the danger area, prevention of damage to property, and the scaling of loose rock from final surfaces. One of the prime objectives in the excavation of rock is to reduce overbreak by controlled blasting techniques. A booklet explaining four major methods of controlled blasting to reduce overbreak was prepared some years ago by the Explosives Department of E. I. DuPont DeNemours and Company, Wilmington, Delaware 19898. Additional information on rock excavation is presented in Articles 4.6K and 11.6K, Use of Explosives.

Allowable tolerances should be specified for final slopes and elevations. These tolerances can be established with the design department.

4. Structure Excavation.

Structure excavation covers many situations, which in turn will generally dictate the methods to be used and the precautionary measures to be taken for the protection of persons and adjacent construction. Excavations left open during nonworking hours are to be enclosed by barricades or temporary fencing. Excavated material should not be stockpiled within a specified minimum distance of the sides of the excavation. The allowable maximum length of excavated trench in advance of pipe laying should be limited to a specified figure, such as 200 feet. When the bottom of a pipe trench is in rock, it is customary to have rock removed to six inches below trench grade. This allows for a bedding course of granular material to provide uniform support of the pipe along its full length. Operations associated with structure excavation include dewatering and support of the excavation.

a. Dewatering. Since most foundations and pipelines are required to be constructed "in the dry," attention must be given to the details for dewatering an excavated space. Except for a ruptured pipeline, water in an excavated space will usually result from a high watertable or from surface run-off. It is customary to require that the watertable be lowered to a specified minimum distance (generally 12 inches) below the bottom of the excavation. When it is necessary to lower the watertable, this can cause settlement of adjacent structures. Limitations may therefore have to be specified on the dewatering method.

Environmental considerations may make it necessary to specify acceptable methods for the disposal of water removed from excavations. In city areas where the pump discharge may have to flow directly into a storm drainage system, it may be necessary to require settling basins to limit the amount of solid particles carried to the storm drains. There should also be a requirement for the Contractor to flush the storm drains at the end of the job to remove accumulated solids. Since wellpoints used in dewatering excavations are equipped with screens, the pump discharge is usually free of solids. Dewatering cofferdams generally presents no environmental problems because the bottom has usually been first sealed with concrete.

If the accidental flooding of an excavation after it has been dewatered can seriously affect the work, the Specifications should require standby pumping equipment. When considered necessary, it should be specified that the Contractor submit to the Engineer for approval his proposed method and equipment for dewatering his excavation.

b. Additional Depth Excavation. When an excavation has been completed to Plan elevation, the subgrade material may be found unsuitable for supporting the proposed structure or pipe. To provide for such a possibility, the Specifications should include a provision that the Contractor may be ordered to remove additional material at the bottom of an excavation, when in the Engineer's opinion it will not support the new construction.

Additional depth excavation may also require additional sheeting and dewatering, plus backfilling the space with suitable material. The suitable material may consist of a granular material or a low strength concrete, depending on the particular requirements. Material and placement requirements for the low-strength concrete fill should be specified in the Concrete Section. Details of payment for the additional work are presented in Article 10.7F, Measurement and Payment; Paragraph 3e, Additional Depth Excavation.

5. Support of Excavation.

An excavation more than four or five feet deep with vertical sides will normally require additional support to maintain its sides (OSHA requirements). In open country where conditions will permit, the Contractor may be given the option to slope back the sides of his excavation. Sloping back the sides of a pipe trench should be limited to the area above elevation of the top of pipe.

a. Timber Sheeting and Bracing. Timber sheeting and bracing is generally used in excavations to a depth of 15 feet. It is commonly used in supporting the walls of trenches, particularly trenches for sanitary sewers which are often more than ten feet deep. Installed sheeting should project a minimum of six inches above the top of the excavation. Since sheeting and bracing is a temporary installation, it is removed as the backfilling progresses. Where removal of sheeting may disturb an adjacent utility, the Engineer may order the sheeting to be left in place. The Specifications should specify that the top of sheeting to be left in place is to be cut off at a specified minimum distance below ground surface.

The Specifications should also require that when sheeting is to be left in place, the bracing and walers are to remain in place to prevent inward movement of the sheeting.

In lieu of installing sheeting and bracing in a trench, which is a time consuming operation, some contractors may request approval to use a steel trench box inside which the pipelayers can work in safety. This works well in areas having no underground intersecting utility lines. Its use in City streets is usually prohibited because underground house connections and other utilities that lie in the path of the box will be disturbed and possibly damaged.

b. Open Cofferdams. Open cofferdams are commonly used in the support of excavation for the construction of bridge piers and abutments in water. Being of a temporary installation, cofferdam design becomes the responsibility of the Contractor. The design is submitted to the Engineer for approval. Allowable tolerances in location and alignment of the cofferdams, plus design criteria, should be provided in the Specifications. When no longer required for the Work, cofferdams should be removed to at least the level of the stream bed.

c. Soldier Piles and Lagging. Vertical sides of deep excavations are frequently supported by walls constructed of soldier piles and lagging, restrained by tiebacks. The Contractor designs the support system in accordance with design criteria outlined in the Specifications. The specialty subcontractor designing and installing the support system should be required to meet the minimum experience qualifications specified in the Specifications. Tieback testing procedures should include a pullout proof test on each installed tieback to verify that it will carry its design load.

When deep excavations are adjacent to existing structures, such as those encountered in the construction of underground stations for rapid transit systems, the continued safety and use of the existing structures must be ensured. A preconstruction survey of the conditions of these adajacent properties is normally required to be conducted by the Contractor, in the presence of the Engineer, prior to execution of the Work. Movement of ground and structures is then monitored throughout the construction period through use of a system of markings and instrumentation. The Contractor should be required to hire a registered professional engineer to conduct the pre-construction survey and supervise the installation and monitoring of instruments and reference points. Photographs should be taken by a commercial photographer. The maximum allowable movements in existing construction should be specified.

6. Removal of Existing Pavement, Sidewalk and Curb.

When a portion of existing pavement is to be removed, it should be specified that the joint to be left in the existing pavement is to be saw cut to produce a neat, straight line at the surface. The saw cut in concrete pavement should be not less than two inches deep. If concrete pavement is to be broken up by use of a ball or hydraulic punch, its use should not be permitted within five feet of existing structures or existing joints that are to remain.

It should be specified that removal of an existing sidewalk or curb is to be made to the nearest joint in the sidewalk or curb.

D. Backfill.

Material for backfill may be suitable material from on-site excavation or material from borrow pits off the site. It can also be a select material furnished from outside sources. Material from excavations which is to be used for backfilling should in general be free of rubbish and organic matter, contain no oversized stones (say, six inches maximum), and have a limited percentage passing the No. 200 sieve (say, 50 percent). Select material is just what its name denotes. The material may have to be processed by selective screening and blending to meet specific requirements. The select material may be a granular material with a specified maximum percentage passing the No. 200 sieve. It can also be an impervious material with a specified minimum percentage passing the No. 200 sieve. Trench backfill placed around the lower half of the pipe is generally a granular material. When backfill is to be placed against the waterproofed surface of a buried structure, the Specifications should state that stones in backfill material that is to be placed within 12 inches of the structure shall be limited to a maximum size (generally 3/4 inch).

Requirements for the placement and compaction of backfill material will vary depending on space limitations, size and shape of structure, and the intended purpose of the backfill.

1. It should be specified that backfill shall not be placed against new structures until they have developed sufficient strength to withstand the stresses placed upon them by the backfilling operation.

2. The placement of backfill material should generally be limited to a specified maximum thickness of loose layers.

3. When backfill is to be placed around a structure, it should be specified that the material is to be spread equally around all sides of the structure. This is required in order to equalize the pressure on the structure.

4. Backfill around pipe laid in a trench should be placed on both sides of the pipe simultaneously. Unequal placement of backfill can cause lateral movement of the pipe with possible opening of the joints. Placement and compaction of backfill in a pipe trench generally follows a prescribed procedure. The first lift is placed to about one-third the height of the pipe. The backfill should be rammed under the haunches of the pipe using hand shovels, to ensure that all voids are filled. The second lift is placed to about threequarters of the height of the pipe. Compaction by manually operated tampers is generally specified for these two lifts. The rest of the trench backfill is compacted by normal methods.

5. No matter how well the Contractor complies with requirements for placing and compacting backfill, there will be some settlement before the backfill stabilizes. When backfilled areas in roadways have to be opened to traffic, temporary

pavement should be specified. The requirements for temporary pavement should be presented in the Section on Pavement.

E. Embankment.

Material for embankment will initially come from on-site excavations. Should there be insufficient suitable material available from excavations, the remaining material will have to be obtained from sources off the site. Following are some checklist items to be considered:

1. Requirements for preparation of the ground upon which embankment is to be placed.
2. Maximum loose thickness of each layer of embankment material spread; also required moisture content and minimum in-place density.
3. If borrow material is required, specify whether the material is to be supplied by the Owner or by the Contractor. Also, requirements for final treatment of borrow area after the removal of material has been completed.
4. Allowable tolerances for final slopes and elevations.
5. Soil erosion controls.
6. Frequency with which in-place density tests will be taken by the Engineer.

Information on procedure and minimum in-place density, is presented in Articles 3.2.3F, End Results, and 7.3B, Test Method Standard.

F. Measurement and Payment.

1. General.

a. Since it is difficult, if not impossible, to predict the necessary limits of most items involving earthwork, payment is generally based on the actual quantities of work performed. One exception is the use of a predetermined Fixed (Plan) quantity to establish the final payment quantity for a unit price item, should there be no change to the Plans. For example, the Method of Measurement clause for a unit price item having a Fixed (Plan) quantity would read: "The quantity of Roadway Excavation to be paid for will be the Fixed (Plan) quantity in cubic yards shown in the Unit Price Schedule, computed by the method of average end areas using data obtained from preconstruction surveys made by the Engineer and the excavation dimensions shown on the Plans." A Fixed quantity in the Unit Price Schedule would be identified by adding the word "Fixed" immediately under the quantity listed in the Estimated Quantity column. Additional information on the subject of Fixed quantity is presented in Articles 4.8C and 11.8C, Fixed (Plan) Quantities.

b. The Contract unit price for an excavation item normally includes the costs of disposal of the material, whether it be reused in the Work or be disposed of at off-site locations. It is important to specify that the suitability of excavated material for reuse in the Work will be determined by the Engineer.

c. When the Contractor is given the option and elects to slope back the sides of his excavation, payment generally remains unchanged from the design concept of vertical sides and support of excavation.

2. Control of Soil Erosion.

Since the extent of work required for controlling soil erosion cannot be anticipated in advance, payment is established on a unit price basis for each type of control provided.

3. Excavation.

a. Test Pits. If minimum plan dimensions of test pits are specified, payment can be established on a vertical foot basis measured from ground surface to bottom of pit. If pit sizes will be variable then payment can be established on basis of the volume of material removed. The Contract price will generally include the costs of sheeting and bracing, dewatering, backfilling, and surface restoration (including temporary pavement).

b. Roadway Excavation may encompass more than one payment item.

(1) The removal of topsoil is an operation separate and distinct from the normal excavation operation. It should therefore be paid for separately. Stripping Topsoil can be paid for on a volume (cubic yard) or area (acre) basis. The Contract unit price should include the cost of stockpiling the topsoil on the site for reuse in the Work or the cost of its disposal off the site if it is not to be reused.

(2) If little or no rock is anticipated, a pay item description of Roadway Excavation, Unclassified, would be appropriate. If rock is determined to be present, separate payment items of Roadway Excavation, Common (Earth), and Roadway Excavation, Rock, should be provided. Boulders over one cubic yard in volume should be measured and paid for as rock. The Contract price for rock excavation should include the costs of scaling loose rock from final surfaces.

c. Structure Excavation. Excavation performed for relatively large structures, such as bridge foundations and buildings, is generally paid for on the basis of a unit price per cubic yard for the volume of material removed from within specified limits. Trench excavation for pipelines may be paid for separately or it may be included for payment in the pipe items and related structure items, depending on local practice or on the choice of the designer.

When trench excavation is paid for separately on a cubic yard basis it may, as in the case of sanitary sewer construction, consist of several payment items, each classified according to the depth below ground surface. For example, pay item descriptions may read:

Trench Excavation at Depth of 0 Feet to 10 Feet
Trench Excavation at Depth of 10 Feet to 15 Feet
Trench Excavation at Depth of 15 Feet to 20 Feet

This is established to more realistically reflect the cost of trench excavation. It is more costly to remove one cubic yard of soil from a sheeted trench at a depth of 18 feet than it is to remove it at a depth of 10 feet.

Trench width payment limits are normally shown on the Plans and are generally equal to the outside diameter of the pipe plus 12 inches to 24 inches, depending on the pipe size. Excavation payment limits for manholes, inlets, and similar structures are generally shown on the Plans. The volume to be paid for would lie within vertical planes located 12 inches outside the limits of the foundation or base of the structure. When structure excavation is not paid for separately and is included in the Contract prices for the various pipe and related structure items, it should be so specified in the Basis of Payment subsection.

If there should be a possibility of encountering rock, a separate payment item for Rock Excavation should be established. Also, a separate payment item should be established for Granular Bedding Course to be placed in trenches for uniformly supporting the pipe.

Temporary Fencing or Barricades placed around excavations left open during nonworking hours should be paid for separately on a linear foot basis. Otherwise the Contractor may not be very diligent in providing this safety protection.

d. Dewatering is rarely paid for separately. The costs for dewatering an excavated area and maintaining it in a dry condition are generally included for payment in the Contract price for the related excavation or cofferdam item.

e. Additional Depth Excavation ordered by the Engineer is paid for separately on a cubic yard basis. In-place measurements are taken before and after the removal of unsuitable material. The Contract price per cubic yard generally includes the costs of disposing of the removed material, necessary additional sheeting, and additional dewatering. Since the space is usually backfilled with concrete or a select granular material, this work is measured and paid for separately. It should be stated that payment for Additional Depth Excavation will be made only for material ordered by the Engineer to be removed.

4. Support of Excavation.

a. Temporary sheeting for bracing the walls of pipe trenches should be paid for separately on a square foot basis. Trench cave-ins which have resulted in numerous fatalities have been caused principally by the absence of sheeting and bracing in the trench. Specifications which require the Contractor to include his costs of temporary sheeting in the Contract prices of pipe items encourage the dangerous practice of attempting to do work without the protection of sheeting and bracing. By establishing a separate payment item for this temporary work there should be no incentive for the Contractor to omit it. The method of measurement for Temporary Sheeting has to be clearly outlined in the Specifications. For example:

> "Temporary sheeting installed in accordance with the Contract requirements will be measured for payment as follows: 1) The area in square feet of continuous or tight

sheeting will be the product of the vertical length of sheeting measured from six inches above the ground surface to six inches below the bottom of the excavation, and the total horizontal length of sheeting measured along both sides of the trench; 2) Where the sheeting is not tight, the horizontal length will be equal to the sum of the widths of the individual sheets measured along both sides of the trench."

The Specifications should indicate that if the Contractor is granted permission to slope back the sides of his excavation, payment will still be made on the basis of a sheeted trench having vertical walls.

b. Sheeting Left in Place. A payment item should be provided for those situations where the Engineer may find it necessary to order the Contractor to leave sheeting in place. Since the top of sheeting left in place has to be cut off to a specified minimum distance below the surface of the ground, the Specifications should define the upper and lower payment limits. The Basis of Payment should specify that the unit price per square foot for Sheeting Left in Place includes the cost of cut-offs and their disposal.

c. Open cofferdams are generally paid for on a lump sum basis for each cofferdam constructed. The lump sum price should include the costs of dewatering and necessary removal of the cofferdam.

d. Soldier Pile Walls. The support system consisting of soldier piles, lagging, and tieback anchors for a deep excavation is normally paid for on a lump sum basis. The lump sum price includes the costs of designing, furnishing, installing, testing, maintaining, and removing the excavation support system. The costs of performing the pre-construction survey may be paid for separately or be included in the lump sum price.

5. Backfill.

Requirements for material to be used in backfilling excavated spaces will vary, depending on its function.

a. Backfilling with On-Site Excavated Material. The cost of disposing excavated material is generally included in the excavation price. Consequently there would be no separate payment when backfilling with material that comes from on-site excavations.

b. Select Granular Material used for backfilling additional depth of excavation, as a bedding course, and for backfilling around underground pipelines, is paid for on a cubic yard basis. Payment quantities are determined from in-place measurements taken within the limits shown on the Plans or ordered by the Engineer.

Truck load measurements for determining payment quantities are not very accurate because of the bulking factor in loose material. If truck measurements must be used, it should be specified that loaded trucks are to be struck off to the top of the truck body. Heaped truck loads are difficult to measure.

6. Embankment.

Embankment constructed with material that comes from on-site excavations would not be paid for separately if these costs are included in the Contract price for the excavation item.

Embankment constructed with material obtained from sources off the site will be measured and paid for separately as Borrow Excavation. For determining payment quantities, measurements should be taken of the material in its original position at the borrow pit rather than in its final position. Earth disturbed in the excavation process can require years of consolidation before it settles to its original volume. If embankment is to be constructed of material coming from both on-site excavations and from borrow pits off the site, then measurements for determining the payment quantity for Borrow can be taken only at the borrow pit. Original cross sections should be taken at the borrow pit after the area has been cleared and the topsoil removed. The Contract price per cubic yard for Borrow Excavation should include the costs of obtaining the borrow pit, clearing the site, removing topsoil, hauling and placing the material, and final clean-up and treatment of the borrow pit.

10.8 Underground Pipelines

A. Introduction.

Most of the pipeline work involved in engineering construction concerns storm drains, water mains, and sanitary sewers. Many requirements and problems in underground pipeline construction are common to all types of pipelines. Also, there are construction requirements and problems that are peculiar to each particular type of pipeline.

B. Common Requirements.

1. Requirements on excavation, sheeting and bracing, and backfilling, should be referred to the appropriate Technical Section.
2. Requirements for pipe joints should not be overlooked. They are important in pressure lines and in sanitary sewers, where leakage can create problems.
3. Pipe in the bottom of the trench may be supported on natural soil, on a bedding of granular material, on a concrete cradle, or be completely encased in concrete, depending on soil conditions in the trench.
4. When pipe is to be laid in an embankment area, it should be specified that embankment must first be placed to at least two feet above the top elevation of the pipe, before the pipe may be laid. Pipe should always be laid in a trench, never on the surface.
5. Where gravity flow is involved, it should be specified that the pipe is to be laid in the upgrade direction with the bell end pointing upgrade.

6. Backfill is not to be placed in the trench until the laid pipe has been inspected and approved.

7. Pipe crossings are often required under existing highways and railroads. When traffic at these locations cannot be interrupted by an open trench operation, the pipe is installed by jacking, augering, or tunneling under the highway or railroad. In the case of storm drains, if the carrying pipe is large enough, the pipe itself may be jacked under the facility. Otherwise, a sleeve is installed or a passageway is tunneled. The open space between sleeve and carrier pipe is sealed at each end of the crossing to prevent the entrance of material that can undermine the highway or railroad tracks. A working pit is required at each end of the crossing. Limits of the crossing including the pits, are shown on the Plans. The Contractor is required to submit for approval complete details of his proposed method of installation including working pits, and indicate the type of casing he proposes to use.

8. When pipe is to be laid in a paved area, temporary pavement or sidewalk should be specified following completion of the backfilling. Permanent pavement or sidewalk should be delayed as long in the Contract period as possible to allow the maximum time for backfill to settle. The Contractor shall maintain the temporary pavement or sidewalk.

C. Common Problems.

1. The laying of pipe in traveled areas can cause a disruption in the movement of traffic and may also present hazards. The Specifications should provide adequate requirements for the maintenance and protection of vehicular and pedestrian traffic. Signing, safeguards, and other required arrangements should be specified to be completed and in place before construction may begin.

2. When excavating pipe trenches in populated areas, the possibility of encountering an uncharted underground utility or other obstruction is always present. The Specifications should include provisions for resolving unanticipated subsurface conditions encountered by the Contractor (see Articles 4.3C and 11.3C, Differing Subsurface Conditions).

Another common problem when trenches are being dug in populated areas is the practice of many contractors to store and stockpile materials on private property, particularly grassed lawns. The Contractor's responsibility for respecting private property and repairing any damage, must be spelled out (see Articles 4.6L and 11.6L, Protection and Restoration of Property and Landscape).

3. It has been common practice among designers to include the cost of temporary sheeting in the related pipe item or excavation item. The only thing to be said in its favor is that it eliminates the field measurements and office computations necessary to determine payments to the Contractor, if it were a separate payment item. However, as pointed out earlier in Article 10.7F, Measurement and Payment; Paragraph 4, Support of Excavation, the consequences of the

dangerous practice by contractors of not installing sheeting, which this method of payment seems to encourage, far outweighs any minor savings in field administration costs. Therefore, separate payment should be established for temporary sheeting required in pipe trenches.

4. Settlement of backfill. No matter how well earth backfill is compacted, some settlement will occur. Unlike a roadway embankment operation where compaction is well organized and controlled, the compaction requirements for trench backfill are generally not rigidly enforced. Many owners, particularly the smaller municipalities, have maintained that enforcing compaction requirements for the full depth of the trench would increase construction costs. Consequently, compaction is enforced only for the backfill from bottom of trench to top of pipe. Above this level the backfill gets only minimal if any compaction. Surface traffic and the weather are depended upon to provide whatever consolidation is deemed necessary. This situation will sometimes present problems after a heavy rainfall. Therefore, temporary pavement should be a must before the permanent pavement is laid.

D. Storm Drains.

1. This work can involve new construction or the relocation of existing facilities. When existing storm drains are to be modified or relocated, it should be determined whether surplus castings of existing structures are to be salvaged for the Owner.

2. Leakage tests should not be specified for storm drains; it is unnecessary. In some wet areas the design may even call for the joints to be open in order to facilitate the collection of water.

E. Water Mains.

1. This item will usually involve relatively minor work such as extending existing water lines, adding hydrants, or relocating existing facilities. Here again, it should be determined if the removed material and equipment are to be salvaged.

2. Since these are pressure lines, trench depths are minimal. The depth of trench needed to provide the minimum cover over pipe is normally dictated by the frost protection requirement for the area.

3. Leakage tests and other requirements of the Utility Owner that apply to the Contractor, should be specified.

4. If the Contractor is to disinfect the new lines, these requirements should be specified.

5. An excellent reterence and source of information are Standards published by the American Water Works Association (AWWA) 6666 West Quincy Avenue, Denver, Colorado 80235.

F. Sanitary Sewers.

1. When this type of work constitutes a minor operation of the Contract, it will generally involve relocation or modification of existing lines and manholes. A good reference and source of information is ASTM C 12, Recommended Practice for Installing Vitrified Clay Pipe Lines.

2. When sanitary sewer work consists of the construction of an entirely new collection system in a community where none existed before, it can develop into a multicontract project. When this type of construction is performed in an established community of paved streets and numerous underground utilities, the problems generated and controls required are unique to this type of work. The author discovered this when he became involved in one of these projects by assuming the responsibility of Resident Engineer, after construction had been underway for approximately six months. Article 11.3F, Maintenance and Protection of Traffic, presents a description of this project with its related problem of controlling traffic in local streets. Another problem generated on this project was noise pollution. Its disrupting effect is illustrated in the case history presented earlier in Article 4.6H, Public Convenience and Safety. Some items to consider in preparing the Specifications are:

a. Permits will have to be obtained from the proper authorities for permission to open State, County, or Municipal roadway pavement. The Contractor's responsibility in obtaining permits should be specified.
b. Sewer laterals are generally laid along the centerline of the street. Since this usually requires closing the street to vehicular traffic, construction schedules of the various prime contractors have to be closely coordinated. It should be specified that the Contractor is to make available during working hours an emergency lane for use by firefighting equipment and ambulances. He should also be required to provide during non-working hours a travel lane for the use of residents on the street. Before work is to begin in a street, residents should be given 48 hours advance notice.
c. Streets shall be maintained in a dust-free condition at all times.
d. For manholes located in easements on private property, a locking type of cover should be specified.
e. For manholes located in streets, it should be specified that contact surfaces of frames and covers are to be machined to ensure a non-chattering fit.
f. Many municipalities will require the Contractor to maintain new pavement for a period of two years following completion of the Contract. This is to take care of the inevitable settlement that will occur because of inadequate compaction of the trench backfill.
g. The Contractor should be required to patrol his work area immediately after a rain to check for trench settlements in traveled areas. There have been instances where a resident has backed his car out of the driveway

only to have the rear wheels sink into a backfilled trench that had settled as the result of a rain.

h. The Contractor should be required to remove all surplus materials upon the completion of work in a street area. Failure to do this results in complaints from residents on whose property the materials have been stored.

i. Completed lines are to be tested for leakage by the Contractor in the presence of the Engineer. When the groundwater table is above a specified level with relation to the sewer to be tested, an infiltration test is to be performed. When the groundwater table is below that level, an exfiltration test is to be performed by filling the line with water to provide an average hydrostatic head of five feet or at least a head of one foot at the upper end of the line. Test duration may vary from four hours to eight hours.

G. Measurement and Payment.

1. General.

a. Most pipe items are paid for at a unit price per linear foot, and pipeline structures at a unit price per each structure. The Contract price may include the costs of all related work. On the other hand, certain items of related work like excavation, sheeting, and select backfill, may be paid for separately. Whichever arrangement is decided upon should, for consistency, apply to all pipeline and similar work in the Contract.

b. Concrete cradles or concrete encasement is generally paid for separately on a cubic yard basis or on a linear foot basis, whichever is applicable.

c. Pipe crossings under highways and railroads are paid for on a lump sum basis for each crossing.

2. Storm Drains.

Drainage structure castings (frames, covers, and grates) are sometimes paid for separately on the basis of a unit price per pound of casting.

3. Water Mains.

a. The Contract unit price per linear foot of pipe will also include the costs of leakage tests and sterilization.

b. Water main accessories such as fire hydrants and valves are generally paid for on a per each basis. The unit price per hydrant would include the costs of the shut-off valve and branch line from the main.

4. Sanitary Sewers.

a. The Contract unit price per linear foot of pipe will also include the costs of leakage tests and temporary pavement. If the Contract unit price should include the cost of all related work, this arrangement would require multiple payment items for the same diameter sewer pipe, classified according to the depth of pipe invert below ground surface. Pay item descriptions would read as follows:

12-Inch Diameter Vitrified Clay Pipe, Extra Strength
0 to 10 Feet in Depth
12-Inch Diameter Vitrified Clay Pipe, Extra Strength
10 to 12 Feet in Depth
12-Inch Diameter Vitrified Clay Pipe, Extra Strength
12 to 14 Feet in Depth

b. Manholes which have to accommodate to the same variable depth of installation as for sewer pipe, would be paid for by the number of vertical linear feet of each type manhole constructed and accepted. Measurements to the nearest tenth of a foot would be taken from the bottom of the base to the top of the manhole cover.
c. Pipe fittings such as wyes and tees would be paid for on a per each basis for each size wye and tee.
d. Permanent pavement would be paid for separately.
e. Controlling dust in the streets is an on-going important maintenance item. Establishing this for separate payment would assure better cooperation from the Contractor and thus reduce the number of complaints from residents. Payment would be on a per ton basis for calcium chloride used, or on an hourly basis for the use of a water sprinkler truck.

5. Additional Information.

Additional information on Measurement and Payment is available in: Article 3.2.4, Method of Measurement; Article 3.2.5, Basis of Payment; Article 4.8B, Measurement of Quantities; Article 4.8D, Scope of Payment; Article 8.14B, Payment Items; Article 10.2K, Method of Measurement; and Article 10.2L, Basis of Payment.

10.9 Roadway Pavement

A. Introduction.

The subject of roadway pavement covers both temporary and permanent pavement. Temporary pavement is needed for detour roadways, over backfilled excavations, and for any other temporary purpose. Permanent pavement, both flexible and rigid, includes subbase course, base course, binder course, and surface or wearing course. The best available source of information on local practices and materials for pavement construction are once again, the Standard Specifications of the local State Department of Transportation.

B. Guidelines.

1. Plan details of the proposed pavement should be studied and the indicated pavement courses specified.
2. Requirements for subgrade preparation should be specified.
3. When an asphaltic concrete base course is to be placed on an aggregate subbase or soil subgrade, a prime coat should be specified.
4. When an asphaltic concrete binder or surface course is to be placed upon a previously laid course of asphaltic concrete or portland cement concrete base, a tack coat should be specified.
5. Thickness acceptance of pavement is based on test cores taken from the finished pavement. Frequency of cores to be taken and the basis of acceptance, should be specified.
6. When trenching is required in paved streets, it should be specified that an initial sawcut of at least two inches deep be made in the pavement.

C. Measurement and Payment.

1. Costs of temporary pavement placed over backfilled areas are generally included for payment in the Contract price of related items. Temporary pavement for a specified detour could be paid for on a square yard basis with the Contract price covering the cost of all pavement courses including subgrade preparation. If the detour is clearly defined between specified limits it could be paid for on a lump sum basis. Depending upon job conditions, removal of the detour and its restoration could be included in a lump sum price for the detour or paid for under an excavation item.

2. The cost of subgrade preparation for permanent pavement is frequently included in the Contract price for the subbase or base course. When paid for separately, it is on a square yard basis or sometimes on a basis of roadway stations of 100 feet (this would also include the subgrade for shoulders).

3. Aggregate subbase course and aggregate base course may each be paid for on a square yard basis for a specified course thickness. If the course will be of a variable thickness, then the unit of measurement should be the cubic yard.

4. Bituminous concrete courses may each be paid for on a square yard basis for a specified course thickness. If a course is to be of a variable thickness, then the unit of measurement would be the ton.

5. Portland Cement Concrete Pavement of a specified thickness is paid for on a square yard basis. If it is of a variable thickness, such as in bridge approach slabs, the unit of measurement would be the cubic yard.

6. Prime coat and tack coat would each be measured for payment by the gallon of material used.

7. A procedure for payment adjustment has been established by some owners for pavement that has been found by test cores to be deficient in thickness beyond allowable tolerances but still within a limit that would allow the pavement to be useable. This procedure is outlined in the AASHTO Guide Specifications for Highway Construction, Section 524, Basis of Payment for Portland Cement Concrete Pavement. It may also be specified in the local State DOT Standard Specifications.

10.10 Bearing Piles

A. Introduction.

Materials commonly used in bearing piles are timber, steel, and concrete. Bearing piles may be divided into two general classifications according to the manner in which they develop their capacity to support loads. End bearing piles transmit most of their load through their tips to a firm substratum. Friction piles develop their capacity to support loads from the friction developed between the pile surface and the soil through which they are driven.

Average length of piles determined from the quantities in the Proposal, are for estimating purposes only. Before permanent piles are ordered, test piles are usually driven and load tests performed to establish their safe load-bearing capacity. The ordered lengths of piles are established by data obtained from test piles and pile load tests.

The most basic pile hammer in use is the falling weight called a gravity or drop hammer. Other types of hammers are air, steam, vibratory, diesel, and hydraulic. Single acting hammers are essentially gravity hammers in which the weight is lifted by pressurized air or steam instead of a hoist line.

Because of the hammer's high impact velocities which, if exerted directly on the pile's head, would damage it, the head of the pile is protected by a pile cap that rests directly on the pile. The pile cap contains a cushion block, usually of wood, to serve as a shock absorber.

Sometimes piles have to be driven in water or to a depth where the top or butt of the pile will be below the surface and below the reach of the hammer. In such situations, an auxiliary member called a follower is usually interposed between the pile and the pile hammer to transmit the driving energy to the pile. The follower must be sufficiently rigid to resist bending or buckling during driving.

B. General Guidelines.

1. The type of piles to be provided, and their general locations, should be specified.

2. Pile design bearing capacity should be stated.
3. Where applicable, specify that in areas to receive bearing piles, the excavation is to be completed before piles may be driven.
4. Driving criteria as applicable should be specified:

   a. Minimum tip elevation.
   b. Minimum pile penetration.
   c. Final blow count, which is generally expressed by the number of blows per inch for the last six inches. If the Contract includes test piles and load tests, the final blow count is established after test piles have been driven and load tests successfully completed.
   d. Driving sequence for pile groups. Driving generally starts at the center and works toward the perimeter.
   e. If pile driving is to take place in close proximity to concrete placement, it should be specified that piles are not to be driven within 60 feet (or other established distance) of concrete that is less than seven days old.

5. The Contractor should be required to submit data on the driving equipment he proposes to use.
6. When driving becomes difficult in granular types of soil, the Contractor should be permitted or authorized to use a water jet to facilitate pile penetration. The Contractor should be required to submit details of his proposed jetting operation. Specify the conditions and limits under which jetting will be permitted. Jetting should be limited to a specified distance (generally three feet to six feet) above the anticipated final tip elevation of the pile. The final three to six feet of pile penetration is to be driven without use of a water jet.
7. Pre-augered or predrilled holes may sometimes be necessary in sensitive areas or where driving may be difficult. This would not apply to open-end pipe piles.
8. The locations of test piles are usually indicated on the Plans. When multiple test piles and load tests are required for the same type of pile, it should be specified that the load test must be successfully completed before the Contractor may drive and load the next adjacent test pile.
9. Allowable tolerances should be specified:

   a. Deviation from Plan location, at cut-off elevation—generally three inches.
   b. Deviation from the vertical for plumb piles—generally 1/4 inch per foot.
   c. Deviation from specified slope for batter piles—generally 1/4 inch per foot.

10. It should be specified whether splices will or will not be permitted. If permitted, specify the maximum number of splices allowed in one pile, and include splice requirements.
11. Where steel pipe and steel shell piles are involved, the Contractor should provide a suitable drop-light for inspecting the inside of each pile throughout its length.
12. If pile cut-offs are acceptable to be spliced to other piles, this should be so stated. The allowable minimum length of pile cut-off acceptable for reuse should be specified. The disposition of pile cut-offs not reused should be specified.
13. It should be indicated whether a pile rejected by the Engineer must be pulled or whether it may remain in place. Rejected pipe or shell piles allowed to remain in place may be filled with sand. Concerning damaged piles, the Contractor should be required to submit his proposed corrective measures to the Engineer for approval.
14. When reference is made to a "marine" environment, it is intended to refer to salt water.
15. A useful reference is the Section on Bearing Piles in Division II Construction, of the Standard Specifications for Highway Bridges, adopted by the American Association of State Highway and Transportation Officials (AASHTO).

C. Test Piles.

Test piles may be specified for different reasons. In some instances they are specified to confirm boring information and required pile lengths. At other times in a search for the most economical pile, several different types of test piles will be driven and load tested. In the driving of a test pile, the behavior of the pile for each foot of penetration is observed and the information noted. All test piles are not given load tests. Load tests for specific test piles are designated by the designer.

It should be required that test piles be identical in manufacture to the permanent piles. Test piles are to be driven by the same equipment proposed for driving the permanent piles. Ordered lengths for test piles are generally established ten feet longer than the estimated in-place length of permanent piles. Test piles are normally driven in permanent pile locations. Because it is the simplest formula, the Engineering News pile driving formula is most often used in driving the first test pile.

D. Load Tests.

A pile load test is a dependable method for approximating the actual bearing capacity of a pile. It is far more reliable than pile driving formulas for determining the safe loading capacity of a pile. To describe it simply, the load test is a method by which a load is applied to the pile in increments while observations are made of the resulting settlement.

The total test load to be applied to a test pile is generally established at twice the design bearing capacity of the pile. The safe allowable load of a pile so tested is normally considered to be one-half the test load under which the net settlement of the pile is not more than 1/4 inch. Sometimes it may be desired to continue loading a test pile until failure of the pile occurs.

Two methods of load testing are generally used. One is the platform method in which a loading platform is supported by the pile and the test load applied directly to the pile. The other is the anchor-pile method where the load is applied to the pile by jacks acting against an anchored reaction frame. The Contractor should be required to submit to the Engineer for review, details of his proposed method for performing the load test, including arrangement of equipment.

Requirements governing performance of load tests are presented in ASTM D 1143, Standard Method of Testing Piles Under Static Axial Compressive Load. This Standard can be modified as necessary to reflect the particular requirements of the Contract.

The minimum time should be specified after a test pile has been driven, before the test load can be applied. The soil surrounding a driven pile must be allowed time to settle and stabilize so that it can perform its function in supporting the pile.

If cast-in-place concrete piles are specified, determine if the piles to be test loaded have to first be filled with concrete before the test load is applied. If they do, it should be specified that concrete in the test pile must be at least three days old before applying the test load.

E. Timber Piles.

Timber bearing piles are generally of Southern yellow pine or Douglas fir. Piles for permanent structures are normally peeled or debarked. Minimum tip and butt dimensions should be specified for piles to be furnished. Material requirements for timber piles can be found in ASTM D 25, Standard Specification for Round Timber Piles.

Piles that are to be situated permanently below the groundwater table normally do not require a preservative treatment. However, timber piles located in seawater require protection against marine borers and other wood-destroying agents. When portions of piles are to be situated above the groundwater table, the piles should be treated. The type and quality of preservative treatment should be specified. Standards for preservative treatment of timber piles can be found in the publications of ASTM; The American Wood Preservers Association, 1625 I Street, N.W., Washington, D.C. 20006; and the American Wood Preservers Institute, 1945 Gallows Road, Vienna, Virginia 22180.

Steel driving points should be specified when needed to protect the tips of timber piles during driving.

Splicing a timber pile is difficult and costly. Close attention therefor should be given to the procedure for establishing the ordered lengths of timber piles and by whom. Some Specifications state that the Engineer will give the Contractor a list of the lengths of piles to be ordered. Other Specifications make this the responsibility of the Contractor.

F. Pipe Piles.

Pipe piles are either open end or closed end. Open end piles are utilized where driving vibrations and the resulting subsidence may be an item of concern, and where obstructions may be encountered. When hard driving is anticipated with possible damage to the tip of the pile, an open end cutting shoe is specified for attachment to the tip. Closed end pipe piles are usually fitted with a conical steel point. As an open end pile penetrates the soil, material inside the pile is removed intermittently. A plug of soil is maintained in the bottom of the pile until the pile reaches its final penetration. The open end concept also permits some type of socketing arrangement when it is desired to dowel the pile into bedrock.

Pipe piles are easily spliced by means of drive-fit sleeves. Sleeves may either be flush on the inside or on the outside of the pipe. Piles can also be installed by jacking, where headroom is limited. Some guidelines are:

1. Material requirements for pipe piles can be referred to ASTM A 252, Standard Specification for Welded and Seamless Steel Pipe Piles.
2. After a pile has been cleaned out, the open top should be temporarily covered to prevent entry of foreign material.
3. Before concrete is deposited in a pile, the interior should be inspected and accepted by the Engineer.
4. Water should be removed from inside a pile prior to concreting. If the Contractor is unable to dewater a pile, he should be required to submit details of his proposed concrete tremie operation.
5. Concrete should not be deposited in a pile until all piles within a specified radius (10–15 feet) have been driven. Depositing concrete should be a continuous operation with no cold joints.
6. Piles extending above the ground or water surface require a protective coating.

G. Shell Piles.

The distinction between pipe piles and shell piles is simply the wall thickness. Shell piles are usually corrugated, of light wall thickness, and require the use of an internal support (mandrel) for driving. The mandrel is cylindrical, tapered or step-tapered, and may be expandable. When inserted into the pile it acts as a rigid core that distributes the driving energy along the engaged length of the pile. Virtually all of the rules regarding integrity, straightness, dryness, and end closures of pipe piles apply to shell piles as well.

H. Precast (Prestressed) Concrete Piles.

Precast concrete piles may be of normal concrete with conventional reinforcement or of prestressed concrete. Precast prestressed concrete piles are manufactured in accordance with the applicable requirements for prestressed concrete. The piles are prestressed by tensioning the steel strand reinforcement before concrete is placed in the form. As the concrete hardens it bonds to the tensioned reinforcement. When the concrete reaches a specified compressive strength, the tensioned strands are released. This prestresses the concrete by placing it under compression. Precast piles are generally manufactured to predetermined lengths and shapes in off-site fabricating plants. They are delivered to the job by truck, rail, or barge. If the piles to be precast are too long (generally over 75 feet) for safe transportation to the job, a casting yard is set up on the jobsite. Some guidelines are listed in the following paragraphs.

1. The storing, transporting, and handling of precast concrete piles should be done in a manner to avoid excessive bending stresses, cracking, spalling or other injurious results. Information on pile support locations, and pick-up points and devices, should be included with fabrication details that are to be submitted by the Contractor.

2. Off-site fabricating plants should possess a minimum of experience (generally five years) in the manufacture of precast concrete.

3. Cement in the concrete for piles that are to be located in seawater should be Type II moderate sulfate resistant cement. The portions of piles to be exposed to splash zones and tidal water, should be given a protective coating.

4. The splicing of precast piles is generally not permitted. Extensions or build-ups are time-consuming and expensive. If a pile extension or build-up is deemed necessary, the Contractor should submit his construction procedure to the Engineer for approval. The allowable maximum height of pile build-up should be specified (normally three feet). When constructing a pile build-up, the top of the pile in place is cut away to expose the reinforcement for a length of 40 diameters. Sometimes an alternate method consisting of grouting dowels in holes drilled in the top of the pile is permitted. Pile build-up requirements for prestressed piles will generally follow those for precast concrete piles with conventional reinforcement, except that the strand reinforcement would be exposed.

5. When 54-inch prestressed concrete cylinder piles are being specified, it may be desired to require the Contractor to provide facilities for the Engineer

to descend inside the driven pile for a visual inspection. The interior of piles in place may be filled with sand and topped with a cast-in-place or precast concrete cap.

I. Steel Piles.

Steel piles are generally of structural shapes having a width and depth of the same dimension. They are often referred to as H-beams or H-piles. A steel pile is an end bearing pile and is usually specified when hard driving through boulders or other obstructions is anticipated. Steel piles are often specified to be driven to refusal on bedrock. Refusal is generally considered to be reached when five blows of the hammer produce a total penetration of not more than 1/4 inch. Pile accessories are available such as hardened steel tips to protect the tips of piles in hard driving, and splicers for facilitating the splicing of piles.

Steel piles are vulnerable to corrosion, particularly if they are part of seawater installations or if they are being driven through fills of organic materials, cinders, or other potentially harmful material. The need for protection against corrosion must be investigated. Corrosion protection is generally provided through cathodic protection or with a coal tar epoxy coating. Input material for specifying cathodic protection is normally provided by the electrical designer.

Coal tar epoxy is commonly used for protecting steel piles against corrosion. It is applied in two coats in the shop or in the field to a total minimum dry film thickness of 16 mils. The coating of piles to be in seawater should cover the parts of the pile that will be in the tidal zone or splash zone, and extend at least five feet below low water. Coated piles should not be stored in direct sunlight for longer than a month. Coated piles have to be handled carefully so that the protective coating is not punctured or removed. Surfaces to be coated have to be properly prepared before application of the coating. The Contractor's attention should be called to the manufacturer's safety requirements dictated by the flammability and potential toxicity of coal tar epoxy products. Information on specifying a coal tar epoxy coating can be found in Volume 2 of the Steel Structures Painting Manual, produced by the Steel Structures Painting Council (SSPC), 4400 Fifth Avenue, Pittsburgh, Pennsylvania 15213.

J. Measurement and Payment.

1. General.

a. The installation of bearing piles includes both fixed and variable costs. Some of these costs cannot be equitably reflected if they all are to be included in one Contract price per linear foot of pile. The Contract should be evaluated to determine which costs represent a significant enough expense to warrant consideration for separate payment.

b. The mobilization of pile driving equipment will generally represent the largest fixed cost. This cost may be appreciable if the piles are to be driven in water and barge mounted equipment is required. Payment for this fixed cost would be established on a lump sum basis. The Contract price should include the costs of moving from one location of operations to another, and of demobilization.

c. Pre-augered or pre-drilled holes for piles are measured for payment by the linear foot, from ground surface to bottom of hole. The Contract price for each size hole would include the costs of drilling equipment, temporary casing to keep the hole open, and dewatering of the hole, if necessary.

d. Pile points for pipe piles or hardened steel tips for H-piles may be measured and paid for on a per each basis. Steel driving tips for timber piles are relatively inexpensive and would not warrant separate payment.

e. Splices may be measured and paid for on a per each basis. Pile build-ups or extensions on precast concrete piles would be measured and paid for on a linear foot basis. The Contract price would include costs of preparing the top of the pile.

f. Coal tar epoxy coating may be measured and paid for on a linear foot of pile basis. Cathodic protection for steel piles would be paid for on a lump sum basis.

g. Establishing the method of payment for pile cutoffs requires some thought. Wood and precast concrete piles, which are not readily spliceable, have to be ordered in specific lengths. If the ordered lengths are established by the Engineer, separate payment should be provided for the pile cut-offs. Cut-offs would be measured and paid for on a linear foot basis.

h. Completed piles are measured for payment on a linear foot basis. The quantity to be paid for is the number of linear feet of each type and size pile incorporated in the Work and accepted by the Engineer, measured from pile tip to actual cut-off elevation. The Contract unit price will usually include costs of the following related work, when applicable:

(1) Jetting.
(2) Use of a follower.
(3) Drop light for inspecting interior of pile.
(4) Rejected piles.
(5) Repair of damaged piles.
(6) Pile cut-offs; this would have to be determined on an individual Contract basis.

2. Test Piles.

Even though test piles may become permanent piles, they should be paid for separately, since the cost per linear foot of driving a test pile is greater than the cost of driving a production pile. To compensate the Contractor for the waste in cut-offs of wood and precast concrete test piles, some Specifications will allow the measurement for payment to extend to five feet above the cut-off elevation. Other Specifications will establish a payment measurement equal to the total length of test piles ordered by the Engineer or indicated on the Plans.

3. Load Tests.

Load tests should be paid for on a per each basis because the actual number of required load tests may be difficult to determine in advance. The Contract price should include the costs of setting up the operation and the removal of material and equipment upon completion of the test.

4. Timber Piles.

If the Contract calls for both treated and untreated piles, two separate payment items should be established. The Contract price should include the cost of steel driving points.

5. Cast-in-Place Concrete Piles.

The Contract price for both pipe and shell piles would include the costs of dewatering, concrete, reinforcing steel and dowels. The Contract price for open end pipe piles would also include the cost of cleaning out inside of the pile.

6. Precast (Prestressed) Concrete Piles.

The length of piles to be measured for payment would also include build-ups. The Contract price may or may not include the cost of protective coating, depending on the determination made.

The Contract price for 54-inch prestressed concrete cylinder piles would include the costs of cleaning out inside of the pile, dewatering, and filling the pile. Pile build-up may be paid for separately or paid for at the same Contract unit price as the 54-inch cylinder piles. Providing facilities to enable the Engineer to descend inside a pile for visual inspection would be paid for separately as it may not apply to every pile. An item titled "In-Place Inspection of 54-Inch Diameter Piles" would be paid for on a per each pile basis.

7. Steel Piles.

Protective coating may be included in the Contract price per linear foot of pile or it may be paid for separately.

10.11 Steel Reinforcement

A. Introduction.

Reinforcement for concrete includes both steel bars and wire fabric; plain or deformed. Material requirements are normally referenced to ASTM or AASHTO

Standards. Reinforcing bars are available in Grades 40 and 60 representing yield strengths of 40,000 and 60,000 psi. Grade 60 is generally specified.

Epoxy coated steel reinforcement is being specified more frequently for bridge concrete decks and other locations where the reinforcement may be subject to corrosion, due principally to the effects of roadway de-icing salts.

B. Guidelines.

1. The local State DOT Standard Specifications are a reliable source of information for specifying steel reinforcement.
2. Specify the required Grade (40 or 60) of reinforcing bars.
3. Chairs for supporting reinforcement on slab forms are normally of metal. If resulting rust stains on the exposed bottom surface of the slab will be objectionable, specify that these chairs are to have plastic tips or be of stainless steel. Reinforcement can also be supported on precast mortar blocks of sizes necessary to maintain the reinforcement in proper position. Metal chairs may be used when they rest on structural steel members.
4. When reinforcing bars are to be spliced it is customary to specify that the splices be staggered. The length of splice lap should also be specified.
5. When the spacing of bar intersections is less than 12 inches it is generally only necessary to tie alternate intersections.
6. Epoxy-coated reinforcement requires special handling, coated tie wire, and coated supports or chairs, all to prevent damage to the epoxy coating. Require that all damaged and uncoated areas be patched in the field.
7. Some reinforcing steel placed in the Work may have to remain exposed to the weather for long periods of time. This can occur at construction joints where connecting work is to be accomplished under another contract. It should be specified that reinforcing steel to remain exposed for more than 30 days shall be protected with a brush coat of neat cement grout.
8. Specify that reinforcing steel in place shall be inspected and approved before placement of concrete begins.

C. Measurement and Payment.

Steel reinforcement is normally paid for at the Contract unit price per pound. The quantity to be paid for is computed using the nominal unit weights of bars listed in the referenced material Standard. The quantity of wire fabric to be paid for is the computed weight in pounds obtained by multiplying the unit weight per square foot and the area of fabric placed in the Work. If an appreciable quantity of a specific size of wire fabric is required, the unit of measurement can be the square foot. Consideration is given to splices and laps when computing payment quantities.

Epoxy-coated reinforcement requires a separate payment item. When both coated and uncoated reinforcement are in the Contract, payment item descriptions would read:

Reinforcing Steel, Epoxy Coated
Reinforcing Steel, Uncoated

The weight of epoxy coating is not included for measurement. The same theoretical unit weights for uncoated bars are used in determining the payment quantity for epoxy coated bars.

Reinforcing steel in small concrete structures like manholes, inlets, and headwalls, or in concrete sidewalk and concrete curb, is generally not measured and paid for separately. Payment is included in the Contract price for the structure or construction item itself. This should be clearly specified to prevent any possible double payment. Preventing double payment can be assured with a statement in the General Conditions. This is further discussed in Article 11.8D, Scope of Payment.

Suggested guidelines and details of measurement and payment for Reinforcing Steel are illustrated in earlier Article 3.4, Sample Technical Section.

10.12 Cast-In-Place Concrete

A. Introduction.

Concrete is one of the most commonly specified items of construction in the industry. It encompasses three major categories; cast-in-place concrete, precast concrete, and prestressed concrete. We are concerned principally with cast-in-place concrete and deal with the more basic commonly used procedures and practices. It will not include the consideration of special materials such as super plasticizers and expansive cements, nor special mixes for fiber reinforced concrete, lightweight concrete, and very high strength concrete.

Human effort and decisions play a prominent role in the many procedures and operations necessary to produce the finished product. Thus, the need for requirements governing ingredients, mixing, conveying, placing, finishing, curing, and protection of the concrete. Specifying these requirements demands a working knowledge of the basic fundamentals of concrete construction. A good source of information and reference in addition to available State DOT Standard Specifications is the American Concrete Institute (ACI) Manual of Concrete Practice.

B. Mix Proportions.

Concrete mix proportions indicate the weight or proportionate volume, of each basic ingredient in the mix; namely portland cement, coarse and fine aggregates, and total mixing water. The most important and expensive ingredient in the concrete is cement. There are five basic types of portland cement, each with its own characteristics. Their specific uses and other related information can be found in ASTM C 150, Standard Specification for Portland Cement. Proportioning by volume is rarely used because the volume of sand is affected by its moisture content. The weights of mix ingredients are established to produce a specific absolute volume of concrete; generally one cubic yard. If it is desired to establish mix proportions for a batch that will produce concrete having an absolute volume other than one cubic yard, the weights are adjusted proportionately. Mix proportions are established by either Owner or Contractor, depending on Contract requirements. When the Contractor has to design the mix, he should be required to retain for this work the services of an approved testing laboratory. After the mix proportion has been approved by the Engineer, the Contractor is not to vary the proportions without prior approval of the Engineer. Responsibility for designing a workable concrete mix will of course, rest with the party establishing the mix proportions.

A mix proportion is designed basically to produce concrete that will attain a specified minimum compressive strength at age 28 days. Some Specifications may in addition require a specified minimum cement content for each Class of concrete. Many Contracts have a need for concrete of different strengths. A commonly used arrangement to differentiate strengths is to designate each concrete mix with the word "Class" followed by a figure equal to the 28 day strength. Thus when referring to concrete having a 28 day strength of 3,000 psi, its designation would read "Concrete Class 3000." Another arrangement substitutes letters for the figures. Thus we can have Concrete Class A, Concrete Class B, and Concrete Class C.

C. Overdesign Factor.

The acceptability of concrete in place is usually determined by results of compression tests performed on test cylinders. The strength of concrete is influenced by numerous variables, including characteristics of the ingredients; human variability in mixing, conveying, placing, and curing of the concrete; and those variables involved in the casting, curing, and breaking of the test cylinders themselves. In order to compensate for the possible loss in strength that may occur from these variables, some Specifications will require that an overdesign factor be considered in designing the mix. When an overdesign factor of 15 is specified, it means that a mix for Concrete Class 3000 has to be designed for a strength 15 percent higher, or 3,450 psi. More information on this subject will be found in ASTM C 94, Standard Specification for ReadyMixed Concrete, and in ACI 214, Recommended Practice for Evaluation of Strength Test Results of Concrete.

D. Heat of Hydration (Mass Concrete).

Heat of hydration is heat generated by the chemical reaction of cement with water. The amount of heat generated is directly proportional to the quantity of cement in the concrete mix. It has been estimated that for each bag of portland cement in the mix there will be a resulting rise of approximately 10 degrees Fahrenheit in the temperature of the concrete, from the heat of hydration. The heat of hydration, the high temperature of ingredients, and the summer heat, can combine to produce an excessively high internal temperature after the concrete is in place. If this internal heat cannot readily escape to the outside, as in the case of mass concrete, then when the concrete finally cools down, the resulting shrinkage in volume can cause cracking of the concrete.

High internal temperature in mass concrete can be minimized or controlled in several ways, such as:

1. Using cement of low heat generation. Different types of cement have different rates of heat generation. Type III high early strength cement is the fastest heat generator, while Type IV which is the slowest, may not be readily available. Type II has a moderate heat of hydration.
2. Minimizing the cement content by using the largest permissible size of coarse aggregate in the mix.
3. Replacing part of the cement with a pozzolan such as fly ash. This subject is treated with more detail in Article 10.12E. Concrete Admixtures, Paragraph 5.
4. Precooling the ingredients, principally the mixing water and aggregates.
5. Maximizing the time interval between successive lifts or adjacent placements, to allow more time for the escape of interior heat.
6. Concrete placement during cooler evening hours.
7. Post-cooling of the concrete in place, with embedded pipe systems.

E. Concrete Admixtures.

Admixtures may not be cure-alls, but when properly selected and added to a well designed mix, they can help in producing concrete suitable for most all conditions. They can increase the strength of concrete, its workability, and its resistance to the weather; accelerate or retard initial set; reduce the heat of hydration; and improve its pumpability. A listing of applicable Standards for admixtures can be found in ASTM C 94, Standard Specification for Ready-Mixed Concrete.

1. Air entraining admixtures are among the more widely used admixtures. They are best known for protecting concrete against the harmful effects of alternate freezing and thawing cycles and against damage caused by deicing salts. They also improve the workability and durability of concrete.

2. Water reducing admixtures are next among the more commonly used admixtures. By reducing the amount of mix water needed, they lower cement requirements and improve workability. Reducing the mixing water reduces the slump and allows the concrete to be consolidated to a greater density.

3. A retarding admixture delays concrete setting time and keeps the concrete in a plastic state long enough to enable it to be placed and finished more effectively. It is particularly helpful during hot weather in a long continuous placement operation, and when long hauls from a mixing plant are involved.

4. Accelerating admixtures accelerate the initial set of concrete and its early strength development. These admixtures are used primarily in cold weather to prevent frost damage.

5. The use of fly ash as a pozzolan in concrete is gaining favor more and more. Fly ash is the finely divided residue resulting from the combustion of powdered coal. The use of fly ash in concrete has advantages and some disadvantages. On the positive side: a) Replacing part of the Portland cement with fly ash can be economical, since fly ash is a waste product; b) Fly ash reduces the heat of hydration, which is desirable in mass concrete; c) Fly ash reduces the water requirements of a mix; d) Fly ash increases concrete workability and reduces bleeding and segregation; and e) Fly ash will improve the watertightness of concrete.

Two main concerns on the negative side are described in the following paragraphs.

a. Fly ash supplied from different generating stations can vary in carbon content and affect the entrained air content in the concrete. The higher the carbon content the lower the entrained air.

b. The strength of fly ash concrete in its early stages will be somewhat less than the strength of concrete without fly ash. This reduced strength may continue up to an age of 90 days depending on the percentage of cement replaced by fly ash. Beyond this age, the strength of fly ash concrete will equal or may even become greater than that for concrete without fly ash.

Additional information on the subject of fly ash can be obtained from the American Coal Ash Association, the American Fly Ash Company, and the National Ash Association, all listed in Appendix B, Sources of Information for the Specification Writer, under number 2, Concrete and Reinforcement.

F. Batching Plant and Mixing.

To produce concrete of uniform quality, the ingredients for each batch must be measured accurately. As mentioned earlier, weight measurements of solid materials are more accurate than volume measurements. It is also easier to adjust weight measurements to reflect changes in aggregate moisture content.

In order to be able to stay within allowable established tolerances for the weighing of materials, Standards have been developed for concrete plants. Plants are certified as complying with these Standards. Concrete plant Standards have been adopted by the Concrete Plant Manufacturers' Bureau and approved by the National Ready Mixed Concrete Association, 900 Spring Street, Silver Spring, Maryland 20910. A copy of these Standards is available.

Concrete batching plants fall into three general categories listed below.

1. Manual batching in which all operations of weighing and batching the ingredients are controlled by hand. This type of equipment will naturally have a limited production capacity.
2. Semiautomatic batching, in which each of the various operations is controlled by a manually activated pushbutton.
3. Fully automatic batching, in which the weighing and batching operations are electrically activated by a single starter switch.

Cast-in-place concrete can be manufactured by any of the three different procedures of mixing described below.

1. Central-mixed concrete, in which the ingredients are mixed completely in a stationary mixer. Stationary mixers can include mixers located off the site in a ready mix plant, and stationary mixers or paving mixers situated on the site. Concrete mixed off the site can be delivered in a truck agitator, a truck mixer operating at agitating speed, or a special nonagitating truck.
2. Shrink-mixed concrete, which is mixed partially in a stationary mixer and then completed in a truck mixer.
3. Transit-mixed concrete, which is completely mixed in a truck mixer.

G. Joints.

Joint locations are shown on the Plans. There are three basic types of joints in concrete construction. They are construction, expansion and contraction joints, and each serves a particular purpose, explained below.

1. Construction joints occur wherever it is necessary to interrupt or limit concrete placement. Since there is no provision for movement across this type of joint, the reinforcing is continuous through the joint. Bonding fresh concrete to hardened concrete requires some preparation of the hardened surface. A more detailed description of this preparation is presented below.
2. Expansion joints create a complete separation between adjoining parts of a concrete structure to allow relative movement both horizontal and vertical, such as that caused by temperature change and by the settlement of slabs on ground.
3. Contraction joints in concrete are constructed to produce a weakened plane and thus control the location of cracking. This eliminates the unsightly random cracks that would normally appear. Contraction joints may be formed, sawed, or tooled. Joint penetration generally does not exceed two inches. Contraction joints are also referred to as control joints.

The bonding of fresh concrete to hardened concrete can involve different preparations, depending on the results desired. Normal preparation generally involves wetting the surface of the hardened concrete for a specified time before the placement of fresh concrete. It can also include spreading a coating of mortar or grout to the hardened concrete immediately before placing the fresh concrete.

In some concrete construction it is essential to provide good bonding conditions at the surface of horizontal construction joints. This would hold true for horizontal construction joints in wall construction below the surface, where watertightness is desired. The preparation of a bonded joint, as this is called, basically involves removal of surface mortar and laitance to expose the coarse aggregate. This can be accomplished either before or after the concrete has attained its final set. The surface of hardened concrete is prepared by sandblasting and washing with a water hose. Bonding fresh concrete to hardened concrete can also be accomplished with the application of an epoxy adhesive compound, provided that the appropriate surface preparation is made. Preparing the surface of concrete before it attains its final set can be accomplished by either of two acceptable methods, as follows:

1. At an appropriate time between initial set and final set of the concrete, approximately four to twelve hours after placement depending on cement content and curing conditions, the surface can be cut with a high velocity air-water jet to expose the coarse aggregate. This operation is commonly referred to as "green cutting." The timing of this operation is dictated by the hardening rate of the concrete. If it is performed too soon, the high pressure stream of water will undermine and loosen the coarse aggregate of the unhardened concrete. And if it is performed too late, the stream of water will have no effect on the surface of the already hardened concrete.

2. A liquid surface retardant can be sprayed on the surface of the concrete shortly after it has been placed and screeded. This compound delays the setting of the surface of the concrete to a depth of 1/8 inch to 1/4 inch, for a period of two to three days. On the second day, the unhardened surface mortar can be easily removed with a pressure water hose to expose clean coarse aggregate.

H. Concrete Finishes.

Practically all surfaces of concrete require some form of finishing. This can vary from a simple screeding operation to an elaborate procedure. Each type of finish serves a particular purpose. Concrete surfaces may be formed, or they may be unformed as in the top surface of slabs and horizontal construction joints. The requirements for finishing concrete surfaces to be exposed to view are naturally more exacting than those for surfaces to be hidden from view.

Generally, for those surfaces to be covered and hidden from view, the only requirement is that honeycombed areas be patched and form tie holes be filled, to exclude entrance of water. An exception would be a surface that is to receive a membrane type of waterproofing. In this situation, the surface of the concrete would have to be free of fins and other projections capable of puncturing the waterproofing membrane.

Concerning the surfaces of slabs:

1. A roadway riding surface would be screeded and given a broomed finish with broom markings transverse to the direction of traffic. Base slabs for a bituminous concrete roadway surface course would be screeded and given a scratched finish made with a rake.
2. Sidewalks would be screeded and given a wood or cork float finish. Sidewalk ramps are sometimes given a broomed finish.
3. Floor slabs are screeded and generally given a steel trowel finish. The surface of a floor slab for an industrial building would probably also be treated with a sealer-hardener compound to eliminate dusting and increase its resistance to wear.

Formed surfaces that are to be exposed to view would require any one of many types of finish. To begin with, there may be special requirements for the forms such as dressed lumber, plywood, or a form lining. After the forms are removed, the required finish may consist of one of the following:

1. An ordinary finish, in which fins and irregular projections are removed, voids patched, and form tie holes filled.
2. A rubbed finish, which would include in addition to the work of an ordinary finish, rubbing the surface with a carborundum stone to remove form marks and give the surface a uniform appearance.
3. A tooled finish, which is obtained with the use of air tools such as a bushhammer, pick, or other approved tool. In this type of operation the tooling process removes the surface mortar and fractures the aggregates at the surface of the hardened concrete, producing an attractive multicolored and textured surface.
4. Sandblast finish.

I. Concrete Curing.

The quality control exercised by the Contractor in selecting, measuring and mixing the ingredients, and in transporting and depositing the mixed concrete, would all be to no avail if proper conditions for curing the concrete were not provided. Development of the strength of concrete during early stages of the hardening process is related to the effectiveness of the curing program. An effective program will prevent excessive evaporation and loss of the moisture needed for hydration of the cement. It will also provide the proper curing temperature that will induce a favorable rate of hardening of the concrete. Where the ratio of exposed surface to volume of concrete is high, such as in slabs, protection against evaporation becomes critical. Loss of moisture can also result from a dry subgrade before concrete placement.

Various materials, methods and procedures are available for the curing of concrete. A normal curing period is of seven days duration. Preventing excessive evaporation and loss of water is accomplished by: 1) Application of water by ponding, sprays, steam, or by saturating such cover materials as burlap, cotton mats, or sand; and 2) Covering the surface with sheets of plastic or by the application of a membrane-forming curing compound.

In addition, a suitable curing temperature will ensure that the concrete will reach its design strength at the specified age. Curing temperature takes on greater importance in cold weather, when it is necessary to protect concrete against freezing. This protection will generally consist of enclosing the area and providing moist heat.

J. Guidelines.

1. To repeat an earlier recommendation, the ACI Manual of Concrete Practice makes available to the specification writer complete up-to-date information on the subject of concrete. A design office should not be without it.

2. When more than one Class of concrete is required, specify where each Class is to be used. This can be presented in the Materials subsection, along with definition of the different Classes of concrete.

3. When uniformity of color in the concrete is desired it should be specified that all cement shall be of the same brand and come from the same mill.

4. Mix proportions.

a. Determine who designs the mix proportions, the Owner (Engineer) or Contractor, and indicate it in the Specifications.

b. If it has not already been stated, it should be specified that mix proportions are to be designed for material measurement by weight.

c. If the Contractor is to design the mix proportions, consider specifying a minimum cement content for each Class of concrete or use of an overdesign factor, as described earlier in Article 10.12C, Overdesign Factor.

d. Concrete that is to be placed under water by tremie methods requires additional considerations in its mix design. The mix has to be plastic and have good flowability, because it cannot be consolidated by vibrators. To promote this flowability the slump should be in the range of six to nine inches, and coarse aggregate should consist of rounded gravel rather than crushed gravel or stone. Also, the mix should be richer in cement to offset the higher slump and possible washing out of cement due to water action. It is customary to require a minimum cement content of seven bags per cubic yard. A retarding admixture is also beneficial in preventing cold joints, by retarding initial set.

5. When heat of hydration has to be taken into consideration, specify the

maximum allowable temperature of the concrete when it is being placed in the forms.

6. Admixtures.

a. Air entrainment is not a requirement for all concrete. When weight is an important factor such as in tunnel linings of the sunken tube type of tunnel and in the counterweights of movable bridges, air entrainment should not be specified. Concrete that is not exposed to freezing and thawing, or to salt water, will generally not require an air entraining admixture.

b. The use of calcium chloride in an accelerating admixture for cold weather concreting should be prohibited. It is harmful to many embedded metals because it tends to promote corrosion.

c. Where a portion of the cement has been replaced with fly ash, maximum replacement has generally been in the range of 30–35 percent of the cement. Since the introduction of fly ash slows strength development of the concrete, its use in structural concrete should be given careful thought. For additional information refer to earlier Article 10.12E, Concrete Admixtures, Paragraph 5.

7. Batching plant and mixing.

a. To be considered acceptable, concrete plants should be certified as complying with the Standards adopted by the Concrete Plant Manufacturers' Bureau.

b. To reduce or eliminate the human errors inherent in manually controlled batching, weighing and mixing operations, a semi-automatic or fully automatic concrete plant as described earlier in Article 10.12F, Batching Plant and Mixing, should be specified.

c. The gradation of coarse aggregate in a lean concrete mix specified for a large volume of concrete must be closely controlled, otherwise problems in consistency and workability can develop. Gradation can be controlled by requiring a finish screening just before the coarse aggregate goes into the batching bins. The consequences of not specifying finish screening for the coarse aggregates in concrete for a dam are demonstrated in the following case history from the water reservoir project described earlier in Article 4.3C, Differing Subsurface Conditions:

> Lean concrete for the interior of the gravity dam was designed for a cement content of two bags plus 30 percent fly ash per cubic yard, using six inch maximum size coarse aggregate. Several times a day during the placing of this lean mix and without any warning, a four-cubic yard bucketful of discharged concrete would be difficult to consolidate, requiring additional time and effort. The concrete moved so sluggishly under vibration that it appeared to be unaffected by the large two-man operated vibrators. This caused numerous production slowdowns. Adjustments were made to the mix proportions but they gave only temporary relief. The Contractor complained about his increased costs and he had reason to, for the mix had been designed by a concrete testing laboratory retained by the Owner.
>
> Up until this time there was no concrete technician assigned to the mixing plant, for economy reasons. The writer was told by his supervisor that since the plant was

fully automatic, a technician was unnecessary. However, the workability problem soon changed this line of thinking and a technician was assigned to the plant. A sample of the three-inch to six-inch size aggregate was taken from the hopper above the mixers at the concrete plant and a sieve analysis performed. The sample was found to contain almost twice the allowable percentage of finer particles. The Specifications for this size aggregate called for zero to fifteen percent passing a three-inch square opening; the sample showed 24 percent passing. Also, the allowable percentage passing a two-inch square opening was zero to five percent; the sample showed eleven percent passing. Discovery of the excess fine material pinpointed the cause of the problem.

The screening plant for the coarse aggregate was located at the quarry, eight miles distant from the mixing plant at the dam site. After the aggregate had been screened to Specifications at the quarry, it was rehandled seven times before it entered the mixer. Particle breakdown from each rehandling of the larger aggregate produced additional fines. These fines would accumulate at the bottom of stockpiles (being washed down by the rain), at the bottom of truck loads during transit from the quarry to the mixing plant area, and at the bottom of the storage bins and hoppers. Since the fines were not uniformly distributed, it was impossible to predict when a load of unworkable concrete would be encountered.

The problem was controlled by making adjustments in the screening procedure at the quarry, to compensate for the additional fines. Had a secondary or finish screening been required at the top of the batching and mixing plant, the excess fines would have been eliminated together with the workability problem and the grief and investigative efforts that they generated.

8. Joints.

a. Three basic types of joints are described in Article 10.12G, Joints. Prior approval of the Engineer should be required before the Contractor may place a joint at a location not shown on the Plans.

b. A joint interrupts the continuity of the concrete. If watertightness is required, waterstops should be specified for vertical joints and a bonded surface preparation specified for horizontal joints. If a liquid surface retardant is used, it should be applied in accordance with the manufacturer's instructions.

c. When expansion joints are involved, a compressible premolded filler should be specified.

d. Where the intrusion of water or the entrance of other substances into a joint is undesirable, specify a joint sealant. A joint sealant in floor slabs will also provide protection of the corners at edges of joints.

9. Formwork.

a. The Contractor should be required to submit shop drawings of his proposed formwork to the Engineer for approval. When desired by the Engineer, specify that the Contractor shall also submit his formwork design calculations.

b. Do not require the Contractor to erect his forms "true to line and grade"; this is practically impossible. One consequence of this specified requirement is described in Article 8.11, Tolerances. Forms are to be erected and maintained within specified allowable variations from the lines, grades, and dimensions, shown on the Plans. Tolerance guidelines can be found in ACI 347, Recommended Practice for Concrete Formwork.

10. Concrete delivery.

a. Specify that no concrete shall be placed until the space to receive it, along with the formwork and all embedded items, have been inspected by the Engineer and accepted.

b. Each load of concrete delivered to the site is to be accompanied by a delivery ticket signed by the batch plant inspector. A listing of information normally furnished on a delivery ticket can be found in ASTM C 94, Standard Specification for Ready Mixed Concrete.

11. Concrete placement.

a. It should be specified that all concrete is to be placed "in the dry," unless otherwise specified or approved. This is to ensure that the Contractor will remove water from within the forms before placing his concrete.

b. When applicable, it should be specified that the Contractor shall comply with the requirements for hot weather and cold weather concreting.

c. Specify that concrete is to be consolidated by internal vibration, as opposed to external form vibrators or spading. If a placement operation should be critical, consider requirement for the addition of a standby vibrator in good working condition, in case of a breakdown.

d. It may be necessary to specify a sequence of concrete placement. This is often required when concrete is being placed for bridge decks, for arches, and for large floor areas. When structural steel beams are to be encased, specify that the concrete is to be initially deposited on one side of the beam until it fills the space under the member and is visible from the other side. When this occurs, the concrete can then be placed on both sides of the beam simultaneously. This sequence is necessary to prevent pockets of air being trapped under the beam when concrete is initially deposited on both sides of the beam simultaneously.

e. Placing concrete by tremie method is used to seal bottoms of open cofferdams so that foundations in water can be constructed "in the dry." The concrete is normally batched and mixed in a floating plant situated adjacent to the cofferdam. It is customary to specify a minimum production capacity for the plant. It is also required that before the Contractor may begin, he have available at or near the site of the tremie seal a quantity of each material sufficient to complete the operation. Another customary requirement is that the Contractor provide and maintain on the site two complete mixers, each with its own separate power plant; the second mixer to be available for emergency use. Sufficient reserve equipment must be on hand to assure continuity of the operation.

f. When referring to the placement of concrete use proper wording. Concrete is not "poured." It is placed or deposited. If a mix is wet enough to be poured, it does not meet Contract requirements.

12. The breaking of test cylinders made with the same concrete mix that went into the forms is one method of verifying adequacy of the mix design to produce

the required concrete strength. The casting, curing, and breaking of concrete test cylinders is a quality assurance activity, and should not be assigned to the Contractor as his responsibility. This must remain a function of the Owner's representative. A clarification of quality assurance is presented in Article 8.12, Quality Control and Quality Assurance.

The frequency and number of test cylinders to be taken by the Engineer should be specified. Cylinders are initially cured on the site in curing boxes provided by the Contractor. After they have hardened sufficiently they can then be transported to a testing laboratory where the curing period is completed and they are broken at various ages for strength determination. Test results are made available to the Contractor.

Some test cylinders are cured under field conditions, along with the concrete in place. This is done to determine when it is safe to strip certain forms. This procedure can also be used to check the adequacy of the Contractor's curing program.

13. The required finish for different parts of the structure should be specified. Mortar to be used for patching purposes should consist of the same cement and fine aggregate, and be proportioned in the same ratio, as was specified for the concrete being patched. Unenforceable terms such as "smooth finish" or "finished in accordance with industry standards" should not be used.

14. In specifying curing requirements, the use of a membrane-forming curing compound should be prohibited for surfaces that are to receive additional concrete or where a bonded coating such as paint, dampproofing, waterproofing, or roofing is to be applied.

15. Cement grout is normally specified for filling spaces that are to provide support under stationary equipment and under steel bearing plates. Frequently it is necessary to specify that the grout be introduced from one side only, in order to prevent the formation of trapped air pockets. The procedure to be followed would be somewhat similar to that described earlier for encasing steel beams. An additional requirement that will help prevent the formation of air pockets would be to specify that the grout be rodded after it is in place. Do not specify that the grout, which is in liquid form, is to be consolidated by vibrating. An explanation for grout designation is presented in Article 9.3, Proper Use of Terms.

K. Measurement and Payment.

1. Method of Measurement.

Cast-in-place structural concrete is measured and paid for on a cubic yard basis, with the exception of standard drainage and other small structures, and possibly some concrete superstructures. The Specifications, however, should not simply state that concrete will be measured by the cubic yard. It is necessary to outline how the measurements will be made. Otherwise the Contractor may claim payment for every cubic yard of concrete delivered. If the concrete work has been completed in accordance with the Plans, the payment quantity may be computed from the Plan dimensions. If, due to field conditions, the Engineer orders changes to the Plan dimensions (generally for footings), then the revised dimensions as ordered by the Engineer will be used in computing the payment volume.

Since quantities thus obtained constitute gross volume, adjustment for voids, openings, and embedded items in the concrete have to be considered. This can be handled by specifying that deductions will be made for openings and recesses each having a volume greater than a specified figure, generally three cubic feet. Space occupied by embedded structural steel may also be considered for deduction. It should also be specified that no deductions will be made for chamfers and embedded reinforcing steel.

To avoid controversy when specifying the method of measurement, do not make reference to payment for the "actual" volume of concrete placed. This introduces the quantities shown on the concrete delivery tickets which will invariably be more than computed theoretical quantities. One exception to computing the volume would occur when irregular cavities or sink holes have to be filled, and the space dimensions are difficult or impossible to measure. The practical method for measurement in this situation would be to use the volume figure on the delivery ticket accompanying the load of concrete.

2. Basis of Payment.

The Contract price per cubic yard for each specified Class of concrete includes in addition to the costs of labor, materials, and equipment necessary to complete the work in accordance with the Plans, Specifications and orders of the Engineer, the costs of formwork, admixtures, expansion joints, waterstops, bonded joints, joint sealant, and protection for hot and cold weather concreting, as applicable.

Payment for the concrete in drainage structures such as manholes and inlets would be included in the Contract price for each structure.

Payment for the concrete in bridge superstructures and individual small structures other than drainage structures would be included in the Contract lump sum price for the bridge superstructure or in the Contract price for each individual small structure.

The cost of cement grouting to provide support under equipment or steel bearing plates would be included in the Contract price for the related equipment or structural steel.

10.13 Structural Steel

A. Introduction.

Structural steel in engineered construction contracts generally involves bridges. There are also times when structural steel is specified for buildings associated with the construction of rapid transit projects, airports, port facilities, and water treatment facilities.

The basic structural steel for construction is defined by ASTM A 36, Specification for Structural Steel. This Standard covers carbon steel for rolled structural shapes, plates, sheet piling, H-piles, and bars of structural quality for use in the welded and bolted construction of bridges and buildings, and for general structural purposes. Other grades of structural steel are available for special use.

Mention should be made here of one of the other grades of steel that is being incorporated in many bridge designs. This steel is commonly referred to as weathering steel. It requires no painting because it develops a natural oxide coating of its own as it weathers. The coating has about the same thickness as a heavy coat of paint. The fully matured coating is dense, tightly adherent, and relatively impervious to additional atmospheric corrosion. The savings in initial painting costs and subsequent maintenance can be substantial. Weathering steel in a bare, uncoated condition, however, is not recommended for some environments because the tight protective oxide will not form properly. These include atmospheres containing corrosive industrial fumes, locations subject to salt water spray and locations exposed to de-icing salts. During the early months of weathering, water run-off can produce staining of adjacent surfaces if precautionary steps are not taken. In bridge construction this staining can occur on piers and abutments, prior to construction of the deck.

Useful information on structural steel is available to the specification writer. Some of these sources of information are (for mail addresses refer to Chapter 7 or Appendix B):

1. Standard Specifications for Highway Bridges, published by the American Association of State Highway and Transportation Officials (AASHTO).
2. Specifications for Steel Railway Bridges, published by the American Railway Engineering Association (AREA).
3. Manual of Steel Construction, published by the American Institute of Steel Construction, Inc. (AISC).
4. Structural Welding Code, published by the American Welding Society, Inc. (AWS).
5. Steel Structures Painting Manual, published by the Steel Structures Painting Council (SSPC).

B. Fabrication.

Before beginning any work in the shop, the Contractor is generally required to notify the Engineer. This is to give the Engineer time to make the necessary arrangements for organizing shop inspection.

Connections made in the shop are usually welded. Welders, welding operators, and tackers are required to be prequalified for the work by tests prescribed in the AWS Structural Welding Code. Records of qualification tests are submitted to the Engineer. The extent of quality assurance testing that will be performed by the Engineer is outlined in the Specifications. This information defines the types of welds to be tested, the percentage of each type of weld that will be subjected to testing, and the test method that will be used (radiographic, ultrasonic, or magnetic particle).

Field connections of main members are first pre-assembled in the shop for adjustment and fit. Complicated structures, such as those having curved girders or extreme skew, may be completely preassembled. Connecting parts pre-assembled in the shop are match marked. Each member is identified with an erection mark corresponding to the mark shown on the approved erection diagram.

C. Field Erection.

Erection drawings and other related information required from the Contractor are to be submitted to the Engineer for approval. This includes the sequence of erection, falsework details including design information, erection equipment to be used, and setting details for anchor bolts and bearing assemblies. Field connections are usually bolted using high-strength bolts.

Sometimes the construction of a large bridge is accomplished under separate substructure and superstructure contracts. In this situation, the superstructure contractor is usually required to deliver the anchor bolt assemblies with setting templates to the Engineer who will then turn them over to the substructure contractor for installation. Delivery by the superstructure contractor must be synchronized with the construction schedule of the substructure contractor. On some projects, anchor bolts are installed by the superstructure contractor in predrilled holes and set in nonshrink grout.

Masonry areas that are to support steel bearing plates have to be properly prepared to a required elevation. Concrete for these areas is generally placed 1/4 inch high and finished to grade by bushhammering or grinding. Steel bearing plates must have full and even bearing on the supporting area. This is accomplished in one of several ways, some of which are:

1. Placing a single thickness of sheet lead or alternating three layers of red lead paint and canvas duck on the concrete, or
2. Setting an elastomeric bearing pad directly on the concrete, or
3. Setting the bearing plate to grade on steel shims or wedges. After steel has been erected and nuts on the anchor bolts tightened, space under the bearing plate is then filled with a nonshrink grout or dry-packed with a stiff mortar.

When expansion bearings have to be set in position, an adjustment to the Plan location must be made to reflect the change in length of the member or span because of the difference between existing temperature and the standard temperature used in the design (normally 60 degrees Fahrenheit).

High-strength bolts used in connections have to be tightened to a specified minimum tension for the size bolt used. Bolts can be tightened by one of the following methods: 1) Turn-of-nut tightening; 2) Tightening by use of a calibrated wrench; or 3) Tightening with the use of a direct tension indicator. Use of this method also simplifies the inspection procedure. Two direct tension indicators are:

a. The load indicator washer, which is a hardened round washer with protrusions pressed out of the flat surface. The washer is placed under the head of the bolt. As the bolt is tightened the clamping force partially flattens the protrusions and reduces the gap between bolt head and load indicator washer. When the gap is reduced to a prescribed dimension, the bolt is properly tightened.

b. The load indicator bolt, which is a high-strength bolt that meets the standards for ASTM A 325 bolts. The end of the bolt contains a 12-point spline. A load control groove separates the spline from the end of the bolt threads. A special wrench holds and tightens both the bolt and the nut. The wrench applies a clockwise turning force to the nut and a counterclockwise turning force to the spline. When the bolt reaches the proper tension, the wrench twists off the spline end. Inspection is simple; it is only necessary to check that the spline has been twisted off.

Information on bolt tightening methods and inspection procedures can be found in The Specification for Structural Joints Using ASTM A 325 or A 490 Bolts, included in the AISC Manual of Steel Construction mentioned earlier.

D. Painting.

The painting of structural steel is accomplished in two phases. The initial phase takes place in the shop after fabrication. It consists of the surface preparation and application of a prime coat of paint. Proper surface preparation is necessary if satisfactory results are to be obtained. Machine finished surfaces are not painted but protected against corrosion by a rust-inhibiting coating which is removed just before the member is erected.

The second phase of painting begins after the steel has been completely erected and inspected. The first step is the paint touch-up of damaged areas of the shop coat, and prime painting the field bolts and welds. Paint for this operation is the same as the paint used for the prime coat in the shop. Following the field touch-up and prime painting, come the finish coats. The required number of finish coats are determined principally by the severity of the environment to which the steel will be exposed. It can vary from one finish coat for interior steel exposed in a dry environment to three finish coats for exterior steel exposed in a severe environment. Film thickness is important. The total dry film thickness normally required to obtain good performance will range between five and eight mils. As explained earlier, weathering steel does not require painting.

Sharp edges such as those on angles, rivet heads, bolts, and junctions of lap joints, require a stripe coating before the full prime coat is applied. Where the steel is to be subjected to particularly severe exposures, the recommendation is that this procedure be mandatory for all coats of paint.

The final coat of paint on bridge structures is usually not applied until all concrete work has been completed, to prevent splatter from the concrete operation.

E. Guidelines.

1. When a bridge is to be constructed under separate substructure and superstructure contracts, each contractor's responsibility as it applies to the furnishing and setting of anchor bolts should be defined.
2. To ensure that the work as laid out will fit correctly, specify that steel tapes to be used by the Contractor be first calibrated with the U. S. Bureau of Standards.
3. Steel members are to be stored above the ground on platforms, skids, or other supports.
4. When being shipped, girders and beams are to be positioned with their webs vertical.
5. When weathering steel is called for and a uniform appearance of the oxide coating is desired, specify that the surface of the steel be given a sandblast cleaning in the shop to remove mill scale.
6. Erecting structural steel should not be referred to as "installing structural steel."
7. To prevent the staining of concrete from weathering and unpainted structural steel, the use of temporary polyethylene coverings should be required on concrete foundations, piers and bearing areas, until the roadway deck is constructed.
8. In the erection of structural steel, the allowable deviations from plumb, level, and alignment, should be specified.
9. Painting.

    a. The paint coating specified should be suitable for the environment in which the steel is to be exposed.
    b. Require the submittal of paint samples for approval of the paint to be used.
    c. The extent of surface preparation, and the methods to be used, should be specified. Galvanized surfaces also require treatment before paint application.
    d. Require the protection of pedestrian, vehicular, and other traffic, which may be affected by the painting operation.
    e. Specify the required minimum total dry film thickness of the coating.

f. Specify method of paint application; brush, spray, or roller. In populated areas brush application is usually required, to eliminate damage from paint spray.
g. Require contrasting shades to distinguish between successive coats.
h. Some bridge designers may desire to leave unpainted those surfaces of the upper flanges of members that are to be embedded in the concrete deck. This should be reconsidered because severe rust staining of concrete piers and abutments can develop before the deck is constructed.
i. The preparation of surfaces and the application of paint may be referred to applicable requirements in the Steel Structures Painting Manual issued by the Steel Structures Painting Council.

10. When construction of a movable bridge is involved, require the Contractor to furnish the services of a trained man for thirty days to supervise operation of the movable span, make adjustments, and instruct the Owner's operator.

F. Measurement and Payment.

1. Method of Measurement.

Structural steel may be measured for payment at a unit price per pound or paid at a lump sum price. For purpose of payment measurement, structural steel is considered to include castings, plates, anchor bolts and nuts, shoes, rockers, rollers, pins, bearing plates, rivets, permanent bolts, welds, expansion dams, roadway drains, scuppers, and sometimes bridge railing.

Weight measurements for payment may be obtained by two different methods; from computed weights or from scale weights.

When the weight of structural steel to be paid for is computed, payment is based on the computed weight of metal as shown on the approved shop drawings. Unit weights in pounds per cubic foot are established for each type of metal involved such as cast aluminum, cast iron, and steel. The weight of required rivet heads, bolt heads, nuts and washers, and welds to be paid for, is estimated. Deductions are made for copes, cuts, and open holes.

When the weight of structural steel to be paid for is based on scale weights, each fabricated member is weighed on a shop scale before shipment to the site. When the painting of structural steel is to be paid for under a separate payment item, and if the shop coat has been applied to the member before it is weighed, four-tenths of one percent of the weight of the member is deducted from the scale weight, to compensate for the weight of shop paint.

Structural steel is often paid for on the basis of a lump sum price wherein no measurement for payment is made. The lump sum price bid applies only to work completed in accordance with the Plans and Specifications. If subsequent changes are ordered by the Engineer, the lump sum price must be adjusted accordingly. The adjusted price will be equal to the lump sum price bid, multiplied by the ratio of the total estimated weight of metal actually incorporated in the Work to the total estimated weight of the metal as shown on the Plans.

In comparing the three methods of measurement, payment on a lump sum basis, where it is feasible, seems to be the most desirable.

2. *Basis of Payment.*

Payment for structural steel will generally include the costs of:

a. Shop and field painting, unless it is being paid for separately on a lump sum basis.
b. The setting of anchor bolts.
c. Grouting and dry packing under bearing plates, assemblies, and bearing pads, plus steel shims and wedges.

10.14 Incidental Construction

A. Introduction.

There are many items of construction limited in scope and in dollar value, which are necessarily part of a project. A brief coverage is presented of some of the items most frequently encountered.

B. Manholes, Inlets, and Catch Basins.

These minor structures may be constructed in place or may be precast and set in place. Most common materials involved in their construction include concrete, reinforcing steel, concrete block, brick, ladder rungs, and casting frames, grates, and covers. Sanitary sewer manholes constructed of brick or block usually require the outside surface to be plastered with a coat of cement-sand mortar to minimize or prevent infiltration. Pipe penetrating the walls of these structures are normally required to be set flush with the inside face of the structure. Basic requirements governing excavation and disposal, backfilling, concrete, and reinforcing steel, should be referred to their respective Technical Sections in the Specifications. They should not be repeated.

Another item of work related to these structures involves adjustment of the tops of existing structures to meet the new grades in a highway reconstruction project.

Inlets and catch basins are usually of a standard depth and are measured and paid for on a per each basis. Manholes, on the other hand, may have to be constructed to different depths. Payment for manholes is usually accomplished under two payment items. Standard manholes of a specified depth, such as six feet, would be measured and paid for on a per each basis. When manholes exceed this depth, the excess depth is paid for at a unit price per vertical linear foot for the number of vertical linear feet that the manhole depth exceeds the standard depth. Measurement is taken from the top of the manhole cover to the bottom of the base. The item would be titled "Manhole, Additional Depth."

The Contract prices generally include the cost of all work necessary to complete the item. This would include excavation, disposal of surplus and unacceptable material, backfilling (except for select material that is paid for separately), reinforcing steel, ladder rungs, castings, plaster coat, and surface restoration. Pavement replacement is sometimes paid for separately.

Existing structures whose tops have to be adjusted to grade would be paid for on a per each basis for each type of structure. A payment item would read, "Adjust Manhole Top to Grade."

C. Dampproofing and Waterproofing.

These treatments are generally applied to exterior walls and roof slabs of underground structures and to bridge deck slabs. Surfaces to be treated have to be clean and dry. After the concrete has cured, a waiting period of ten days is usually necessary to allow the concrete to dry.

Dampproofing normally involves the application of two to four coats of asphalt or coal tar pitch. Measurement and payment is generally made at a unit price per square yard for the area covered.

Waterproofing is specified where watertightness in a structure is critical. The installation procedure is more involved and the requirements more stringent. Waterproofing may include:

1. Membrane waterproofing, in which multiple layers of fabric are embedded in bitumen, or a single membrane is bonded to the concrete.
2. Bentonite waterproofing, in which bentonite panels are placed directly against the concrete surface.

In membrane waterproofing, the concrete surface must be free from projections or holes that can cause the waterproofing membrane to be punctured. Holes or voids in roof slabs or bridge decks should be filled with mortar. When bentonite panels are specified, the Contractor should be cautioned against any premature wetting of the panels. Waterproofing applied to horizontal surfaces should be temporarily protected against damage until permanent protection is placed. This applies to areas that may be subjected to traffic. Temporary protection may consist of planks or sheets of plywood or hardboard. Permanent protection on horizontal surfaces can consist of a layer of concrete or cement-sand mortar reinforced with wire fabric; asphalt plank or asphalt impregnated fiber board laid in a hot mopping. Vertical surfaces are protected with insulation board. In all situations, waterproofing materials are to be applied in accordance with the manufacturer's written instructions.

Membrane Waterproofing and Bentonite Waterproofing are each measured and paid for at a unit price per square yard for the area covered. Permanent protection course is paid for separately at a unit price per square yard for each type of protection course provided. The unit price would include the cost of wire fabric reinforcing. Temporary protection course is not paid for separately, but is included in the Contract price for the waterproofing item.

D. Slope Protection.

Slopes subject to erosion require a covering of erosion resistant material. This may include crushed stone, riprap, precast concrete paving blocks, cast-in-place concrete, and burlap sacks containing dry-mixed concrete. The cover materials most commonly used are crushed stone and riprap.

If the material in the slope is of a gradation in which the fines may wash down through voids in the riprap, a filter course or bearing course should be specified. This will generally consist of crushed stone graded from a maximum of 1-1/2 inches to a No. 4 size.

The slope must first be constructed to Contract requirements. A toe trench is next excavated to key the bottom course of the riprap slope protection. A crushed stone filter course is then spread and followed with the placement of riprap. The maximum and minimum dimensions of individual pieces of riprap will depend on the requirements of the particular situation. Riprap is generally graded uniformly from maximum to minimum size. When specifying placement of the riprap, it should first be determined if placement should be made using a skip or clamshell bucket, or whether dumping and spreading will be permitted. End dumping generally causes segregation and results in the larger pieces nesting together. When applicable, it should be specified that riprap is to be placed to the full course thickness, not in layers. Also specify allowable tolerance in course thickness of the riprap in place. Some slopes may only require a cover of crushed stone of three to four inch maximum size, uniformly graded.

Filter Course and Riprap may each be measured and paid for by the square yard, cubic yard, or ton. When the item is being paid for by weight, the weigh scales should be certified and the Engineer should receive certified copies of the weight tickets. The Contract price for Riprap includes the cost of excavating the toe trench and necessary backfilling.

E. Curb.

The most frequently used materials for curb construction are portland cement concrete, bituminous concrete, and granite.

Portland cement concrete curb is commonly referred to in the industry as simply concrete curb. Concrete curb can have different dimensions and shape variations. When this occurs in the same Contract, separate payment is made for each curb classification. Concrete curb is constructed in sections of a standard length, ranging between ten and twenty feet. Generally, a space of 1/8 inch is provided between sections. This is generally accomplished by inserting steel plates 1/8 inch thick in the forms and withdrawing them after the concrete has hardened. Expansion joints are constructed at normal intervals. When the curb is adjacent to concrete pavement, curb joints line up with joints in the pavement. Forms have to be removed early enough to permit the face of the curb to be finished properly. Backfilling should take place immediately after finishing is completed. This helps to protect the curb against displacement and damage. When white concrete curb is specified, it is necessary to use white portland cement and a light colored sand in the mix. Steel faced curb will sometimes be specified for areas that will be subjected to heavy truck traffic. The steel facing, which usually consists of a rolled channel, acts as the face form for the curb.

Bituminous concrete curb is generally specified for temporary construction. Granite curb, because of its greater durability, is specified sometimes for areas in the northern part of the country. Granite curb will usually be supported on a concrete base or on granular bedding.

Curb is measured for payment by the linear foot. The measurement should be made along the exposed face of the curb. Specifying how the measurement will be taken is essential for the elimination of any possible disagreement. Some Specifications go one step further and state that the measurement will be taken at the gutter line. Depressed curb at driveways and street corners will normally be measured for payment as standard curb, unless it constitutes an appreciable quantity.

The Contract price per linear foot for each type and classification of curb will generally include the costs of reinforcing steel, joints, excavation, and backfill. The Contract price for Steel-Faced Curb will also include cost of the steel facing. When granular bedding or concrete base is standard for a particular type curb, this cost may also be included in the Contract price for the curb.

When Bituminous Concrete Curb is used for temporary construction, its Contract price should also include the cost of its removal, if this cost is not accounted for elsewhere.

F. Sidewalk.

Sidewalks are normally constructed of portland cement concrete for permanent use and of bituminous concrete for temporary use.

Portland cement concrete sidewalk is commonly referred to simply as concrete sidewalk. In the construction of concrete sidewalk which is normally four inches thick, a granular bedding course is usually specified. Sand is generally preferred for a bedding course. To control cracking, contraction joints are specified at a five foot spacing. As in concrete curb construction, the contraction joints are formed by inserting steel plates inside the sidewalk forms. When concrete hardens sufficiently, the plates are pulled out leaving an open space. Expansion joints are provided at normal intervals. In addition, when structures such as light standards, fire hydrants, and manholes are located within the sidewalk area, expansion joints around these structures are specified to prevent undesirable cracking in the sidewalk. Another item that should not be overlooked is the requirement that the subgrade be wetted down immediately before concrete placement is to begin. A dry subgrade will draw moisture from the concrete; moisture that is needed for complete hydration of the cement. It should also be specified that the Contractor backfill immediately after removal of forms. This is to protect edges of the sidewalk from being damaged while the concrete is still green.

If sidewalk is of one thickness, the item description can simply read, "Concrete Sidewalk" or "Bituminous Concrete Sidewalk." If more than one thickness is specified, it should be reflected in the item description, which would read "Concrete Sidewalk 4″ Thick."

Measurement for payment is normally by the square yard for concrete sidewalk and by the square yard or ton for bituminous concrete sidewalk. If the bedding or base course is to be measured and paid for separately, the units of measurement would be the same. In determining the quantity for payment, some Specifications will deduct the area occupied by a structure situated within the sidewalk, if that area is greater than one square yard.

The Contract price per square yard or ton for each type and thickness of sidewalk will generally include the costs of excavation, base or bedding course, and backfilling, unless these items are being paid for separately.

G. Fencing.

The most common types of fencing specified, are right-of-way (R.O.W.) fencing and chain link fencing.

The purpose of highway R.O.W. fencing as its name suggests is to delineate the limits of the highway right-of-way. Installation of R.O.W. fencing may require some necessary clearing and grubbing. Fencing will generally follow the contour of the ground. When the fencing traverses open country it will normally consist of preservative treated wood posts or galvanized steel posts, and galvanized steel woven wire fence fabric. Wood posts are set in holes excavated in the soil and then backfilled. Steel posts are driven in place. In rock, posts are set in drilled holes and grouted in place. When a highway traverses populated areas the R.O.W. fencing will generally be chain link fence.

Height of chain link fencing is variable, depending on its usage. Common heights are five feet and eight feet. The eight feet height is usually used for security purposes. Chain link fencing materials are generally of galvanized steel, aluminum coated steel or vinyl coated steel. Posts are set in concrete footings and should not be disturbed for at least seven days. Security fencing will generally contain barbed wire attached to extension arms along the top of the fence. Aluminum surfaces of posts that are to be in contact with concrete should first be given a prime coat of zinc chromate or asphalt paint.

Fence gates are provided for pedestrians and for vehicles. Vehicular gates may be single or double leaf. Locking devices should be specified where required.

Where an electrical transmission or distribution line crosses the fence, grounding of the fence is required.

Fencing is measured for payment by the linear foot for each type and height of fence. Measurement is made along the top of fence from outside to outside of end posts with deductions made for gates. Gates are measured for payment on a per each basis for each type and size of gate specified.

The Contract price per linear foot of fence will include the costs of any required clearing and grubbing, concrete footings for posts, and barbed wire overhang, as applicable. Electrical grounding of fence is generally either paid for separately or is included for payment with other electrical work. A payment item description for vinyl coated chain link fence five feet high would read, "Chain Link Fence Vinyl Coated, Five Feet High."

The Contract price per each gate will include the costs of locking devices. A payment item description for a vehicular gate ten feet wide would read, "Gate, Chain Link Fence, Vinyl Coated, Ten Feet Wide."

H. Topsoiling and Seeding.

This operation used to be performed just before completion of the project. However with increasing emphasis on the environment and erosion control of raw surfaces of the soil, topsoiling and seeding is now usually accomplished, time of year permitting, as soon as a unit or portion of the project has been satisfactorily completed. Topsoiling is not always specified when seeding is required. Earth slopes of cuts and embankments are generally given a cover crop seeding, without topsoil.

When topsoil exists on the site, it is stripped and stockpiled prior to beginning a normal excavation operation (see Article 10.7C, Excavation; Paragraph 3, Roadway Excavation). When it is to be furnished from sources off the site, the Contractor is required to give the Engineer advance notice of the source. Topsoil to be suitable for use should be free of subsoil, refuse, roots, stones two inches and larger in size, brush, weeds, and other materials considered detrimental to the growth of grass. The topsoil should be tested to determine its pH value and establish the rate of limestone to be added. Areas to receive topsoil are cultivated with a disk harrow to break up the surface crust and provide a better bond with the topsoil. Thickness of topsoil layer may vary from three to six inches.

The item of seeding involves the furnishing and spreading of limestone, fertilizer, grass seed, and mulch. Certified test reports of materials are required, particularly for the grass seed that is to be supplied. Seeding will also include surface preparation of the areas to be seeded, if topsoil is not specified. This would include removal of rubbish, brush, weeds, and stones two inches and larger in size. The soil is then cultivated to a depth of three inches to prepare a seed bed.

The time for sowing seed is usually limited to early Spring or early Fall. The spreading of materials may be done hydraulically or in the dry condition by mechanical means. In the hydraulic method the seed, fertilizer, and limestone are mixed with water to produce a slurry which is sprayed on under pressure, at a specified rate. In the dry method, materials are applied with mechanical spreaders. The limestone is spread first, followed by fertilizer and seed.

To protect seeded areas and foster seed germination, hay mulch is spread to a thickness of one to two inches. Where large areas are involved the mulch may be blown into place. Mulch is anchored in place with the use of asphalt which is sprayed onto the mulch as it leaves the blower. When mulch is spread by hand it is tacked down with sprayed-on asphalt.

The Contractor has the responsibility to protect and maintain the seeded areas including necesary mowing, until final acceptance of the Contract. In addition, areas that do not produce a uniform stand of grass have to be refertilized and reseeded.

Topsoiling is usually measured for payment by the cubic yard, computed from in-place measurements. Seeding is measured for payment by the square yard or acre, computed from in-place measurements.

There are two payment item descriptions for topsoiling, depending on source of the material. They are:

Placing Topsoil from Stockpiles
Furnishing and Placing Topsoil

The Contract price per cubic yard includes the cost of surface preparation and cost of loading and hauling topsoil from stockpiles.

The Contract price per square yard or per acre of Seeding, includes the costs of surface preparation when no topsoil is specified, fertilizing, liming, mulching, and maintaining the seeded areas including refertilizing and reseeding where necessary.

Chapter 11

Presenting the General Conditions

11.1 Introduction (See Article 4.1)

Published Standard General Conditions offer many advantages. To present a few:

A. They are prepared in consultation with legal personnel knowledgeable in the use of construction Contract Documents.
B. They have been reviewed by people with years of experience in the construction industry; and.
C. Most important of all, they have over the years withstood the test of the courts.

Accordingly, modifications to a published Standard Document should be made only where necessary to reflect specific requirements of the particular Contract. The final document should then be reviewed by the Owner's or Engineer's attorney.

The specification writer will be better prepared to work on the General Conditions after he has devoted time to preparing the Technical Sections and has become familiar with the specific requirements of the Contract. He will then be better able to present the special features and requirements that have to be highlighted and brought to the attention of bidders.

Chapter 4, which explains the Articles of the General Conditions, should be reviewed in conjunction with the use of this Chapter. Articles of the General Conditions discussed herein have the same titles and follow the same order as those presented in Chapter 4.

11.2 Definitions and Terms (See Article 4.2)

A. General (See Article 4.2A).

To one studying a foreign language, the use of a dictionary to define the meanings of words is necessary if the language is to be understood. Words that Engineers use in a Specification may seem to be a kind of foreign language to some readers.

The specification writer should make certain that the General Conditions include a Definitions Article.

B. Abbreviations (See Article 4.2B).

Plans of many contracts contain a Legend defining abbreviations indicated on the Plans, like NIC (Not in Contract). Abbreviations defined in the Legend shown on the Plans should not be repeated in the Specifications.

C. Definitions (See Article 4.2C).

This Article can play an important part in enabling the reader of Specifications to correctly interpret the meanings of words and terms. Words or terms not ordinarily used should be defined in this Article.

11.3 Scope of Work (See Article 4.3)

A. Intent of Contract (See Article 4.3A).

Attention is directed to AASHTO Article 104.01 referred to in Article 4.3A, which includes the statement: "The Contractor shall furnish all labor, materials, equipment, tools, transportation, and supplies required to complete the work in reasonably close conformity with the plans, specifications, and terms of the contract." Being presented in the General Conditions, this requirement becomes applicable to all the Work of the Contract. It is therefor considered unnecessary to begin each Technical Section with the repetitive statement: "The Contractor shall furnish all materials, labor and equipment necessary to construct (install) the following work. . . ." Article 10.2H, Description, makes similar reference in the fourth paragraph, to the use of repetitive wording in the opening statement of each Technical Section.

When items of material or equipment are to be furnished to the Contractor by the Owner or by other contractors, the information should be presented in this Article of the General Conditions. It should also be presented in the Notice to Contractors (see Article 12.2A, Notice to Contractors).

B. Changes (See Article 4.3B).

Contractors may claim reimbursement for additional work on the basis of verbal orders. Verbal orders are sometimes difficult to confirm and frequently lead to disputes. This Article in the General Conditions of the Contract should be reviewed to see that it clearly indicates that consideration for adjustment in costs or Contract time will be given only for changes authorized in writing by the Owner or Engineer.

Also, this Article should indicate that the basis for adjustment in cost or Contract time is to be agreed to in advance of the issuance of a Change Order.

C. Differing Subsurface Conditions (See Article 4.3C).

This Article is sometimes titled Differing Site conditions. Bidders are instructed In the Bidding Documents to visit and inspect the site of the project to satisfy themselves as to the problems and difficulties involved in performing the proposed Work. A contractor would be hard put to substantiate a claim for encountering an unanticipated condition on or above the ground surface after having supposedly made this inspection of the site. The author therefore believes that Differing Subsurface Conditions would be a more realistic and descriptive title of the situation.

Changes resulting from differing subsurface conditions should not be confused with the changes discussed in Article 11.3B, which are due to modifications initiated by the Owner or his Engineer. When differing subsurface conditions are encountered, the Contractor has to initiate action by notifying the Engineer.

Some suggested guidelines:

1. Where underground construction is involved, it is almost inevitable that some unanticipated condition will be encountered. A changed conditions clause should be provided in which the Owner will assume the risk relating to unknowns in subsurface conditions. The more the gamble is eliminated from a contract, the lower will be the bid.
2. A changed conditions clause should specify the procedure to be followed when the Contractor encounters a differing subsurface condition. The Contractor must notify the Engineer before such encountered conditions are disturbed; otherwise he may not be entitled to receive consideration.
3. All known existing utilities should be shown on the Plans. If they are not shown, they do not exist as far as the bidder is concerned. He generally does not have the time to research and contact each Utility Company.
4. Do not introduce a disclaimer of responsibility for reliability of the subsurface information presented; particularly if this information was used in designing the project. A bidder has no choice but to rely on this information. Also, disclaimers should not be made for factual data.

D. Variations in Estimated Quantities (See Article 4.3D).

When applicable, this provision should be included in the Specifications.

E. Extra Work (See Article 4.3E).

This subject is adequately covered in Article 4.3E and in the referenced AASHTO Article at the end of Chapter 4.

A procedure for verifying the Contractor's daily field records for determining payment on force account work is presented in Article 4.8F, Payment for Extra and Force Account Work.

F. Maintenance and Protection of Traffic (See Article 4.3F).

Some project Specifications present the requirements of this Article as a Technical Section in the Division of Miscellaneous Construction. Requirements necessary to maintain and protect vehicular and pedestrian traffic through or adjacent to the construction site should be fully presented. As stated in Article 4.3F, the requirements can vary greatly from one project to another.

When the impact of construction on the movement of traffic is not recognized in the Contract Documents, the results can sometimes be chaotic. The author was involved in just this type of situation back in 1964–65. He had accepted new employment with a consulting engineering firm as resident engineer on the construction of sanitary sewers in a town in New Jersey. The project had been underway for some months and he was to replace the present resident engineer. The Work included installation of approximately 200,000 L.F. of sewer pipe ranging in size from six inches to thirty inches in diameter. This work was distributed among five prime contractors who were already working on the site. The pipe was being laid in paved streets, and each particular street block being worked in had to be closed to traffic except for provision of an emergency lane for ambulances and fire trucks. Each of the five contractors was maintaining two separate pipe laying operations. This meant that ten different street areas were being closed to traffic at the same time.

Concerning maintenance and protection of traffic, the Specifications contained only two paragraphs of general requirements. The first paragraph required the Contractor to obtain the necessary permits before laying pipe in a State or County road, and comply with the traffic control requirements of the particular Agency. The second paragraph concerned the maintenance and safeguarding of traffic in local streets. One general sentence covered these requirements and read as follows: "The Contractor shall erect such temporary guards, fences, warning signs, lights, and signals as may be necessary or required to protect all traffic on the streets and roads." Although traffic was continually being detoured around the many construction areas, there were no instructions governing the control of this traffic. The result was chaos and a constant headache for the author, who not only had to placate irate citizens, but also establish with the Contractors some standard form of traffic control.

The following guidelines are offered:

1. Study the Contract Plans dealing with traffic control.
2. If a detour is shown on the Plans, requirements for its construction, maintenance, removal, and the restoration of disturbed areas, should be presented.

3. If the construction site is situated within a populated area and if clean streets and roadways are to be maintained, it may be necessary to require that:

   a. All trucks or other vehicles leaving the construction site to enter paved public roadways are to be cleaned of mud and dirt from their wheels and exterior body surfaces.
   b. Material spillage on traveled roadways is to be removed immediately.
   c. Traveled roadways are to be maintained free of dust.

4. Where applicable, require sufficient watchmen to patrol the area hourly during nonworking hours, and replace flares and other non-operating lighting units.
5. Spell out detailed instructions that may be required for controlling traffic on designated streets.
6. Require that flagmen be experienced.
7. Present any required restrictions governing the movement of construction equipment on public roads and streets.
8. The Contractor should not be permitted to store materials or park his vehicles on any active traffic lanes, roadway shoulders, or sidewalks.
9. When streets are to be closed, prior notification is to be given to public and school bus companies, and to local public agencies, particularly fire and police departments. This is in addition to the notification that is to be given to affected businesses.
10. When the removal of existing sidewalks may affect access to businesses, entrances, and properties, temporary walkways are to be provided.
11. Measurement and Payment.

    a. Payment for Maintenance and Protection of Traffic is generally made on a lump sum basis. If the project should involve protection of other types of traffic, this payment item description would be modified to read "Maintenance and Protection of Highway Traffic."
    b. Separate payment is usually made for detours that are designated on the Plans.
    c. Controlling Dust is generally paid for separately, by the ton of calcium chloride used, or by the truck-hour for the use of a water sprinkler truck.
    d. Temporary Barriers and Barricades are sometimes each paid for on a linear foot basis.
    e. Flagmen or Watchmen are sometimes each paid for on a man-hour basis.

12. A good reference on this subject is the local State DOT Standard Specifications.

G. Rights In and Use of Materials Found on the Work (See Article 4.3G).

If there should be a possibility that some of the material to be excavated on the project can meet the requirements for select backfill, provide for this possibility in the Specifications. The method of measurement and basis of payment for the operations involved should be clearly specified.

An illustration of this type of situation is presented in the following case history taken from the water reservoir project described earlier in Article 4.3C, Differing Subsurface conditions. Each end of the dam penetrated the side slopes of the high ground adjacent to the river. To prevent water stored in the reservoir from escaping around the ends of the dam, the Specifications called for impervious material to be used for backfill. It was specified that 20 percent to 50 percent of this backfill material had to pass a No. 200 sieve, with no individual stones larger than six inches in any dimension. Preliminary soils investigation of the project indicated the existence of such material in the areas of excavation, but the extent was unknown. To cover the possibility of an insufficient supply of select material on the site, a pay item for furnishing select material from sources off the site, "Borrow Excavation for Dam," was included in the Contract. In excavating for the foundation of the dam, material removed from the west bulkhead area was found to satisfy the requirements for impervious backfill. Screening tests showed 31 percent to 48 percent passing the No. 200 sieve. The Contractor stockpiled this excavated impervious material on the site until such time that it would be needed for backfilling around the ends of the dam.

Three payment items were provided in the Contract for this operation, as follows:

1. Earth Excavation for Dam—The unit price included the costs of excavation and disposal of the material. Excavated material which satisfied the requirements for impervious backfill was stockpiled on the site for later use.
2. Borrow Excavation for Dam—The unit price included the costs of obtaining and delivering select material from sources off the site, for use in backfilling the dam.
3. Backfilling of Dam—The unit price included the costs of placing and compacting select material for backfilling the dam, regardless of its source.

H. Cleaning Up (See Article 4.3H).

Maintaining work areas and storage areas free of waste materials, rubbish, and debris takes on greater importance when the Work is being carried out in a populated area. It sometimes requires a firm resident engineer to get an uncooperative contractor to comply with this requirement.

I. Value Engineering Proposals by Contractor (See Article 4.3I).

In some Specifications this provision may be titled Cost Reduction Incentive. Being a relatively new concept, many Standard Specifications may not have this provision. The Federal Highway Administration may have pioneered this concept as it was included in their Standard Specifications for Construction of Roads and Bridges on Federal Highway Projects, 1974.

J. Temporary Utility Services (See Article 4.3J).

If water or electrical power is available on the site, check the Plans to see that its location is indicated.

When two or more prime contractors are to occupy the same area, additional instructions will be necessary to define their respective responsibilities for providing, using, and maintaining temporary utility services.

K. Warranty of Construction (See Article 4.3K).

A warranty is made by a party to the Contract, and is a statement that the particular work will conform to the Contract requirements. A guarantee is made by a third party, binding him to the terms of another's (second party) contract. It is assurance by the third party that the second party will provide equipment or services of the quality required. One example of this is a performance bond (see Article 5.3E, Requirements of Contract Bonds) in which a third party (surety) guarantees the performance of a second party (contractor). Should a specific item of equipment be called for by the Owner, it is suggested that the Contractor not be required to furnish a warranty for a period greater than the guarantee period offered by the manufacturer.

Some contractors have attempted to interpret the one year warranty as meaning that one year after final acceptance of the Work, they could no longer be held responsible. This warranty is not intended to relieve the Contractor of his responsibility for latent defects discovered after expiration of the warranty period. The statement of Paragraph F, in the illustration quoted in Article 4.3K, should eliminate any misunderstanding on the part of the Contractor.

It is suggested that the warranty period be established to begin with the date of final acceptance and not the date of substantial completion. Since most equipment warranty periods begin when the item is placed in beneficial service, use of the date of substantial completion in warranty periods could cause confusion and create problems.

Some contracts require longer warranty periods for specific items of work. This would include roofing, waterproofing, and items of equipment. In this situation, the one year blanket period would have to be modified. This could be accomplished by adding the clause: "except where longer warranty periods are specified for certain items in the Contract."

L. Disposal of Material Outside the Work Site (See Article 4.3L).

When disposal sites on private property have to be obtained by the Contractor, it is important that he be required to furnish the Engineer with a copy of a written agreement with the property owner. Upon completion of the Work and as a prerequisite for final payment, the Contractor should be required to furnish a copy of an acceptable release from the owner of the disposal site.

The burning of material is generally subject to regulations of the local Air Pollution Control Code.

When material is to be spoiled on a site situated within view of a public road, it may be necessary to include requirements governing the placement and final treatment of the deposited material. This could include prohibiting any disruption to the existing drainage pattern, prohibiting an unsightly appearance, and requiring that the area be seeded.

11.4 Control of Work (See Article 4.4)

A. Introduction (See Article 4.4A).

The control of work as presented in the Articles of this subsection outlines the responsibility and authority of both the Contractor and the Engineer. Stated simply, the Contractor exercises control by the actual performance of the Work, while the Engineer exercises his control through approval or disapproval of the Work.

B. Authority of the Engineer (See Article 4.4B).

It is essential that the authority and responsibilities of the Engineer in administering the construction contract be clearly defined.

The Contractor's work is subject at all times to inspection by the Engineer. Should a standard set of General Conditions still continue using the word "supervision" in describing this function of the Engineer, substitution of the word "inspection" is recommended. Using "supervision" in relating to the authority of the Engineer can only expose him to unnecessary liability. The Contractor is responsible for supervision of the Work, not the Engineer.

Some contracts state that in the event of a dispute between the Contractor and the Owner, the sole arbiter will be the Engineer, who represents the Owner. Since many of the points in question may refer to an error or omission on the Engineer's part, one may tend to question how his decisions in these matters can be impartial. The Engineer should not be required to render decisions in matters involving his own negligence. Concerning questions related to interpretation of the Plans and Specifications, no one is better qualified to answer them than the Engineer who prepared the documents. However, when confronted with a question that concerns the adequacy of the Contract Documents, he may be ethically unable to render an impartial decision since his own performance is being questioned.

When the Engineer has been given authority to suspend the Contractor's operations (see Article 4.4B, last paragraph), many courts have held that, given this "right," the Engineer also has a corresponding "duty" to exercise it. Thus, when an unsafe condition results in personal injury or property damage, the Engineer is being found negligent by the courts because he should have recognized the unsafe condition and used his given authority to stop the work. As a result, many Specifications are now omitting this authority of the Engineer to stop the Contractor's operations. They are also clearly stating that the Engineer will not be responsible for safety at the site because providing safe working conditions is a responsibility of the Contractor. The Engineer still retains authority to reject defective work or work that does not conform to Contract requirements (see Article 11.4N, Removal of Unacceptable and Unauthorized Work).

C. Construction Manager (See Article 4.4C).

If a Construction Manager is designated as the Owner's authorized site representative, his authority and responsibilities should be defined in a similar manner as those for the Engineer. When both Construction Manager and Engineer are designated as site representatives of the Owner, a clear delineation of their respective authority and responsibilities is a must.

D. Plans and Contractor's Drawings (See Article 4.4D).

Some contractors may withhold the submittal of their drawings until the last minute, in anticipation that the Engineer will feel compelled to accept materials, equipment, or installation methods not in full accordance with the Contract requirements, in order not to delay progress of the job. On the other hand, other contractors may submit their drawings in volume at the start of the job, and the Engineer unknowingly finds himself processing drawings that are less critical to the Contractor's schedule. Before this development can be corrected, letters are received from the Contractor complaining that his drawings are not being processed on time and delays to the project are resulting. To prevent either of these situations from developing, it should be specified that within a designated number of days after Award of Contract, the Contractor shall submit to the Engineer for review, a schedule of specific target dates for the submission of drawings required of him by the Contract Documents. It should also be required that all drawings for interrelated items are to be submitted at approximately the same time.

An unreasonable length of time taken by the Engineer in processing drawings submitted by the Contractor can affect the Contractor's scheduling and may result in claims for delay. To eliminate this potential source for trouble, specify that the Contractor shall allow a maximum time of_____ days after each submittal for processing by the Engineer.

Some contractors may attempt to bypass responsibility for initially reviewing the drawings of their subcontractors for conformance to Contract requirements by transmitting them directly to the Engineer. By so doing, they hope to get the Engineer to perform this review for them. This practice can be prevented by including a requirement that all drawings, particularly those submitted by subcontractors, are to be reviewed, coordinated, and approved by the Contractor prior to their submittal to the Engineer. The drawings shall bear the Contractor's certification indicating compliance with this requirement. The Specifications should indicate that drawings submitted without this certification will not be processed and will be returned to the Contractor. Also, if the drawings contain any variations from Contract requirements, the Contractor is to inform the Engineer in writing, describing such variations and the reasons for them. It should also be required, where appropriate, that working drawings of the Contractor's design are to contain the seal and signature of a licensed professional engineer.

If it is feasible, have the Contractor submit drawings that are reproducible, rather than prints. Multiple copies of prints require marking up duplicate sets of prints, which can result in omissions and errors.

The Specifications should also inform the Contractor that an item of work which requires his submittal of a drawing will not be permitted to begin until that drawing has been approved by the Engineer.

There have been many discussions on the relative liability to the Engineer in the use of the word "reviewed" or "approved," when processing drawings submitted by the Contractor. Some maintain that the Engineer is equally liable, regardless of which word is used. This would be a matter to be resolved with the Owner's or Engineer's legal staff.

E. Conformity with Plans and Specifications (See Article 4.4E).

It is recognized that some types of work which do not conform completely to Contract requirements may be acceptable, depending on the extent of its deviation. This would occur more commonly in roadway pavement construction. The thickness of an individual pavement course may sometimes be just beyond the allowable minimum, yet be capable of satisfactory performance. When feasible, an adjustment in the Contract price can be considered. More information on this subject is provided in the seventh paragraph of Article 10.9C, Measurement and Payment.

F. Coordination of Plans and Specifications (See Article 4.4F).

The Specifications should provide a means for resolving discrepancies that may occur between Contract Documents. This can be accomplished by presenting an order of precedence establishing which document will prevail. For example:

In the event of any discrepancy between the Contract Documents, the following sequence in descending order of precedence shall govern:

1. Written Agreement.
2. All subsequent agreed-to modifications.
3. Addenda.
4. Special Provisions.
5. Plans.
6. Supplemental Specifications.
7. Standard Specifications.

Other clarifications to consider are: 1) On the Plans, written figures shall govern over scaled dimensions; and 2) In case of a difference between small and large scale plans, the large scale plan shall govern.

G. Field Record Drawings (See Article 4.4G).

These drawings are sometimes called "as-built drawings." Use of this term may infer that the information shown accurately portrays the finished construction. This will not always be true, because it is not possible in the day to day activities to observe and record all changes. If there should be inaccuracies in the "as-built drawings," liability may be involved. The use of "field record drawings" would indicate that the drawings are simply the Engineer's or the Contractor's record of the construction.

When the Contractor is to prepare the record drawings he should be reminded that changes in the work of his subcontractors are also to be recorded.

H. Cooperation by Contractor (See Article 4.4H).

A Contractor's Superintendent who is not qualified for the Work can be disastrous for a project. Before starting work, the Contractor should be required to submit for approval the name, qualifications, and experience, of his proposed Superintendent. If the Contractor should desire to change his Superintendent during the life of the Contract, he should be required to first notify the Engineer in writing and provide for approval the necessary information on his proposed new Superintendent.

If the Work is to be performed in a populated area, it may be desired to have the Contractor furnish the Engineer with the name, address, and telephone number of a person in the Contractor's organization who is authorized to obtain the necessary labor, materials, and equipment outside of Contract normal working hours, in case of emergency.

I. Cooperation with Utilities (See Article 4.4I).

The Specifications should spell out work that will be performed by each Utility Owner or Operator, and the estimated time for its performance. It should be a requirement that the Contractor shall at all times permit free, safe, and clear access to the affected facilities by personnel of the Utility Owners or Operators for the purpose of inspection, maintenance, repair, and construction of new facilities.

The Contractor should be required to notify all affected Utility Owners or Operators prior to his performing work on or near their facilities.

The Contractor should also be required to protect and maintain in service, throughout the life of the Contract, all existing drains and utilities shown on the Plans that are not to be abandoned under this Contract.

When any utility structures, facilities, or equipment are damaged by the Contractor, he is to immediately notify their owners. The Contractor may also be required to bear the expense of the damage repair.

J. Cooperation Between Contractors (See Article 4.4J).

If the site of the Work is to be occupied by more than one prime contractor, this should be mentioned in the Specifications. Otherwise, the Contractor may have prepared his bid on the basis of having sole occupancy, in which case he could suffer unanticipated delays and other damages, and would therefore look to the Owner for redress.

Other contracts of the project that are under construction or are scheduled for award during the period of this Contract should be listed. The information should include Contract number and title, estimated date of Notice to Proceed, and estimated time of completion, for each prime contract. If a part of the Work of this Contract is dependent on the availability of facilities to be constructed by other contractors, the estimated dates that these facilities are scheduled to be completed and made available to this Contractor should also be listed.

The rights, responsibilities, and limitations of each prime contractor should be defined in the Specifications. The Specifications should also indicate who is to have the responsibility for coordinating the work of the various prime contractors. This is a responsibility of the Owner which should be assigned to his site representative, and not to one of the prime contractors. The joint schedule of operations described in Article 4.4J is further discussed in Article 11.7E, Prosecution and Progress.

K. Construction Stakes, Lines, and Grades (See Article 4.4K).

Human error in the layout of construction is a factor that cannot be overlooked. A correct layout of construction can save time and money for both Contractor and Owner. One way of assuring a correct layout is to have it checked by an independent source. This can be accomplished by requiring the Contractor to perform the detailed layout of his Work from established control points. This gives the Engineer the opportunity to verify the correctness of the Contractor's layout. If the Standard Specifications should indicate that the Engineer is to perform the detailed stakeout for the Contractor, discuss the possibility for modifying this arrangement and making it a responsibility of the Contractor. The value of an independent check is illustrated by the following case history.

Early in his field career, the author was serving as a construction inspector on a bridge project located on the East Coast. The Specifications required the Contractor to perform the detailed construction layout. The Contractor, however, convinced our Resident Engineer that it would be mutually economical to combine both the Contractor's and the Engineer's survey parties into one party, thus reducing the number of required survey personnel by half.

One of the approach ramps to the bridge was to be supported on a series of concrete piers spaced 50 feet on centers. In laying out the piers for this ramp, an error of one foot in station was made in the layout for one of the piers. Instead of two consecutive spans being 50 feet each, one was 49 feet and the other 51 feet. Because of the combined survey party, there was no independent check of the layout. Construction of the concrete pier proceeded to the degree of completion that made it ready for structural steel. When the steel erector raised the first stringer to rest on this pier, the error was discovered. As a result, the concrete pier shaft had to be demolished down to the footing and reconstructed. Since the Contractor and the Engineer were equally responsible for the error, they both had to share the costs of rectifying the mistake.

Construction stakeout performed by the Contractor is generally not paid for separately; its costs are distributed among the various payment items in the Contract. However, some Specifications designate this work to be paid for separately on a lump sum basis.

L. Engineer's Field Office (See Article 4.4L).

Some Specifications present these requirements in a Technical Section under the Division of Incidental Construction. Other Specifications include these requirements in the General Conditions. Since these facilities are not part of the permanent construction, the author favors its presentation here in the General Conditions.

The size and extent of the facilities required will depend on the type of project and its magnitude. To determine requirements for a particular Contract, the field engineer or the agency that is to administer the Contract should be consulted. Items to be considered include:

1. Minimum floor area in square feet to be provided.
2. Engineer's office to be separate from any building used by the Contractor. Office to be situated at a location approved by the Engineer.
3. Specify whether the office furniture and equipment are to be new or may be in good, serviceable condition.
4. Fire and theft insurance to be provided by the Contractor for the protection of personal property in the field office, unless this is provided by the Owner.
5. Exterior windows and doors are to be lockable and equipped with insect screens.
6. Parking space for a specified minimum number of vehicles.
7. Watchman services to be provided during nonworking hours and janitor services during working hours.
8. Construction in or over water will require that the Contractor furnish suitable water transportation for Engineer's field personnel.

When Engineer's Field Office is specified in a Technical Section, payment is made separately on a lump sum basis. Fifty percent is to be paid on the first payment estimate following completion of the office facilities, and the balance pro-rated on a monthly basis for the remainder of the Contract time. Otherwise, there will be no separate payment and the costs are to be included among the various payment items in the Contract.

M. Inspection of the Work (See Article 4.4M).

The duties and authority of the inspector appear to be well defined in AASHTO Article 105.10, referred to in Article 4.4M. The restricted authority given the inspector is in line with his limited background and knowledge of the details of the overall Contract. Basically, he only has authority to reject work that does not conform. He is not authorized to give approval, modify any of the Contract requirements, nor direct any of the Contractor's personnel. Nevertheless, the important role that the inspector plays is pointed out in the third paragraph of Article 4.4M. Also, the significance of the Engineer's authority to direct the Contractor to uncover work at any time before final acceptance is well demonstrated in the case history presented in Article 4.4M.

It should also be specified that if the Contractor plans to perform any work during hours other than his normal work schedule, he shall notify the Engineer so that an inspector may be present. The importance of timely inspection is illustrated in earlier Article 3.2.3G, Inspections.

N. Removal of Unacceptable and Unauthorized Work (See Article 4.4N).

It is interesting to note in Article 4.4N that referenced AASHTO Article 105.12 makes no mention of the Engineer's authority to suspend an operation of the Contractor for his failure to conform to the Contract. One consequence of being assigned this authority was presented in earlier Article 11.4B, Authority of the Engineer. Two other options equally effective, yet less vulnerable to damage claims, are made available to the Engineer; namely (1) that unacceptable work will not be paid for, and (2) if the Contractor fails to comply with an order to remove unacceptable work, the Engineer may have it removed by others and deduct the cost from monies due the Contractor.

O. Load Restrictions (See Article 4.4 O).

There is not much that can be added to this Article except that if there are any load restrictions that apply to existing bridges and culverts or to proposed structures during their construction, they should be spelled out here.

P. Maintenance of the Work During Construction (See Article 4.4P).

The Contractor must be made aware that he is responsible for protecting and maintaining the Work until its final acceptance. This responsibility would cease for any portion of the Work taken over by the Owner prior to completion of the Contract. The costs for protecting and maintaining the Work are generally not paid for separately, but are included in the Contract prices for the various payment items.

Q. Failure to Maintain Project (See Article 4.4Q).

The authority given the Engineer as explained in Article 4.4Q is necessary if continued maintenance of the Work by the Contractor is to be assured.

R. Acceptance of the Work (See Article 4.4R).

AASHTO Article 105.16, referred to in Article 4.4R, permits partial acceptance of the Work. Many public works contracts will not accept the Work in piecemeal fashion; it must be 100 percent completed. However, as explained in subsequent Article 11.6P, Possession and Use Prior to Completion, it may be desirable for the Owner to take partial possession of the Work.

S. Claims for Adjustment (See Article 4.4S).

Including a provision for early filing of claims and a procedure for their resolution, can help to minimize time consuming and costly arbitration or litigation. This provision is outlined in Article 4.4S and referenced AASHTO Article 105.17. Note the requirement that the Contractor give notice of his intention to file a claim before starting the contested work. This is an important requirement.

When a defective Specification is responsible for a claim it may be deficient because of:

1. The allotment of insufficient time for its preparation and review (Chapter 14); or
2. Failure to use the proper language (Chapter 9); or
3. Improper assignment of risks (Article 8.13, Identifying and Controlling Risks).

T. Automatically Controlled Equipment (See Article 4.4T).

This Article deals with the malfunctioning of automatically controlled equipment and specifies the maximum allowable time that manually operated equipment may be used in its stead. It is believed that this subject can be handled more realistically by presenting it in the Technical Section that involves the use of the particular automatically controlled equipment.

11.5 Control of Materials (See Article 4.5)

A. Introduction (See Article 4.5A).

Selection of materials is the responsibility of the Designer. It should not be left to the Contractor. Specifying the requirements for materials should be done intelligently and realistically.

When it is desired to present a general requirement that is applicable to all materials in the Contract, the requirement can be presented in this subsection of the General Conditions.

B. Source of Supply and Quality Requirements (See Article 4.5B).

It is advisable to include the following requirements and instructions, or their modified versions, as they apply to the particular Contract:

1. The Contractor shall furnish all materials to be used in the Work, unless otherwise provided.

2. All materials to be furnished shall be new, unless otherwise specified.
3. Sources of supply once approved are not to be changed without the prior approval of the Engineer.
4. The Contractor should be advised when a particular item of material or equipment requires expeditious handling to assure timely delivery.
5. The Contractor should be alerted to specific items that are limited to a domestic source of supply, such as steel products.

C. Local Material Sources (See Article 4.5C).

When the Owner designates a material source of supply such as a borrow area, he becomes responsible for the adequacy of that source to the extent specified. Instructions to the Contractor concerning protection of the environment and final treatment of the borrow area may be required. Should the Contractor locate a local source of material on private property, the requirements for furnishing copies of the written agreement and acceptable release should be similar to those specified in Article 11.3L, Disposal of Material Outside the Work Site.

D. Samples, Tests, Cited Specifications (See Article 4.5D).

The AASHTO Article referred to in Article 4.5D presents many important requirements and instructions for the Contractor.

Acceptance testing should be both performed and paid for by the Owner. This is too critical a function to be assigned to the Contractor or performed by a testing agency hired by the Contractor.

E. Certification of Compliance (See Article 4.5E).

A certificate of compliance for some materials and manufactured items, as explained in Article 4.5E, has practical value for both Owner and Contractor.

F. Plant Inspection (See Article 4.5F).

If the Contractor is to supply basic testing equipment for the plant laboratory, it should be so enumerated.

The Contractor's cost of providing plant laboratory facilities is normally not paid for separately. This cost is included in the Contract prices for the related items of work. For example, the cost of providing the plant laboratory at a concrete batching plant would be included for payment in the Contract prices of the concrete payment items.

G. Field Laboratory (See Article 4.5G).

If a field laboratory is to be provided by the Contractor:

1. Present specific requirements. Many of them will parallel those specified for the Engineer's field office.
2. List testing equipment to be supplied by the Contractor.
3. Specify parking space, if any is required.
4. Specify the maximum number of times that the Contractor may have to move the field laboratory during the life of the Contract.

Payment for Field Laboratory may be made separately on a lump sum basis. Where appropriate, the Field Laboratory can be included for payment in the Contract price for Engineer's Field Office Facilities.

H. Foreign Materials (See Article 4.5H).

Determine whether the use of foreign materials is permitted. If certain materials and manufactured items must be of domestic origin and manufacture, they should be so specified.

Should Federal funds be involved in the purchase of foreign materials, a check should be made for any Federal regulations that may establish Buy American requirements. Also, if foreign materials are to be shipped by ocean vessel, a check should be made for any requirement that the shipment must be in vessels flying the U.S. flag.

Additional requirements presented in AASHTO Article 106.07, referred to in Article 4.5H, should be noted.

I. Storage of Materials (See Article 4.5I).

When areas on the Plans are designated as storage areas for the Contractor's use, they should be specifically mentioned in this Article. Indicate when these areas will be made available, and for what length of time.

J. Handling Materials (See Article 4.5J).

When special handling is required for a particular material, the details should be presented in the Technical Section dealing with that particular material.

K. Unacceptable Materials (See Article 4.5K).

Enforcing the requirement that unacceptable and rejected material be removed immediately from the site can at times become frustrating when the Contractor is not cooperative. To facilitate the Engineer's efforts in this respect, some Specifications will include a provision that, should the Contractor neglect to remove unacceptable and rejected material from the site within a reasonable time, further payments may be withheld until this requirement has been met. The alternative may also be specified that the Engineer will have this material removed by others and charge the cost to the Contractor.

L. Owner Furnished Material (See Article 4.5L).

The Contractor should not be held responsible for the adequacy and acceptability of material that is to be furnished by the Owner. The responsibility belongs to the Owner. Conditions that may have a bearing on this transaction should be specified, such as:

1. If the material is to be delivered by the Owner, require the Contractor to give sufficient advance notice (15 to 30 days) of the desired delivery date.
2. If the material is to be picked up by the Contractor, give location of pick-up point and other related information.
3. Require the Contractor to inspect the furnished material and report, within _____ days of receipt, any noted damage.

11.6 Legal Relations and Responsibility to the Public (See Article 4.6)

A. Introduction (See Article 4.6A).

This subsection in particular is steeped in legal requirements. Changes or modifications should be carefully considered and reviewed by legal counsel.

B. Laws to be Observed (See Article 4.6B).

Most Specifications require that the Contractor shall keep himself fully informed of all Federal, State, and Local laws, ordinances, regulations, orders, and decrees of bodies having any jurisdiction affecting the Work or those engaged in it. This is a mighty big order. In addition to this, the Contractor is required to protect the Owner and his representatives against any liability that may arise from the violation of any such laws, ordinances, regulations, orders, or decrees, by himself or his employees. The entire risk of compliance appears to rest completely with the Contractor.

By calling the Contractor's attention to known laws and regulations, the Specifications will undoubtedly enable him to better assess the requirements and more realistically estimate the costs involved. One such example of a known law is the legal requirement existing in some States that requires the Contractor to notify the Public Utilities at least two days before beginning any excavation or pile driving, by calling a specified central telephone number. This requirement should be called to the attention of the Contractor.

C. Permits, Licenses, and Taxes (See Article 4.6C).

The Contractor should be required to submit to the Engineer a copy of the permits and licenses he obtains for the Work.

The Specifications should identify here, and include in an Appendix, a copy of all permits obtained by the Owner for the Work. The Contractor should be instructed that he is to comply with all applicable requirements and provisions of these permits.

D. Patented Devices, Materials, and Processes (See Article 4.6D).

If a particular design, product, or process is specified and it is known that its use will require payment of a license fee or royalty, it should be called to the Contractor's attention.

This Article in the General Conditions should be reviewed to ensure that the Engineer, along with the Owner, will be held harmless by the Contractor against any loss for infringement of patent rights.

E. Restoration of Surfaces Opened by Permit (See Article 4.6E).

AASHTO Article 107.04, referred to in Article 4.6E, realistically and very adequately controls a situation that can become troublesome for the Engineer, when attempting to define the on-site responsibilities for pavement restoration. It first alerts the Contractor that he will probably be called upon to restore the roadway pavement as a result of openings made by others. It next defines the method of payment for that work, thus eliminating one more controversial item. There is not much more that can be added except that some Specifications allow consideration of an adjustment in Contract time for delays caused by such permitted work by others.

F. Federal Aid Participation (See Article 4.6F).

Federal requirements that are applicable to the Work of the Contract should be included in the Specifications.

G. Sanitary, Health, and Safety Provisions (See Article 4.6G).

1. When both Federal and State requirements apply, and if there should be a conflict between the two, specify that the more stringent requirements shall take precedence.

2. If the Work is located within the watershed area of a source of water supply, more stringent requirements may be necessary in providing sanitary accomodations for the workmen.

3. A relatively new health concern is appearing on the scene. This is the situation where asbestos or other hazardous materials may be encountered in performance of the Work. Special instructions would then have to be issued to the Contractor. When it is applicable, include a provision alerting the Contractor that should he, in the performance of his work, encounter the presence of asbestos or any material containing asbestos or any other hazardous material, he is to promptly notify the Engineer. He is also to cease doing any work involving the handling of asbestos or other hazardous material, until he receives specific instructions.

4. Do not make the Engineer responsible for enforcing safety requirements, nor give him the authority to stop an operation that he considers to be unsafe (Article 11.4B, Authority of the Engineer). Providing safe working conditions on the jobsite is a responsibility of the Contractor. However, if the Engineer observes an unsafe condition, he is required to notify the Contractor and go on record as having done so.

5. Trench cave-ins caused by the absence of protective sheeting result in numerous injuries and fatalities. When the Contractor is not being directly reimbursed for this temporary protection, he is tempted to bypass it (see Article 10.7F, Measurement and Payment; Paragraph 4a, Temporary Sheeting).

6. The Contractor should be required to report to the Engineer in writing all accidents arising out of or in connection with the performance of the Work, whether on or off the site, which causes property damage, personal injury, or death. The report should give full details, including statements of witnesses. An accident resulting in death or serious injuries or damages should be reported immediately to the Engineer by telephone or messenger. It is essential that the Engineer have a complete file on all accidents.

7. Specify that the Contractor shall have a standing arrangement with a local hospital for the removal and hospital treatment of employees who may be injured or who may become ill.

8. On contracts considered to be highly hazardous or of major proportions, consider requiring the Contractor to employ a full-time safety supervisor responsible to carry out safety and accident prevention programs.

H. Public Convenience and Safety (See Article 4.6H).

This Article and Article 11.3F, Maintenance and Protection of Traffic, are both concerned with public safety and minimizing inconvenience to the public. Consequently, they have many requirements in common. Instructions governing such items as the storage of material; rodent control, particularly during demolition; dust and noise control; maintaining existing mail boxes; accessibility to fire hydrants, street shut-off valves of water and gas, and manholes; watchmen during nonworking hours; construction operations over vehicular and pedestrian traffic; maintaining access to businesses and residences; and maintaining existing drainage facilities all have to be considered.

Requirements covering specific operations are generally presented in the Technical Sections involving the operations. Special attention should be given to excavation operations in populated areas, because unprotected excavations pose a hazard to children in the neighborhood.

I. Railway-Highway Provisions (See Article 4.6I).

Requirements presented by the Railroad Company concern those areas of the Contractor's work and methods of operation that take place in and adjacent to the Railroad right-of-way. Article 4.6I describes some typical Railroad requirements.

All work to be done within or adjacent to the Railroad right-of-way is subject to approval of the Chief Engineer of the Railroad. When certain materials to be provided by the Contractor have to conform to Railroad requirements, the Contractor is to notify the Railroad Company, which will arrange for testing.

The Specifications should present information on all train movements through the site.

The work of Maintaining Railroad Traffic is generally paid for separately as a lump sum item.

J. Construction Over, In, or Adjacent to Navigable Waters (See Article 4.6J).

A copy of the permit obtained by the Owner from the jurisdictional Federal Agency for construction of the permanent structure should be included as an Appendix to the Specifications. The Contractor should be advised that he is to obtain any additional permits that may be required for the performance of his work.

The Contractor is normally prohibited from interfering with navigation in the waterway. There are exceptions, in which case specific procedures will be outlined by the jurisdictional agency. One such situation would involve the dredge excavating a trench for a sunken tube type tunnel in which the discharge pipeline of the dredge would be temporarily blocking the channel. The dredge would be required to signal approaching vessels to slow down or stop, and await further signals from the dredge. The pipeline would then be opened temporarily to clear the channel.

All available information on traffic flow in the waterway should be presented in the Specifications.

K. Use of Explosives (See Article 4.6K).

It should first be determined if the use of explosives will be permitted. If not, the Specifications should so specify. If the use of explosives is permitted, present the general conditions under which it may be used. For example, specify the hours restricted for blasting. Also, determine if the Contractor will be required to provide on the site a thunderstorm monitor and an automatic lightning warning device. Some Specifications may begin this Article with a statement that the use of explosives will be permitted only with the prior written approval of the Owner (or Engineer).

Requirements governing the use of explosives in rock excavation are discussed in Article 10.7C, Excavation; Paragraph 3, Roadway Excavation.

L. Protection and Restoration of Property and Landscape (See Article 4.6L).

When Work is to be performed in a populated area, the Contractor should be reminded that no material or equipment is to be stored or stockpiled on private property without the written permission of the property owner. A copy of this written permission shall be furnished to the Engineer. The unauthorized storing of material on private property is a troublesome item; this particularly becomes a problem when sewer pipelines are being laid in town or city streets (see Article 10.8C, Common Problems; Paragraph 2).

An effective deterrent to the Contractor in this respect is to specify that, should the Contractor cause damage to any public or private property and fail to satisfactorily restore such damaged property within a reasonable time, the Engineer may (after 48 hours written notice) have such restoration work performed by others and the cost deducted from any monies due the Contractor.

When applicable, it may be necessary to make specific reference to protection of the landscape, particularly trees and shrubs, including their roots.

M. Forest Protection (See Article 4.6M).

AASHTO Article 107.13, referred to in Article 4.6M, appears to adequately cover the subject except for any special conditions that would have to be defined.

It should be indicated whether burning will be permitted in a public forest, and under what conditions.

N. Responsibility for Damage Claims (See Article 4.6N).

Engineers normally do not develop enough expertise to properly advise the Owner on the types of coverage and policy limits of property and casualty insurance. In addition, and of greater importance, the professional liability insurance carried by private design professionals does not as a rule include coverage for claims arising out of the Engineer's giving, or failure to give, advice on insurance matters. Accordingly, the Owner provides the insurance requirements after consultation with his insurance consultant. These requirements should be inserted in the Specifications without change. They should, however, be checked for the following provisions:

1. Requirements should be explicit as to the types and limits of insurance to be furnished by the Contractor.
2. The Contractor should not be authorized to enter upon the site of the Work until he has furnished the necessary certificates of insurance to the Owner, with a copy to the Engineer.
3. Insurance policies are not to be changed or cancelled until written notice of a specified number of days (10–30) has been given to the Owner, with a copy to the Engineer.

The cost of providing the required insurance is normally not paid for separately but is included in the various payment items in the Contract. Sometimes, when a payment item for Mobilization has been established in a contract, the Specifications will include the cost of insurance in the Contract Price for Mobilization.

O. Third Party Beneficiary Clause (See Article 4.6 O).

Since the subject matter of this Article is of legal consequence, the author can add nothing that would be of benefit to the specification writer. As a passing observation, in consulting State DOT Standard Specifications it was observed that this particular Article was not included in a majority of Standard General Conditions.

P. Possession and Use Prior to Completion (See Article 4.6P).

This Article deals with possession and use by the Owner of a portion of the Work before completion of the entire Contract. Most importantly, it reminds the Contractor that such possession and use does not constitute acceptance of that portion of the Work nor is it a waiver of any provision of the Contract.

When a project is to be income producing, the Owner may wish to place the project in operation before it is 100 percent complete. This would require a provision for substantial completion (see Article 4.2C, Definitions). Payment for substantial completion would be provided by a semifinal payment estimate.

Q. Contractor's Responsibility for Work (See Article 4.6Q).

As explained in Article 4.6Q, the Contractor is generally held responsible for the protection of the Work until the Contract has been completed and accepted. If specific instructions are necessary for its protection during a particular operation, they should be presented in the Technical Section dealing with that operation.

R. Contractor's Responsibility for Utility Property and Services (See Article 4.6R).

Determine if there is a legal requirement for the Contractor to give prior notice to the Utilities before performing any work involving pile driving, borings, or excavation. If there is, alert the Contractor in this Article. If there is no such requirement, caution the Contractor that he shall exercise care when working in the vicinity of existing utilities, particularly utilities below the surface. Make reference to related specific requirements that are presented in the Section on Excavation. Provide in the Section on Excavation, requirements governing the digging of test pits to locate underground utilities. Separate payment for this work should be specified.

A list of names and addresses of Owners or Operators of Utilities adjacent to or within the project limits, including names of individuals and telephone numbers, should be provided in this Article.

Additional information is presented in Article 10.7C. Excavation; Paragraph 1, Notification to Utilities, and in Article 10.7F, Measurement and Payment; Paragraph 3a, Test Pits.

S. Furnishing Right-of-Way (See Article 4.6S).

When areas within the right-of-way, or structures to be demolished, will not be available to the Contractor upon Notice to Proceed, they should be identified and their dates of availability indicated. When areas such as storage and staging areas are to be made available to the Contractor for limited periods, this should also be spelled out.

The Contractor should be entitled to an extension of time for delays resulting when availability occurs after the dates specified.

T. Personal Liability of Public Officials (See Article 4.6T).

This Article provides liability protection for individual representatives of the Owner, in addition to public officials. Since this Article deals with a legal matter, the author can add nothing that would be of benefit to the specification writer.

U. No Waiver of Legal Rights (See Article 4.6U).

AASHTO Article 107.21, referred to in Article 4.6U, affirms the right of the Owner to recover any overpayment made to the Contractor, even after the Contract has been accepted and final payment made. It is interesting to note that, in its 1983 Standard Specifications for Road and Bridge Construction, the New Jersey DOT provides also for reimbursing the Contractor if it has been determined that he was underpaid by the Owner. In addition, the State of New Jersey limits the period for adjusting an overpayment or underpayment to three years after final acceptance of the Work.

V. Environmental Protection (See Article 4.6V).

Some contracts may require special instructions in addition to those normal requirements presented in Article 4.6V. For example, if work is to be performed within a watershed area that is a source of potable water supply, two special instructions in the Specifications would be:

1. Chemical toilets are to be supplied by the Contractor for use by his employees—no field latrines.
2. The method of curing concrete that is to be in contact with potable water is to be limited to water curing. The use of a chemical membrane forming compound for curing is prohibited.

Another special instruction would concern the treatment of dredged material being removed from the bottom of a harbor or waterway which contains chemical wastes. Requirements governing deoderizing or disinfecting of the material would have to be considered.

W. Minimum Wage Rates (See Article 4.6W).

A Schedule of Minimum Wage Rates is usually included in an Appendix to the Specifications.

X. Equal Employment Opportunity (See Article 4.6X).

Additional requirements may include:

1. Certification of Nonsegregated Facilities, which confirms that the Bidder does not maintain, for his employees, facilities which are segregated on the basis of race, creed, color, or national origin.
2. Established goals and timetables for minority and female participation.

11.7 Prosecution and Progress (See Article 4.7)

A. Introduction (See Article 4.7A).

Time is an important factor in the successful completion of a construction contract. Because of its importance, Contract time should be specified in terms or units that can be easily understood and accounted for. Contract completion can be specified by a specific date; in calendar days; or in working days. Time that is expressed in working days has a built-in potential for disagreement between Contractor and Engineer, in determination of what constitutes a nonworking day. By the definition of a working day as presented in various State DOT Standard Specifications, the determination of a working day can be subject to question because it depends on interpretation by the individual. To illustrate; a typical definition of a working day would read: "A calendar day exclusive of Saturdays, Sundays, and legal holidays, when in the opinion of the Engineer, weather and soil conditions are such that the Contractor can advantageously work more than half of his current normal force for more than five consecutive hours on a major payment item then being performed." If the Contractor should not agree with the Engineer's interpretation, there arises a disagreement. On the other hand, a calendar day by definition (See Article 4.2C, Definitions) requires no interpretation, and the passage of time can be easily monitored. Accordingly, it is recommended that Contract time be specified either by a calendar date or in calendar days.

B. Subletting of Contract (See Article 4.7B).

When the Specifications indicate that specialty items will be considered in determining the maximum dollar value of work that can be subcontracted, the following should be checked:

1. The Specifications include definition of a specialty item (see Article 4.2C), and
2. Specialty items in the Contract are listed in this Article, 11.7B.

If there is to be no designation of specialty items, all references to specialty items should be deleted. When the Standard Specifications provide for or make reference to specialty items, a statement in the Special Provisions advising that there are no designated specialty items in the Contract is recommended.

Some Specifications will impose the same 50 percent limit on subcontractors who wish to sublet to second tier subcontractors, part of the work originally subcontracted to them by the Contractor.

C. Preconstruction Conference (See Article 4.7C).

The preconstruction conference should be specified to be held after Award of Contract and prior to issuance of the Notice to Proceed, to discuss essential items pertaining to the prosecution and completion of the Contract. If no preconstruction conference is planned, there will be no need to specify this Article.

D. Notice to Proceed (See Article 4.7D).

Upon award of Contract, the Contractor may decide to unofficially begin preliminary work on the Contract or order material before receiving the Notice to Proceed. The Engineer may not be aware of this action taken by the Contractor. To prevent any complications that may result from this unauthorized work, the following instruction can be included: "Any preliminary work started or materials ordered prior to receipt of the Notice to Proceed, shall be at the risk of the Contractor." If it is desired to have the Contractor do no work before receipt of the Notice to Proceed, the following instruction can be included: "The Contractor is not authorized to perform any work or order any materials until he has received a Notice to Proceed from the Owner."

The Specifications may also require that physical construction on the site shall begin not later than a specified time after the date stipulated in the Notice to Proceed.

E. Prosecution and Progress (See Article 4.7E).

Some projects will require two or more prime contractors to be working in the same general area (see Article 11.4J, Cooperation Between Contractors). For the Contractor to realistically prepare a progress schedule, he must have information on the schedules of other contractors on whom his work may depend, or whose activities will affect his progress. This information would be presented in following Article 11.7F, Limitation of Operations.

When multiple contracts are to be awarded at the same time, the Specifications should require each contractor to submit a preliminary progress schedule for the Engineer's review. At the preconstruction conference to be attended by all contractors, these schedules may require adjustment for overall project coordination, and agreement by the contractors affected.

Progress schedules should be updated each month. The periodic updating of the progress schedule is important to the construction record of a project, particularly if claims may be involved.

When a CPM type of schedule (see Article 4.7E) is called for:

1. The Contractor is required to first develop a preliminary schedule in bar chart form, to cover the first 90 days of work. This is to be submitted to the Engineer for discussion at the preconstruction conference.
2. The Specifications should require the Contractor to submit to the Engineer a statement of network analysis capability, verifying that he has the "in-house" capability or that he employs a consultant or firm qualified to satisfy the requirements for preparing and maintaining a progress schedule in the form of an activity oriented Detailed Network Diagram.

F. Limitation of Operations (See Article 4.7F).

The Contractor has a right to know, and should be advised, of any situation or requirement that will restrict or limit his operations. The Contractor should be made aware of the following:

1. If the entire work area will not be made available to the Contractor upon Notice to Proceed, the unavailable areas should be specified and the dates of availability given.
2. Space limitations on the site, where certain storage areas may be available to the Contractor for a limited period. The Contractor may have to provide additional storage areas.
3. Hauling vehicles of the Contractor may be restricted to designated streets or routes. Sometimes haul routes are indicated on the Plans.
4. Existence of a high groundwater level.
5. Stage construction in which the Contractor is required to complete the Work in stages and release each stage to the Owner upon its completion. Describe each stage and specify its completion date.
6. Activities of other contractors on the site. List the names of contractors, anticipated dates that they will be on the site, the items of work they will be involved in, and the estimated time to complete these items.

Additional information that may have an affect on a contractor's operations is presented in Article 8.10, Disclosure of Known Information.

G. Character of Workmen; Methods and Equipment (See Article 4.7G).

On some projects it may be necessary to require that the Contractor provide his employees with badges showing the employees' name and number.

In tunnel construction, equipment powered by gasoline, butane, or propane, should be prohibited. Also, fire fighting equipment should not include any toxic chemical fire extinguishers.

It is good practice not to specify the construction equipment to be used by the Contractor, unless there is a good reason for doing so. Some Specifications may require that certain types of equipment be of a specified minimum capacity; particularly equipment for use in the compaction of soils. Recommendations for not specifying construction methods can be found in Article 8.16, Methods and Results.

The Specifications should not specify that the Engineer will approve the Contractor's equipment.

H. Progress Photographs (See Article 4.7H).

When specifying this item, it is best to confer with the design Project Engineer to determine the number of required photographs to be taken each month and the number of prints to be provided for each photograph.

In preparing instructions to go with this Article, review Article 4.7H and consider the following:

1. Photographs are to be taken where directed by the Engineer.
2. Indicate the maximum number of photos that will be required per month and the number of prints of each photograph. Prints should be eight inches by ten inches in size, exclusive of a one inch binding margin to be provided on the left side.
3. Each print to have a title block containing Contract number and name, name of Contractor, date taken, and brief description of view. The back of the print shall indicate name and address of photographer.

A suggested requirement for sanitary sewer pipeline work to be performed in populated areas is to specify that preconstruction photographs show existing conditions of structures, pavement, sidewalks, and lawns, subject to possible damage. Preconstruction photographs should be taken every 150 feet along the centerline of the proposed sewer lines.

I. Suspension of Work by Owner (See Article 4.7I).

This Article deals with a subject of legal consequence. The Owner who desires to shift the financial risk of delay onto the Contractor, would present a no-damage-for-delay clause (see Articles 8.13A, General, and 8.13B, Identifying the Risks). The Contractor would be granted an extension of time only. Such an arrangement is sure to place Bidders on notice that there could be financial trouble ahead. Few problems during construction of a project have a more financially adverse effect on the Contractor than delays.

There should be a provision for reimbursing the Contractor for expense resulting from a suspension of work for convenience of the Owner. Specifying a maximum limit of time for which the Owner may suspend the Work should also be considered.

J. Determination and Extension of Contract Time (See Article 4.7J).

An unrealistic completion date can impose financial burdens and undeterminable risks on the Contractor. Furthermore, the Owner will be paying more, with no guarantee that his project will be completed within the stipulated time.

In line with the reasons presented earlier in Article 11.7A, Introduction, the author can neither see any useful purpose in designating working days to define Contract time.

If all work on the Contract should be completed with the exception of landscape items, on which work is restricted to specific seasons, it should be specified that no time will be charged against the Contract until such work on the landscape items can resume.

Some Specifications grant an extension of time for shortages of materials, if the Contractor furnishes documentary proof that he had diligently made every effort to obtain such materials from all known sources within a reasonable distance of the site of the Work.

K. Failure to Complete on Time (See Article 4.7K).

The amount of liquidated damages to be specified should not be determined by the Engineer. In establishing the amount to be assessed the Contractor, there should be no intent on the part of the Owner to profit from this procedure. For a liquidated damages clause to be enforceable it should specify an amount that is reasonable and not out of proportion to the actual damages. An owner has to be prepared to justify the specified amount, since it can be challenged by the Contractor, if it appears unreasonable.

If the completion of individual stages in a contract is critical, liquidated damages may have to be established for each stage. When specifying liquidated damages for each construction stage, the following instruction should be included: "In the event the Contractor fails to complete two or more overlapping stages within their respective stipulated times, liquidated damages will not be additive. The maximum amount of liquidated damages to be assessed per calendar day will be equal to the largest of the respective amounts."

It may sometimes be to the financial advantage of the Owner to have his project completed and made available before the specified completion time. As an incentive for the Contractor to complete the Work before the time specified, a bonus clause can be written into the Contract providing for payment to the

Contractor of an additional sum of money for each day that the Contract is completed before the time specified for completion.

L. Default of Contract (See Article 4.7L).

This subject appears to be well covered in Article 4.7L and its referenced AASHTO Article 108.09.

M. Termination of Contract (See Article 4.7M).

Termination of Contract, as presented in Article 4.7M and referenced AASHTO Article 108.10, is an action initiated by the Owner. Some Specifications will, in addition, define the conditions under which the Contractor has the right to terminate the Contract. These conditions include: 1) Failure of the Owner to make payment as specified in the Contract (see Article 11.8H, Progress Payments); and 2) The situation where work, which was suspended by the Owner, remained suspended for a period beyond the time specified (see Article 11.7I, Suspension of Work by Owner).

N. Disputes (See Article 4.7N).

The process of settling disputes by mediation or arbitration is described in Article 4.7N.

11.8 Measurement and Payment (See Article 4.8)

A. Introduction (See Article 4.8A).

All work that is to be performed by the Contractor must be accounted for when preparing the measurement and payment subsections of the Technical Sections.

The consideration of certain procedures in establishing progress payments will ease the Contractor's need to borrow money during progress of the Work. These considerations include:

1. Separate payment for mobilizing and setting up plant and equipment (see Article 8.14A, Payments; Paragraph 1, Mobilization of Plant and Equipment, and Article 10.4, Mobilization).
2. Advance payment for specific materials stored on the site (see Article 8.14A, Payments; Paragraph 2. Materials and Equipment Stored on Site, and Article 11.8I, Payment for Material on Hand).
3. Minimizing the amount of the retainage withheld from progress payments (see Article 11.8H, Progress Payments).

B. Measurement of Quantities (See Article 4.8B).

Since these measurements serve to determine the payment quantities, it is essential that the procedures be presented clearly on how these measurements will be taken.

In addition to Article 4.8B, detailed guidelines in specifying the measurement of quantities are presented in Article 3.2.4, METHOD OF MEASUREMENT, and Article 10.2K, Method of Measurement.

C. Fixed (Plan) Quantities (See Article 4.8C).

Establishing this method of measurement for items of construction which require extensive field measurements is advantageous, for reasons presented in Article 4.8C.

Consider including the provision that, should the Contractor believe that a specified fixed quantity is incorrect, he can request in writing that the Engineer check the questionable quantity. The Contractor should be required to accompany his request with calculations, drawings, or other evidence, to indicate why the fixed quantity is believed to be in error. If it is determined that the quantity is truly in error, payment will be made in accordance with the corrected fixed quantity.

D. Scope of Payment (See Article 4.8D).

The instructions presented in AASHTO Article 109.02, referred to in Article 4.8D, are worthy of mentioning again. They outline what the Contract prices cover, and deny the Contractor double payment for the same work. Presenting these instructions in the General Conditions makes them applicable to all Technical Sections. It therefore becomes unnecessary to repeat them in each Technical Section.

Additional guidelines in preparing the Basis of Payment subsection are presented in Article 3.2.5, BASIS OF PAYMENT, and Article 10.2L, Basis of Payment.

E. Compensation for Altered Quantities (See Article 4.8E).

Contrary to the provisions of AASHTO Article 109.03, referred to in Article 4.8E, many Specifications now provide for renegotiation of the Contract price of major items of work, when their final quantities differ sufficiently from their estimated quantities. An explanation of this provision is presented in Article 4.3D, Variations in Estimated Quantities.

F. Payment for Extra and Force Account Work (See Article 4.8F).

Equipment rental rates should be agreed upon in writing before this work is started. Rental rates can be established from the Rental Rate Blue Book for Construction Equipment, published by Nielsen/DATAQUEST Inc., or Contractor's Equipment Ownership Expense, published by The Associated General Contractors of America, Inc., both listed in Appendix B, Sources of Information for the Specification Writer, under 7, General. Equipment which is ordered to remain on the job on a standby basis is generally paid for at 50 percent or 75 percent of the normal rental rate, for such standby time during regular work hours. Some Specifications make a distinction in the hourly rate, between equipment which is rented and equipment which is owned by the Contractor.

AASHTO Article 109.04, referred to in Article 4.8F, fails to specify a value limit on what are to be considered small tools. Many Specifications will define a small tool as having a value (when new) of less than $200 or less than $500, depending on the Owner's standard.

Include the requirement that, at the end of the work day, both the Contractor's representative and the inspector assigned to the force account work are to compare their records and agree on the labor, materials and equipment used that day on the force account work. Agreement to this field record is to be signified by their signatures on the daily report.

An explanation for the need to provide this method of payment is presented in Article 4.3E, Extra Work.

G. Eliminated Items (See Article 4.8G).

When an item of work is eliminated from the Contract, reimbursement to the Contractor generally does not include the loss of anticipated profit. Some Specifications may also exclude reimbursement for overhead.

H. Progress Payments (See Article 4.8H).

1. If the Contract is large enough to be considered for semimonthly payments, specify the minimum value of work which must be accomplished to justify a midmonth payment.
2. The Engineer prepares progress payment estimates. The Contractor should be given the opportunity to review the payment estimate before the Engineer submits it for payment.
3. Since cash flow is of major concern to every contractor, consideration should be given to specifying the time frame in which payments will be made; normally no later than 30 days after the estimate period. The Specifications

should also indicate that, if payment is delayed beyond the specified time period, the Contractor will be paid with interest at the then current prime rate charged by the (name of a specified bank) for each day of delay in payment beyond that specified.

4. Attention should be called to the requirement in Article 11.8J, Lump Sum Breakdown, which states that the breakdown of a lump sum item must be received and approved before any progress payment on it will be made.

5. If the proper bonding is specified (see Article 12.3E, Requirements of Contract Bonds), an unreasonable retainage of monies should be unnecessary. Furthermore, it can be costly to the Owner, for the Contractor often must borrow operating funds to cover the monies retained. The interest on such loans will be reflected in the bid prices, thus increasing the project cost. Retainage requirements will vary with different owners. For example;

a. Progress payments made in full, with no retainage.
b. A retainage of three percent on all progress payments.
c. A retainage of five percent on progress payments until 50 percent of the Work is completed, then no further monies withheld.
d. A retainage of ten percent until 50 percent of the Work is completed, then no further monies withheld.
e. A retainage of ten percent until 50 percent of the Work is completed, after which the retainage is reduced to five percent.

Concerning reduction of retainage after 50 percent of the Work has been completed, some Specifications make this reduction conditional on the Contractor's work being on schedule or within ten percent of schedule. Other Specifications will permit the Contractor to deposit securities in lieu of the retainage of money. This is explained in Articles 4.8K, referenced AASHTO Article 109.08, and 11.8K, all titled Payment of Withheld Funds.

6. When a semifinal payment estimate is prepared, the money retained is generally reduced to one and one-half to two times the estimated cost of the work to be completed.

7. Additional information on progress payments is presented in Article 8.14A, Payments.

I. Payment for Material on Hand (See Article 4.8I).

If the Contract warrants advance payment, specify the eligible materials and equipment, and conditions governing these payments patterned after those presented in Article 4.8I. In addition, specify the percentage of the cost to the Contractor that will be considered for advance payment. This figure may vary from 85 to 100 percent of purchase cost. The Contractor is responsible for paying necessary warehousing, handling and delivery costs, and insurance coverage.

Some Standard Specifications contain the provision that no payment will be made for materials or equipment until they are permanently incorporated in the Work. This provision in the Standard Specifications would have to be modified in the Special Provisions, if advance payments are to be allowed.

J. Lump Sum Breakdown (See Article 4.8J).

Lump sum contracts or contracts containing lump sum payment items must have a mechanism for determining the money due the Contractor at the end of each progress payment period. The Specifications should include a requirement that the Contractor shall submit to the Engineer for approval a breakdown of his lump sum contract price or of each lump sum item in the Bid Schedule. The breakdown shall indicate the amount of money assigned to each principal operation of work, consistent with the Progress Schedule, in sufficient detail to provide a basis for determining progress payments.

To ensure that the Contractor will not procrastinate in submitting this information, specify that lump sum breakdowns must be received and approved before any progress payments on them will be made.

K. Payment of Withheld Funds (See Article 4.8K).

Accepting securities from the Contractor in lieu of a retained percentage will still protect the Owner's interests and at the same time enables the Contractor to utilize his retained monies.

L. Acceptance and Final Payment (See Article 4.8L).

List all documents that are to be submitted to the Engineer as a prerequisite for final payment. These documents would include warranties, operation and maintenance manuals, releases, field record drawings, and negatives of progress photographs.

M. No Direct Payment.

Specifications may present separate payment items for specific requirements in the General Conditions, such as Maintenance and Protection of Traffic; Engineer's Field Office; and Field Laboratory. If there should be no clarification on how payment for costs of the other requirements specified in the General Conditions are to be handled, the Contractor may believe that he is entitled to additional reimbursement for complying with these requirements. To discourage the Contractor from entertaining any such thoughts, it is recommended that a provision titled "No Direct Payment" be added to these General Conditions, and to read as follows:

"Unless otherwise specifically provided, no direct payment will be made for any of the work described and specified in these General Conditions and in the Bidding Documents. The costs thereof shall be deemed to be included in the prices bid for the various items listed in the Proposal."

Chapter 12

Presenting the Bidding Documents

12.1 Introduction (See Article 5.1)

In presenting the Bidding Documents, most State Departments of Transportation follow the arrangement presented in Sections 102 and 103 of the AASHTO Guide Specifications. Project Specifications which do not follow the AASHTO arrangement will generally combine most of this information and present it under the title of Information for Bidders or Instructions to Bidders.

Chapter 5, which explains the Articles of the Bidding Documents, should be reviewed in conjunction with the use of this Chapter. Articles of the Bidding Documents discussed herein have the same titles and follow the same order as those presented in Chapter 5.

12.2 Bidding Requirements and Conditions (See Article 5.2)

A. Notice to Contractors (See Article 5.2A).

In describing the project to help the prospective bidder determine if he possesses the financial and technical ability to do the job, be sure to indicate unusual features. Additional information to be presented should include:

1. Alerting the bidder if a portion of the right-of-way will not be available to the Contractor upon issuance of the Notice to Proceed.
2. The availability of electrical power and domestic water.
3. Access to the site.
4. Availability of subsurface data.
5. Listing of materials and equipment, if any, that will be furnished by the Owner.
6. The time specified for receipt of bids should indicate if it is Standard or Daylight Saving Time, or simply Local Time.
7. Furnishing necessary information if a guided tour of the site is planned for prospective bidders.
8. Whether payment for the set of Contract Documents will be refunded upon its return in good condition, or whether there will be no refund.

The Owner may have spent years planning his project, and the Engineer will probably have spent many months and possibly one to two years preparing the Plans and Specifications. Bidders require time to examine the Plans and Specifications, inspect the site, calculate the quantities of material and equipment required, obtain prices for materials, contact subcontractors, arrange for financing and for Contract bonds, and finally, to prepare the bid itself. Bidders should be allowed sufficient time to prepare and submit their bids; otherwise insufficient time forces them to increase the contingency items.

B. Prequalification of Bidders (See Article 5.2B).

A contractor cannot perform properly if he does not possess the capability to do so. Bonding companies will generally exercise care when providing bonds for contractors. However, a contractor's success in obtaining performance and payment bonds does not in itself necessarily indicate his competency or financial responsibility. Insurance against the possibility of getting an inexperienced or incompetent contractor on the job can be provided by requiring prequalification.

To require a bidder to submit his qualifications at the same time he submits his bid may be considered to be unfair. It is maintained that, under this procedure, both bidder and owner will have to wait until bids have been opened before determining if the low bidder has qualified to be awarded the Contract. And if the low bidder cannot qualify for award, he not only has to bear the expense of preparing his proposal but also suffer the embarrassment of disqualification.

C. Contents of Proposal Forms (See Article 5.2C).

Much of the informational content of the proposal form is prepared by others. The specification writer should review and, where possible, confirm the information.

D. Issuance of Proposal Form (See Article 5.2D).

Reasons that may prompt the Owner to refuse to issue a set of Bidding Documents to a bidder are outlined in AASHTO Article 102.03, referred to in Article 5.2D.

E. Examination of Plans, Specifications, Special Provisions, and Site of Work (See Article 5.2E).

1. Examination of Site. The bidder should be instructed to visit the site and determine for himself the nature and location of the proposed Work, and any site conditions which may affect the Work and its costs. If a tour of the site is scheduled, include all information relating to the tour.

2. Reference Drawings and Information. The Engineer's design is sometimes related to reports, investigations, and drawings, produced by others. Examples

of this are subsurface explorations and data; drawings of existing structures to be demolished, modified, or underpinned and protected; and plans of existing utilities and other underground structures. The Engineer is therefore not in a position to guarantee their accuracy. A hesitancy to incorporate this information as part of the Contract Documents may be understandable. The information should, however, be made available to the bidders. Before starting the Specifications, the writer should determine if there are any available drawings showing existing conditions and construction that may have a bearing on the proposed Work. Contract Documents should disclose all related information affecting the costs of performance, which the Engineer knew or should have known. Failure to do so may impose liability for misrepresentation (see Article 8.10, Disclosure of Known Information). The information should also identify the Plans and Specifications of related and adjacent construction contracts. Disclosure of this information is normally accomplished by:

a. Describing the information on file in the office of the Owner or Engineer, and stating that it is available for inspection and review by bidders. Some Specifications allow bidders to obtain copies of this data at the cost of reproduction.
b. Including reference drawings and boring logs with the Contract Plans, but with a note that they are not part of the Contract Documents and are being furnished only as information in the possession of the Owner (Engineer). The Owner (Engineer) can make no warranty or representation of the accuracy of information supplied by others.

3. Addenda. Bidders should be advised that requests for explanations or interpretations of the Contract Documents must be made in writing, and must be received at least _____ days before the scheduled opening of bids. They should also be advised that neither verbal requests from bidders nor verbal interpretations or explanations from representatives of the Owner or Engineer will be considered binding. The Bidding Documents should also present the minimum time limit for issuing addenda, as _____ days prior to the opening of bids. The number of addenda should, wherever possible, be kept to a minimum. The greater the number of addenda the greater the probabilities of error by bidders, through oversight.

The material in an Addendum should be presented in a manner that will minimize any possibility of its being misinterpreted. For example, rather than presenting only the changed words of a sentence being corrected, it would be much better to restate the entire sentence containing the changed words. To illustrate, instead of stating "Change the last six words of the second sentence in the first paragraph to read: "_____," it would be better to state "Change the second sentence in the first paragraph to read: "_____."

4. Subsurface Data. The bidder should not be made responsible for obtaining subsurface data as part of the bidding activity. The Owner generally has the time

for taking complete borings and having them analyzed. The bidder usually is given a limited time in which to analyze and prepare a bid. It is unfair and often impossible for the bidder to take and analyze adequate borings during the period of his bid preparation. Having the Owner take and analyze one set of borings is much more realistic and practical than having this duplicated by the number of bidders. When the data is supplied by the Owner, all bidders have access to the same identical information.

It should be indicated when the borings were taken and by whom. Also specify where the soil samples, rock cores, and soil test results, are available for examination by the bidders.

Some Contract Specifications remind bidders that the condition of soil samples when inspected may not be the same as when they were taken. If samples are not hermetically sealed, they have a tendency to dry out during storage.

Another explanation presented in the Specifications may state that the observed water levels and other water conditions indicated on the boring logs are as they were recorded at the time of exploration. These levels and other conditions may vary with time, according to the prevailing climate, rainfall, and other factors.

The interpretation of the character of material in the samples represents the opinion of individuals. One other instruction to be considered advises bidders to have the subsurface data and interpretations independently evaluated by someone qualified in this technical field, before using them for bidding purposes.

A careful distinction should be made between presenting factual data, and presenting interpretations or opinions. Disclaimers of responsibility for factual data provided should, in general, not be made.

F. Interpretation of Quantities in Bid Schedule (See Article 5.2F).

1. Introduction. If there is one single document that can quickly present the general make-up of a contract, it is the Bid Schedule. By reviewing this document, one can obtain from the pay item descriptions a fairly good picture of the type of work involved; from the estimated quantities, an idea of the extent of the proposed work; and from the "Unit" Column, how payment will be made. The reader can also determine whether alternative or optional bids are involved.

2. Section Identification. It is a great help to the reader when the Bid Schedule includes a column listing the Section number of the Technical Section in which he can find information on a specific payment item of work. The time saved in not having to thumb through numerous pages of Specifications searching for a particular item of work is well worth the effort expended by the specification writer.

3. "Approximate" and "Estimated" Quantities. Many Specifications use the word "approximate" when describing the quantities appearing in the Bid Schedule. "Approximate" means nearly exact or reasonably close. A variation

of more than ten percent between final quantities and the "approximate" quantities shown in the Bid Schedule can lead to claims for extras. Use of the word "estimated" is recommended. This can be presented by stating that "The quantities shown in the Bid Schedule are estimated, and have been prepared for bidding purposes and for the comparison of bids." The word "estimate" means a rough or preliminary calculation, and allows for a greater variation with the actual quantities than does use of the word "approximate." Quantities prepared for bidding purposes should therefore be referred to as estimated quantities, rather than approximate quantities.

4. Unit of Measurement. Units of measurement listed in the Bid Schedule should be clearly presented with no possibility of being misinterpreted. The author recalls that in reading the Bid Schedule during the final review of a set of Contract Specifications for construction of a rapid transit tunnel, the unit of measurement for Portland Cement Contact Grout was listed as Cubic Foot. One would thus normally assume from the title of the item that the cubic foot measurement applied to volume of grout used. However, in consulting the Technical Section on Contact Grouting, the measurement and payment clauses specified that the portland cement used in the grout was to be measured and paid for by the cubic foot; not the grout itself. To have been presented correctly, the unit of measurement shown in the Bid Schedule should have read Cubic Foot of Cement; not merely Cubic Foot.

5. Unbalanced Bids. Some bidders may, for various reasons, attempt to unbalance their bid prices. By assigning a higher price on an item of work that is to be completed early in the Contract and balancing this with reduced prices on later items of work, a contractor can receive more money in the early stages of the job and thus help to finance his operations. The subject of unbalanced bids is also discussed in Article 5.2I, Irregular Proposals, and Article 10.4, Mobilization.

6. Fixed Quantities. Fixed quantity items in a Bid Schedule should be made identifiable by inserting the word (Fixed) immediately below the related figure in the Estimated Quantity column. Additional discussion of this subject is presented in Articles 4.8C and 11.8C, Fixed (Plan) Quantities.

7. Alternative Bids. In this situation the bidder is required to submit bids on alternate designs. After bids have been opened and evaluated, the Owner will decide which alternative bid to accept. An alternative design, such as in the case of a bridge superstructure, can include many payment items. Also, there will be many items of work common to both alternative bids. The common items would be grouped and listed in the Bid Schedule under the heading of BASE BID. The total of the bid prices for the common items would be labeled TOTAL BASE BID PRICE.

Payment items peculiar to Alternative No. 1 design would be grouped similarly in the Bid Schedule, but under the heading of ALTERNATIVE BID NO. 1. The total of the bid prices for these items would be labeled TOTAL ALTER-

NATIVE NO. 1 PRICE. Payment items for Alternative No. 2 would be similarly arranged.

Bid Price Summary could be presented as follows:

BID PRICE SUMMARY

TOTAL BID INCLUDING ALTERNATIVE NO. 1

TOTAL BASE BID PRICE	$________
TOTAL ALTERNATIVE NO. 1 PRICE.	$________
TOTAL BID PRICE (BASE BID PLUS ALTERNATIVE NO. 1)	$________

TOTAL BID INCLUDING ALTERNATIVE NO. 2

TOTAL BASE BID PRICE	$________
TOTAL ALTERNATIVE NO. 2 PRICE	$________
TOTAL BID PRICE (BASE BID PLUS ALTERNATIVE NO. 2)	$________

8. Optional Bids. In this situation, the bidder is given the choice of bidding on one of two or more presented designs, materials, or pieces of equipment. The arrangement in the Bid Schedule would, in general, follow that presented for the Alternative Bids, with one major difference. In an Alternative bid arrangement, each bidder is required to submit bids for each Alternative listed. The Owner makes the final selection. In an optional bid arrangement, the bidder selects the option on which he will submit prices. He is required to bid on only one of the presented options. This can be made clear in the BID PRICE SUMMARY by inserting the word OR underlined, between the Price Summary for each option.

G. Pre-Bid Conference (See Article 5.2G).

The date to be set for a pre-bid conference should be established to allow bidders sufficient time to study the Plans and Specifications before attending the conference. Necessary changes resulting from questions raised at the conference can be accomplished by the issuance of an Addendum.

H. Preparation of Proposal (See Article 5.2H).

A bidder has a right to expect that the information presented in the Contract Documents will enable him to prepare a complete and accurate estimate, and that he will not be penalized for any deficiencies in these Documents.

Any special instructions to be given to bidders on preparation of the Proposal and alerting them to specific items or requirements can be presented in this Article. Special instructions can include calling attention to:

1. The allowable maximum price that can be bid for the item of Mobilization.
2. The presence of Fixed prices or Fixed quantities.
3. Alternative or optional bids.
4. Acknowledgement of the receipt of Addenda.
5. Required designation of subcontractors.
6. Allowable price adjustment for material or labor costs (see Article 12.2R, Escalation Clauses).
7. Buy American.

I. Irregular Proposals (See Article 5.2I).

Additional irregularities that can result in a bid being rejected would include:

1. If the Proposal was not prepared in accordance with the instructions presented in previous Article 12.2H, Preparation of Proposal.
2. If the total amount of the bid exceeded the bidder's current capacity rating on file with the Owner (State DOT).
3. If the bidder failed to submit the required sworn statement concerning collusion.

J. Proposal Guaranty (See Article 5.2J).

The form of proposal guaranty can vary with the particular requirements of the Owner. The Owner may specify: 1) A bid bond; 2) A cashier's check or certified check; 3) A cashier's check or certified check, or a bid bond; or 4) A cashier's check or certified check, plus a bid bond.

On privately funded projects, a proposal guaranty is generally not required from selected bidders who have been invited to submit bids.

K. Delivery of Proposals (See Article 5.2K).

Some Specifications remind the bidder that his proposal must be accompanied by the proposal guaranty and other specified documents.

Specifications may also require that if the proposal is to be submitted by mail, the sealed envelope containing the proposal must be enclosed in another envelope addressed to the Owner and marked "BID ENCLOSED."

L. Withdrawal or Revision of Proposals (See Article 5.2L).

Withdrawal or revision of a proposal is not always a guaranteed option for bidders. Some Specifications may stipulate that once the proposal has been deposited with the Owner, a bidder may not modify or withdraw his bid. There is generally one exception. When proposals for more than one contract are to be received and opened at the same place and time, an apparent low bidder on one contract will be allowed to submit a written request to withdraw his proposal for the second or succeeding contract, prior to the opening and reading of proposals for the second contract. This makes it possible for a bidder to protect himself against receiving award of more contracts than he is equipped to handle.

M. Combination or Conditional Proposals (See Article 5.2M).

When combination proposals are involved, some Bidding Documents will stipulate that a bidder submitting a bid for combined contracts must also submit separate bids for each individual contract.

N. Public Opening of Proposals (See Article 5.2N).

Sometimes Bidding Documents state that only the bid totals will be publicly read at the Opening of Proposals. Unit prices will not be made available until after verification by the Owner.

When proposals for multiple public works contracts are to be opened and read aloud, some Bidding Documents will specify that the proposals will be opened and read in alphabetical order by County name.

O. Disqualification of Bidders (See Article 5.2 O).

Additional reasons presented by some owners for disqualifying a bidder are:

1. An unsatisfactory performance record, as shown by past work for the Owner.
2. If there are reasonable grounds to believe that a bidder is interested in more than one proposal for the Work contemplated.

P. Material Guaranty (See Article 5.2P).

When information is desired before contract award, on the origin and manufacture of material to be incorporated in the Work, this material should be listed in the Bidding Documents. Material guaranties are becoming more common, as the "Buy American" clause is increasingly being inserted in construction contracts. Variations of this requirement have been presented. For example:

1. The use of foreign materials is prohibited, unless the Contractor can produce evidence that a specific material is not available domestically because of circumstances beyond his control.

2. Specifically named materials, such as structural steel, must be of domestic origin and manufacture.
3. The use of foreign products will be permitted if the bidder can present evidence that the cost of the foreign product represents a saving of more than a specified percentage of the cost of the same product produced in the United States.

Additional information on foreign materials is presented in Article 4.5H, referenced AASHTO Article 106.07, and Article 11.5H, all titled Foreign Materials.

Q. Non-Collusive Bidding Certification (See Article 5.2Q).

Should it be determined after award of the Contract that the Contractor falsified his sworn statement and was guilty of collusion, he or his surety may be held liable to the Owner for a specified sum of money in addition to possible prosecution under State or Federal laws.

R. Escalation Clauses (See Article 5.2R).

Subcontractor associations have maintained that material suppliers will not or cannot provide firm price guarantees beyond a period of three or four months. This can pose severe economic hardships and possible defaults by contractors working under a fixed price contract. Many recommend that cost increases in materials should be shared by both Owner and Contractor.

Including an escalation clause is one way of sharing this risk between Contractor and Owner (see Article 8.13C, Controlling the Risks; Paragraph 2, Escalation Clauses). Increases beyond a specified percentage in the cost of labor, materials, and permanently installed equipment, beginning six months after the date of bid opening would be given consideration. Cost adjustments would be made upward or downward as appropriate, on specifically mentioned major items of material and equipment.

The exact formula for calculating the price adjustment should be prescribed. Formulas may be based on official price quotations, a cost index recognized in the construction field, or on the basis of increase or decrease of actual cost to the Contractor as compared to his estimated costs at the time bids were opened. A satisfactory procedure should be developed on a job-by-job basis.

Specify the percentage of escalation that the Owner will assume. This figure may vary between 50 percent and 75 percent, with no consideration for overhead and profit. Acceptance by the Owner of 100 percent of labor cost increases would weaken the bargaining power of the Contractor with the unions. Acceptance by the Owner of 100 percent of cost increases in materials would remove the Contractor's motivation to avoid or to at least minimize such cost effects.

When the Contract provides reimbursement to the Contractor for steep increases in labor or material costs beyond a specified period of time, this information should also be presented in the Bidding Documents.

12.3 Award and Execution of Contract (See Article 5.3).

A. Consideration of Proposals (See Article 5.3A).

Some Specifications will provide that, where a unit price has been omitted from a proposal, the accepted unit price will be that unit price which is equal to the extended price for that item divided by the estimated quantity for that item.

If the Bid Schedule specifies the maximum price that can be bid for an item such as Mobilization (see Article 10.4, Mobilization), it should be stated that if the price bid for that item exceeds the specified maximum permissible amount, the price will be reduced to the specified maximum amount, and the total contract bid price adjusted downward accordingly.

B. Award of Contract (See Article 5.3B).

The Award of Contract may not always go to the lowest responsible, qualified bidder. Legislation in one particular State gave a five percent preference to contractors based in that State, when they bid on projects that were 100 percent State financed. In one instance, this preference enabled the second low bidder on a major bridge contract to receive the award.

C. Cancellation of Award (See Article 5.3C).

In this respect, it was noted that one State DOT limited its right to cancel a contract before its execution to the existence of a national emergency.

D. Return of Proposal Guaranty (See Article 5.3D).

Requirements governing return of proposal guaranties will vary with different owners. Modifications of the requirement specified in referenced AASHTO Article 103.04 is illustrated by the following examples: 1) The proposal guaranties of all but the three lowest bidders will be returned within ten days after the opening of bids; 2) The proposal guaranties of the second and third lowest bidders will be returned within three days after execution of the Contract; or 3) The proposal guaranty of the successful bidder will be returned within three days after he delivers the Contract bonds and the required certificates of insurance.

Should all bids be rejected, the proposal guaranties of all bidders will be returned within ten days after rejection.

E. Requirements of Contract Bonds (See Article 5.3E).

Surety bonds are generally required on publicly funded projects. Sometimes the Owner will require the Contractor to furnish only one surety bond which will include both performance and payment obligations on a single form. This type of bond is usually discouraged because claims against the bond would be interfered with by the competing interests of the Owner, workmen, and material suppliers. The Surety would be forced to determine the Owner's priority.

Generally, there is no separate payment to the Contractor for furnishing the Contract bonds. The costs involved are to be distributed among the prices bid for the various items listed in the Bid Schedule. When a payment item has been established for Mobilization, some Specifications may include the cost of Contract bonds in the Contract price for Mobilization. Some Owners may provide separate payment by establishing a lump sum payment item for Performance Bond and another for Payment Bond.

A maintenance bond (see Article 3.2.1D, Warranty and Article 4.3K, Warranty of Construction) protects the Owner against defects in materials and workmanship, generally for a period of one year after final acceptance. Some contracts will dispense with a separate maintenance bond by requiring the performance bond to continue in full force and effect for the one year warranty period specified in Article 4.3K.

F. Execution and Approval of Contract (See Article 5.3F).

The time limit in which the low bidder must return the signed Contract can refer to various starting dates, depending on the requirement of the Owner. To illustrate, the number of days specified for return of the signed Contract may start from: 1) The date of award of the Contract; 2) The date that Contract forms have been mailed to the low bidder; or 3) The date that Contract forms have been received by the low bidder.

After the Contract has been executed by all concerned, the Owner will furnish the Contractor, free of charge, additional sets of the Contract Documents; normally six to ten sets, depending on the size of the Contract. The Contractor is also usually advised that if he requires additional sets, he will be charged at the cost of reproduction.

G. Failure to Execute Contract (See Article 5.3G).

There are some Specifications that limit the liability of a low bidder, when he fails to execute the Contract, to the difference in cost between the low bid and that of the second low bid, up to the full amount of the bid security.

Other Specifications may also include a statement to the effect that a low bidder who fails to execute a contract will be excluded from bidding on future projects, for a specified period of time.

Chapter 13

Specification Format and Arrangement

13.1 Introduction

A lot of correspondence takes place from the time that a project is first advertised to the time that it is completed; and it may go beyond, if there is litigation. Much of this correspondence concerns questions and other matters relating to the Specifications. If the writer of the correspondence can easily identify the particular portion of the Specifications that he is referring to, misunderstandings and chances of error can be lessened. Identification can be simplified by assigning numbers and letters to the various Sections, subsections, articles, paragraphs, and subparagraphs. Thus it would become unnecessary to prepare a detailed description to identify material in the Specifications that is being referred to.

13.2 Section Format

A. Five Part (AASHTO) Format.

The five part format is frequently referred to as the AASHTO format (Article 3.2, The Technical Section (Five Part Format)). Specifications for the majority of civil public works projects follow this five part Section format. Owners include the Federal Highway Administration, State Departments of Transportation, and other public agencies. Because of the nature of the work involved in these projects, the Contracts are established as Unit Price Contracts in which most or all of the items of work are paid for on a unit price basis (see Article 1.3A, Fixed Price Contract; Paragraph number 2, Unit Price Contract). Accordingly, the measurement and payment subsections play an important role in the Technical Section. A Technical Section prepared in the five part format is presented in Article 3.4, Sample Technical Section. The author recommends use of the five part Section format for civil public works construction specifications.

B. Three Part (CSI) Format.

The three part format, introduced by the Construction Specifications Institute, is commonly referred to as the CSI format. The three parts of this format are numbered and titled as follows:

PART 1—GENERAL
PART 2—PRODUCTS
PART 3—EXECUTION

Construction Specifications for building projects are prepared in the three part Section format. Building contracts are generally paid for on a lump sum basis in which there is little or no need for measurement and payment subsections (see Article 1.3A, Fixed Price Contract; Paragraph number 1, Lump Sum Contract).

The major differing feature in the two formats concerns the measurement and payment requirements. These two subsections which constitute an integral part of the AASHTO format, are not featured in the three part format.

Specifications for some recent civil projects, notably rapid transit construction, have been prepared in a modified CSI format incorporating measurement and payment clauses for the items of Work. In some project Specifications, a PART 4—MEASUREMENT AND PAYMENT was added. In the Specifications for other projects, a PART 4—METHOD OF MEASUREMENT and a PART 5—BASIS OF PAYMENT were added. In still other project Specifications, the three part format was retained by incorporating the measurement and payment requirements into PART 1—GENERAL.

13.3 Arranging the Sections and Identifying Section Content

A. Arrangement of Sections.

Technical Sections are frequently grouped under Divisions, in which Sections dealing with related work are placed in the same Division.

1. Technical Sections using the five part format in Specifications for civil public works contracts would generally be grouped under six Divisions. Division titles most commonly presented in Standard Specifications of the State Departments of Transportation and the Federal Highway Administration, are:

Earthwork
Base Courses
Bituminous Pavements
Rigid Pavements
Structures
Incidental Construction

Titles of the first five named Divisions indicate the type of work to be considered when grouping Technical Sections within a Division. The Earthwork Division, which concerns work performed in the early stages of the Contract, would also include Sections dealing with other items of work performed early in the Contract, such as Demolition. Sections grouped in the last named Division of Incidental Construction would include:

Maintenance and Protection of Traffic
Storm Drainage
Sanitary Sewers
Beam Guide Rail
Fencing
Sidewalks and Curbs
Waterproofing
Topsoiling and Seeding
Traffic Markings
Traffic Signals and Lighting Systems
Signing

Project Specifications which do not group their Sections by Division, would arrange Sections in the order of execution of the Work. This is illustrated in Article 3.3, Section Arrangement.

2. Technical Sections prepared in the CSI format would be grouped according to the CSI 16 Division Masterformat, as follows:

1—General Requirements*
2—Sitework
3—Concrete
4—Masonry
5—Metals
6—Wood and Plastics
7—Thermal and Moisture Protection
8—Doors and Windows
9—Finishes
10—Specialties
11—Equipment
12—Furnishings
13—Special Construction
14—Conveying Systems
15—Mechanical
16—Electrical

*This Division does not concern itself with technical requirements of the Work. The Sections grouped under this Division deal with matters which, in a civil public works contract, would be presented in the General Conditions of the Contract.

B. Identifying Section Content.

To begin with, each Section should be identified with a Section number and title. The title is generally in capital letters, underlined. To illustrate:

SECTION 10
CLEARING AND GRUBBING

Each Section should consist of five subsections (five part format) numbered consecutively. Subsection titles should be in capital letters, as illustrated in Article 3.2, The Technical Section. Each subsection should be subdivided into Articles, with each Article identified by a capital letter, starting with "A". Major words in the Article title should have an initial capital letter. An Article may or may not be assigned a title, depending on its contents. If an Article contains two or more paragraphs, each paragraph can be numbered in consecutive order. Should a particular numbered paragraph feature numerous items or subparagraphs, each item or subparagraph would be identified by a letter (lower case). Further identification breakdown would require numbers and lower case letters in parenthesis.

13.4 Standard Specifications and Special Provisions (See Article 6.1B, Standard Specifications, Supplemental Specifications, and Special Provisions, and Article 8.4, Using Standard Specifications)

A. Presentation.

Standard Specification Sections applicable to the specific Contract must be cited in the Special Provisions. Sections in the Special Provisions should be presented in the same order as Sections in the Standard Specifications. The number and title of a Special Provision Section must be identical to the Standard Section that it is modifying.

A Special Provision Section should begin with the normal, brief description of the work involved, even to the extent of repeating the same wording of the Standard Section, if it is applicable. Following a description of the work, present the statement: "This work shall be in accordance with Standard Specification Section (number and title), except as modified herein."

B. Modifying the Standard Section.

The specific portion of the Standard Section that is to be modified must first be identified. In modifying the Standard Section, it may be necessary to add to it, delete those portions that are not applicable, or possibly replace some of the existing requirements. For example:

1. When adding a requirement, the following statement, or one similar to it, should first be presented: "In addition to the requirements specified,".
2. When deleting a requirement(s), it can be presented as follows: "The requirement(s) specified is (are) not applicable to the Work of this Contract."
3. Replacing a requirement(s), can be presented as follows: "In lieu of the requirement(s) specified,".

Sometimes, an addition or deletion can be accomplished by revising the Article, subarticle, or sentence. It can be presented as follows: 1) "The requirement specified is revised to read:"; or 2) "The second sentence is revised to read:".

C. Modifying Articles of the Standard Section.

Article numbers and titles in the Special Provisions should be identical to the Standard Section Articles being modified. Format and language used in the Special Provisions should be similar to that of the Standard Section being modified. Articles in the Special Provisions should be presented in the same order as the Articles of the Standard Section. Those Articles in the Standard Section that require no modification should be simply passed over.

D. Making Reference to Articles of the Standard Section.

When reference is made in the Special Provision to a specific Article or paragraph in the Standard Section, it implies reference to all the paragraphs or subparagraphs thereunder, unless specified otherwise.

E. Specifying Work Not Covered by the Standard Specifications.

When the work of an item of construction cannot be specified by utilizing a Standard Section, a complete new Section has to be introduced in the Special Provisions. The new Section should be presented in the same format and language as the Standard Sections.

F. Special Provisions Note.

It is essential to advise the reader that the Special Provisions are to be used jointly with the specified set of Standard Specifications. This can be accomplished by having the first page of the Special Provisions in the Contract Specifications book contain the following instruction:

SPECIAL PROVISIONS

The Work of this Contract shall be performed in accordance with the requirements of those Sections of the (Owner's name) Standard Specifications dated _____ and Supplementary Specifications dated _____ cited herein, except as they may be modified, amended, or supplemented by these Special Provisions.

13.5 Table of Contents

A. The Table of Contents performs an important function in the Contract Specifications Book. It gives the reader a condensed outline of what is inside. In addition, it tells him where he can quickly find what he is looking for. This can be valuable to personnel of both the Contractor's and Engineer's organizations, for when construction gets underway, time is of the essence to these people.

B. There should be only one Table of Contents and it should be located up front, immediately following the inside title page. It should be a complete, detailed Table of Contents covering the entire book. Some project Specifications follow the practice of presenting a General Table of Contents up front. This is followed further along in the book with an individual detailed Table of Contents for the General Conditions. Then further along in the book is another individual detailed Table of Contents covering the Technical Sections. The reader is thus confronted with three separate Tables of Content, each located in a different part of the book. This can lead to confusion and a waste of the reader's time.

C. The Table of Contents should present the assigned Section numbers and titles, and related page numbers. Individual Articles of the Bidding Documents and the General Conditions, should also be listed. To illustrate, an abbreviated version of a Table of Contents is presented:

Section No.	Title	Page No.
	BIDDING DOCUMENTS	
BD1	BIDDING REQUIREMENTS AND CONDITIONS	
A.	Notice to Contractors	BD-1
B.	Prequalification of Bidders	BD-3

Of course, if the format of the Table of Contents has already been established by the Owner, the specification writer is obliged to follow it.

Additional discussion on the arrangement of material is presented in Article 2.4B, Optimum Arrangement for Users, and Article 3.3, Section Arrangement.

13.6 Page Format

A. Page Size and Margins.

Page size is normally 8-1/2 by 11 inches. Page margins are generally one inch on three sides and 1-1/2 inches on the binding side. If the contents are printed on one side of the page, the binding margin will always be on the left side. If the contents are to be printed on both sides of the page, the binding margin will be on the left side for odd numbered pages and on the right side for even numbered pages. Typing is single spaced.

B. Page Identification.

Each page should be properly identified in case it becomes misplaced. The Contract number should be presented in the lower inside or binding corner of the page. The Section number and page number for that Section, should be centered at the bottom of the page. If a book of Special Provisions is being prepared, it may be desirable to have an auxiliary continuous page numbering system incorporating the letters "SP", beginning with "SP-1" on the first page. This would be presented in the lower outside corner of the page. For example, the first page of Special Provision Section 32, for Contract No. A250, would be identified as follows:

A250	32-1	SP-73

Chapter 14

Procedures in the Production of Specifications

14.1 Introduction

The production of a meaningful document requires adequate time, qualified personnel, and a reasonable budget. An added assurance is provided when Specifications are produced in an established Specification Department. When the Specifications are prepared by principals of the firm, project engineers, staff engineers, and design engineers, a disjointed document usually results.

Completion of an acceptable set of Specifications within the allotted time, is no easy task. The quality of the final document is dependent on timely information received from the project engineer and design departments. When this flow of information is interrupted or delayed, production suffers with a resulting increase in costs.

14.2 Preliminary Information

Upon the activation of a new project, the project engineer should make available to the Specification Department such initial information as:

A. Name of Client.
B. Title and brief description of the project.
C. Site location.
D. List of multiple contracts, when applicable.
E. A copy of that portion of the Agreement with the Client that deals with the format, content, preparation and delivery, of documents to be prepared by the Specification Department.

14.3 Meetings with Client

When the construction Contract Documents are to be initially discussed with the Client, the project engineer should arrange to have a representative of the Specification Department also present. Attendance at this meeting will make it possible for the specification writer to become familiar with the Client's special requirements that concern the Contract Documents.

The specification writer should also be present at subsequent meetings which concern the Specifications. This particularly holds true for the meeting to discuss the Client's final review comments of the Contract Documents. The specification writer is best qualified to answer questions pertaining to the Specifications.

14.4 Project Engineer Responsibility

It is recognized that, in many instances, the delay in receiving specification information is due to the difficulty in obtaining decisions from the Client. It is also recognized that project engineers may have multiple assignments which prevent them from devoting full time to the coordination of one project. When the project engineer is providing assistance to the specification writer, it helps assure production of a good set of Specifications. Areas of assistance include those listed in the following paragraphs.

A. Job Meetings. The project engineer has the opportunity to coordinate the work of design departments and ensure that needed information will be conveyed to the Specification Department.

B. The project engineer should not route an original of correspondence from the Client, to the various departments. This information may be received too late by the specification writer, or may even get misplaced. Copies should be made of the correspondence and transmitted simultaneously to each affected department.

C. During the final stage of Specification preparation, time is of the essence. At this stage particularly, the project engineer should promptly pass along information in his possession.

D. If the date established for completion of the Contract Documents is to be rescheduled for a later date, the Specification Department should be so advised.

E. If the project engineer should find it necessary to be out of the office for more than one or two days, it would be helpful to designate an assistant to perform in his absence.

14.5 Job Meetings

The production work of all departments involved should be coordinated by the project engineer. This improves the flow of information to the Specification Department, making it unnecessary for the specification writer to expend nonproductive time expediting the flow of drawings and other information. New and revised Plans, and updated lists of payment items, should be forwarded periodically.

The project engineer should initiate a program of job meetings at which direct answers to questions can be obtained; decisions made; bottlenecks eliminated;

general information disseminated; and any lack of progress discussed, with solutions recommended. Job meetings tend to increase communication, coordination, efficiency in production, and quality control. Attendance serves to familiarize participants with their respective obligations. The specification writer should be invited to every job meeting. Should the business of a meeting have no bearing on the Specifications, the specification writer can excuse himself from that meeting; this action should be his determination.

The frequency of job meetings can be increased as the final stage of design is being approached. Job meetings help minimize unexpected problems, and help meet scheduled dates for a complex project. Too few meetings can result in the tedious and time consuming process of communicating individually with the project engineer and with each design department.

A discussion of special requirements will inform all concerned departments simultaneously. Special requirements would concern an early exploratory boring contract; pile load tests; and Railroads, Public Utilities, or Public Agencies.

14.6 Scheduling

Schedules for in-house production should be established to conform to the submittal dates set by the Client. These dates could involve submittals at various stages of completion, such as:

A. Preliminary (outline) review (35 percent completion)
B. Intermediate review (65 percent completion)
C. Pre-final review (90 percent completion)
D. Final review (100 percent completion)

Preparation of the Specifications should be scheduled to begin when sufficient Plan information will be available. Plans should be at least 50 percent complete before being made available to the specification writer. The project engineer should establish a realistic date for accomplishing this. Each design department should include a list of payment items with the initial delivery of their Plans.

A common oversight in the scheduling process is to establish the same date for final completion of both the Plans and the Specifications. This does not allow time for the specification writer to review the final Plans and incorporate necessary changes into the Specifications, before their submittal to the Client. In order for the specification writer to accomplish this, completion of the Plans should be scheduled for two weeks before required completion of the Specifications. If Plans are not scheduled for completion before the Specifications, the final Plans cannot be reviewed until after the Plans and Specifications have been delivered to the Client. Then, any inconsistencies or discrepancies found between the Plans and Specifications would have to be taken care of by the issuance of an Addendum during the bidding period. If no review is made of the final Plans, inconsistencies discovered during construction will result in Contract claims and extra costs to the Owner. In either case, extensive Addenda or Contract claims will reflect poorly on the design firm.

The Mechanical and Electrical Departments generally prepare the technical input for their Sections and deliver them to the Specification Department. This should be scheduled for delivery early enough (at least two weeks) to allow time for the specification writer to complete these Sections and make any necessary modifications before final submittal to the Client.

Last minute design changes should, where possible, be transmitted to the specification writer not later than one week before submittal to the Client; otherwise, their incorporation in the final Specifications may not be possible.

14.7 Site Inspection

Bidding Documents place the prospective bidder on notice that, to familiarize himself with requirements of the Contract, he is expected to visit the site (see Article 12.2E, Examination of Plans, Specifications, Special Provisions, and Site of Work). The specification writer on the other hand, seldom if ever has an opportunity to visit the site of the Work, because a job budget rarely considers such a provision. An allowance should be provided for the specification writer to visit the site and familiarize himself with the site conditions and potential problem areas. It makes no sense to require a prospective bidder to visit the site and on the other hand have the specification writer, preparing the construction requirements, be himself ignorant of site conditions.

The specification writer should accompany the designers on their site inspection trip. This would enable him to note those conditions that can have an adverse effect on construction costs. It would include such conditions as the limitation of access to the site, overhead and underground utilities, limitations on blasting operations, and navigation conditions. Noting these conditions would make it unnecessary for the specification writer to ferret out this information by frequent questioning of the project engineer.

14.8 Cooperation of Design Departments

Meaningful Plans, in many instances, are not routed to the specification writer until shortly before the Specifications are scheduled for submittal to the Client. Receiving this late information during the hectic days of final completion frequently necessitates overtime involvement.

It is difficult to prepare the Specifications for an item of construction without having details or some idea of what is involved. If the specification writer has to continually seek out the various sources of information, a waste of time and effort results. It also increases the pressure under which he has to work. Late receipt of information from design disciplines leaves little or no time to analyze and evaluate its impact on the total Specifications, thus causing conflict or ambiguity.

Changes to a drawing can generally be accomplished by scrubbing out certain features and redrawing the changed view. Once completed, the reproduction of the changed Plan is simply a matter of hours. This is not so with the Specifications. A required change which may affect only one paragraph directly must be examined critically to determine its effect on the entire document. This can be time consuming. If there is no time available to incorporate the last-minute changes made in Plans or payment items, the expense of preparing an Addendum becomes necessary.

Design departments should adhere to the schedules set up by the project engineer. Timely receipt of detailed Plans, more than any other factor, will help to ensure the scheduled completion of a quality set of Specifications. Updated Plans should be distributed periodically; particularly when the set is being printed for submittal to the Client. In addition, it is suggested that:

A. Payment item lists established by the design departments be transmitted to the specification writer for early agreement.
B. Mechanical and Electrical Departments adhere to the schedules established for completion and delivery of their technical input to the specification writer.
C. Changes made to Plans in the late stage of design, and which affect Specifications, be immediately made known to the specification writer.

14.9 Review Procedures

Before releasing a draft of the Specifications to the Client, an in-house review should be made to ensure that the Specifications reflect the intent of the design and are consistent with the Plans. Consideration should be given to designating time for in-house reviews, particularly since Plan changes are frequently made. Draft copies of Specifications should be distributed directly to involved design departments and the project engineer, for their review.

In-house reviews and the return of comments to the Specification Department should be prompt. This will enable the specification writer to review and evaluate these comments, and correct the Specifications before their transmittal to the Client. Occasionally an in-house review will be bypassed by a design department. This then takes on greater importance, because lines of communication between design departments, project engineer, and specification writer, now leave much to be desired.

14.10 Feedback From Construction

The production of a set of perfect Specifications is a rarity. The errors of commission and omission often present in Contract Specifications, come to light during the construction phase. Upon completion of the project, a summary report is frequently prepared by the resident engineer. The report will include a record of problems on the job caused by errors in the Specifications. Transmittal of this information to the Specification Department will help prevent a repetition of these errors in subsequent Specifications. Additional feedback from the field office that could help improve the quality of subsequent Specifications would include information on change orders, negotiated agreements, and claims for extra work.

Feedback information should be presented as written, documented field data. Verbal comments are generally not only worthless but may actually be misleading. If feedback information from the field may not be forthcoming, consideration can be given to visiting the project for information on problems with the Specifications.

14.11 Control of Quality

A. The courts have repeatedly pointed up recognition of the Specifications as a principal design document. This certainly directs attention to a need for the reappraisal of in-house procedures in its preparation, if errors and omissions are to be minimized.

B. Independent proofreading is one positive method of detecting errors. The specification writer should not proofread his own material, because the person who commits the error is generally unable to recognize his own mistake. Nor should it be proofread by nontechnical personnel. It is best to be assigned to another staff member in the Specification Department. In addition to checking word for word against the input sheets, a proofreader should check the quality of typing, conformance to required format, clarity, and constructibility. When reference is made to a Standard Section or to an industry Standard, the number, title, and applicability of contents of the referenced Standard, should be verified. When questions or corrections develop, the proofreader should confer with the originator.

C. As a precaution to prevent the accidental mix-up of specification pages for different submittal editions, it is suggested that each page contain the date of its edition. The final edition date would be removed from the pages before transmittal to the Client for publication.

D. When preparing an Addendum to modify the Bidding Documents, it is essential that the specification writer have in his possession a copy of the published document being modified. There are times when an Addendum is prepared by others without prior consultation of the Specification Department. The specification writer should be given the opportunity to review the Addendum before it is distributed to bidders. To better control the quality, all Addenda for the Contract should be prepared by the specification writer.

14.12 Reference Files

Maintaining the office reference files complete and up-to-date plays an important part in the operation of the Specification Department. The files can consist of elements as those listed in the following paragraphs.

A. Standard Specifications of Clients. Clients may include:

1. Federal agencies.
2. State Departments of Transportation.
3. Counties and Cities.
4. Public authorities, including owners of rapid transit projects and toll highway systems.

B. Published Contract Specifications. This file would help the specification writer quickly locate a set of Specifications previously prepared for the same Client or for work of a similar category. Classifying completed Contract Specifications would include such methods as:

1. A card index listing completed Contract Specifications by job category. Categories would include such classifications as movable bridges, concrete dams, flood protection, highways, sanitary facilities, and tunnels.
2. A card index listing completed Contract Specifications alphabetically, by Contract title. Each card would also contain the assigned in-house project number.
3. A card index listing completed Contract Specifications numerically, by their assigned project number. Cards would also contain the corresponding Contract title.
4. A card index listing items of unusual construction, for which Specifications have been previously prepared. The card would contain the corresponding Contract title of the Specifications in which the unusual item has been specified. The listing would include items such as blasting, compressed air operations, fly ash concrete, tremie concrete, dredging; pressure grouting, slurry walls, tunnel lining, and underpinning.

A published copy of Contract Specifications is made a part of the finished Contracts file. The copy should not be removed from the Department. These file copies assume greater importance when an Addendum has to be prepared.

File copies should be handled with care, maintained in good condition, and returned to the files when no longer needed.

C. A current set of the National Reference Standards described in Chapter 7, should also be maintained. Copies of the more frequently consulted reference Standards such as ASTM, should not be permitted to leave the Department. Reference Standards less frequently used, such as those of the Steel Structures Painting Council (SSPC), may be borrowed by other Departments. There should be a record of the reference borrowed, by whom, and date borrowed.

D. A file of correspondence folders pertaining to individual contracts should be maintained. The folder should only be removed temporarily from the files when the specification writer is interested in reviewing some material contained therein. If a particular piece of correspondence is important enough to retain, the specification writer should make a copy so that the original document can remain in the folder.

Material in a correspondence file should not be loaned outside the Department. If it is needed, a copy should be made and transmitted.

Chapter 15

Qualifications of the Specification Writer

15.1 Introduction

Paper 8781, Responsibilities of the Engineer and the Contractor Under Fixed-Price Construction Contracts, March 1972, ASCE Journal of the Construction Division, previously described in Article 4.4S, Claims for Adjustment, contained the following statements which still hold true today.

> The proper framing of a set of construction specifications is not easy. Engineering specialists called specification writers, are normally employed for that purpose, and their work requires good judgment, a broad knowledge of the technical aspects of the job and appreciation of the construction problems entailed, plus the ability to express clearly and concisely all of the terms, conditions and provisions necessary to present an accurate picture to the constructor. It is a very large order.
>
> Very often the questionable parts of specifications are injected by those too far removed from the actual application of them to be knowledgeable, and if it were up to the Engineer on the job, different provisions would therefore be chosen.

To many Engineers, specification writing is considered a thorn in their side. They do not look upon it as a fundamental engineering activity. Yet in the past it has been said that more than fifty percent of all construction claims occur because Plans and Specifications are unclear, ambiguous, or contradictory.

The availability of standard documents will not by themselves assure good project Specifications, particularly if the writer is incapable of properly applying them. A determining factor in the preparation of good Specifications is the capability of the specification writer.

Paper 14001, Summary Report of Questionnaire on Specifications (Contractor Returns) and Paper 14799, Summary Report of Questionnaire on Specifications (Owner and Owner Representative Returns) published September 1978 and September 1979, respectively, in the ASCE Journal of the Construction Division, prepared by the ASCE Committee on Specifications and chaired by the author at that time, offer some interesting information. It was a response of the construction industry and presented current practices, problems in producing the Specifications, and recommendations for their improvement. One question common to both surveys asked for qualifications that a specification writer should possess.

An interesting response in the Contractor Returns bears mentioning here. In the question "What Improvements Would You Like to See in Specifications," some contractors replied that Specifications should be written only by technically trained persons; attorneys should not prepare Specifications.

15.2 Education

A qualification considered highly desirable by respondents in the two Summary Reports on Specifications was a technical education. Along with a higher education, enrollment in courses teaching Engineering Construction Specifications and Construction Contract Law were considered to be beneficial.

15.3 Field Experience

ASCE Paper 9192, Summary Report of Questionnaire on Construction Inspection, September 1972, prepared by the ASCE Committee on Inspection and chaired by the author at the time, included an interesting comment on Specifications. Responding contractors indicated that one of the factors contributing to the problems of inspection was the poor quality of the Specifications. They maintained that too many specification writers with little or no experience in construction practices were establishing standards and specifying requirements for the Contractor to meet in performance of the Work.

Both of the ASCE Summary Reports described earlier in Article 15.1, Introduction, indicated that field construction experience was an important qualification for a specification writer to possess. Seventy percent of the 223 Contractors responding and thirty-five percent of the 622 Owners and Private Design Professionals responding, considered this to be a top qualification. Five to ten years of such experience was preferred, with a portion of this in the role of Resident Engineer.

If a specification writer possesses limited or no construction experience on a particular operation, he should consult with a fellow specification engineer, or the Construction Department of the Firm, or read up on the subject. Additional suggestions for handling this situation are discussed in Article 8.9, Know Your Subject.

In line with the recommendations for field experience, a specification writer should be afforded an opportunity during the construction phase of a project to observe the field application of his requirements, particularly as they may apply to unusual items of work, new procedures, or the use of a new material.

15.4 Knowledge of the Work

Possessing a knowledge of the Work was considered in both Summary Reports on Specifications to be relatively important in recommending qualifications for a specification writer. It was also recommended that he have some design experience along with an understanding of the project.

15.5 Ability to Write Good English

An ability to write good English was considered in both Summary Reports on Specifications to be another important qualification.

A. In responses from Owners and Private Design Professionals, it was considered important to possess technical writing ability, fluency in English, and an ability to communicate.

B. In responses from Contractors, the second most common recommendation was an ability to communicate in clear, simple English.

The specification writer should have a command of English that will enable him to state the requirements clearly and concisely, so that both the Owner and Contractor can have a clear understanding of what has to be accomplished. The significance of this capability is detailed in Chapter 9, Specification Language.

15.6 Additional Qualifications

A. A specification writer should have the desire to:

1. Review technical periodicals relating to his work.
2. Send for available related literature.
3. Become active in related professional organizations.
4. Keep up-to-date on construction materials and equipment.

B. In addition to the qualifications already presented from both Summary Reports on Specifications, the following recommendations are also considered worthy of mentioning. They are listed in the order of their descending frequency of mention in the Summary Reports.

1. Responses from Owners and Private Design Professionals:

a. Professional registration.
b. Specification writing experience.
c. Common sense.
d. Knowledge of the Construction Specifications Institute (CSI).
e. Formal training in specification preparation.
f. Familiarity with State Standards or Local Codes.
g. Research ability.

2. Responses from Contractors:

a. Common sense, responsibility, and being a logical, organized individual.
b. Knowledge of Contract law.
c. An obligation to protect the interests of both the Owner and the Contractor.

15.7 Practices That Promote a Scarcity of Qualified Specification Writers

The development of a set of Contract Specifications normally receives considerably less attention than do the design and execution of the construction. It has been maintained that an otherwise highly qualified engineering firm will engage engineers with impressive credentials to perform the design. The individual charged with preparing the Specifications generally lacks the qualifications to produce a set of Specifications adequate for the purpose.

The ASCE Summary Report of responses of Owners and Private Design Professionals presented, in the answers to questions 2, 3, and 4, illuminating information on the then current practices in the preparation of Specifications. Those practices indicated a need for improvement. This is illustrated in the following paragraphs.

Question No. 2: Do You Maintain a Separate Specifications Department?

a. Approximately fifty percent of the 55 responding Federal and State Agencies answered in the affirmative.
b. Of the 508 Private Design Professionals responding, only fifteen percent answered in the affirmative. Smaller engineering firms generally do not maintain a separate Specifications Department.

Question No. 3: If You Do Not Maintain a Separate Specifications Department, Do You Employ Full-Time Specification Writers?

a. Approximately fiften percent of the 29 Federal and State Agencies not maintaining a separate Specifications Department, responded in the affirmative.
b. Of the 437 Private Design Professionals who did not maintain a separate Specifications Department, only six percent employed full-time specification writers.

Question No. 4: If Your Answers to Questions 2 and 3 are Negative, Who Writes Your Specifications?

The responses of Private Design Professionals are listed in the following order of descending frequency of the same answer.

a. Project Engineer.
b. Designer.
c. Principal of Firm.
d. Staff Engineer.
e. Department Head.
f. Outside Consultant.
g. Senior Engineer.

Additional practices that promote a scarcity of qualified specification writers, include:

a. Offering inadequate salaries that fail to interest a higher level of personnel to pursue a career in Specifications.
b. Failure to establish a reasonable percentage of the design budget, for the preparation of Specifications.
c. Failure of engineering management to recognize the relative importance of Contract Specifications.

Appendix A

Samples of Cited Documents and Forms

BALTIMORE REGION RAPID TRANSIT SYSTEM
STATE OF MARYLAND DEPARTMENT OF TRANSPORTATION
MASS TRANSIT ADMINISTRATION

NOTICE TO CONTRACTORS

LEXINGTON MARKET
STATION STRUCTURE
Contract No. NW-02-06 Date 6-10-77

Sealed Bids addressed to the Mass Transit Administration, Office of Contract Administration, 109 East Redwood Street, Baltimore, Maryland 21202, and marked "Bid for Contract No. NW-02-06 ", will be received at the above address until but not after 2:00 p.m. local time, December 15, 1977 Add.2
At that time the Bids will be publicly opened and read aloud at a location at the same address. The Work to be done under this Contract includes construction of a station structure and north ventilation shaft by the cut and cover method; northeast entrance complex; southeast entrance; south ventilation shaft in a construction shaft built by others under a prior contract; protection of existing structures; control of goundwater; relocation, reconstruction, support and maintenance of existing utilities; maintenance and protection of traffic; street decking; paving and other restoration work. The work also includes the construction of the OCC Building, a three story elevated structure, complete as specified.

Estimated Value for this Work is in the range of $ 25 to 35 million.

Bid documents may be obtained on or after June 10, 1977 at the Office of Contract Administration, at the above address, upon receipt of a check in the correct amount made payable to the Mass Transit Administration, which sum will not be refunded. When requesting documents, give street address. Cash will not be accepted. Price for documents are as follows;

1. Bid Documents and one set of half size Contract Drawings not including MTA Standard Specifications, June 1976: $75.00/Set.

2. MTA Standard Specifications, June 1976 (must be used in conjunction with 1. above): $7.50/Book.

3. Each extra set of half size Contract Drawings: $50.00/Set.

4. Safety Manual: $1.00/Each.

5. Preliminary Subsurface Design Study, Phase 1 - Northwest Line, Volumes 1 and 2, dated June 1974 (Preliminary Soils Data for entire Northwest Line which is common to all Northwest Line Contracts) are available separately: $20.00/Set.

NW-02-06
CONFORMED:
2/3/78

NTC 1 of 2

Exhibit A.

6. **Soils and Exploratory Data and Design Summary Report for the Contract excluding 5. above: $100.00/Set.**

Copies of documents are available for inspection at offices of the Administration; Dodge Reports, Baltimore, Maryland, Washington, D.C., New York, N.Y., Chicago, Illinois, and San Francisco, California offices; and the Building Congress and Exchange of Baltimore, Inc.

A Pre-bid meeting will be held on October 4, 1977 at 2:00 p.m. local time in the MTA administrative offices, 1st floor, 109 E. Redwood Street, Baltimore, Maryland, for the purpose of explaining the project. Questions regarding the Work should be directed by mail to Mr. R. Hampton or Mr. R. Chow at the Administration offices at the above address, or by telephone at numbers (301) 383-4442 and (301) 383-4977 respectively. Add.2

Each bid must be accompanied by a Bid Bond, on a form furnished by the Administration in the amount of ten percent (10%) of the bid price. Performance and Payment Bonds in the amount of the contract price will be required.

The Administration reserves the right to reject any and all bids and/or waive technical defects if, in its judgment, the interests of the Administration so required.

Any Contract resulting from bids submitted is subject to a financial assistance Contract between the Administration and the U. S. Department of Transportation.

All Bidders will be required to certify that they are not on the U. S. Comptroller General's List of ineligible Contractors.

The Mass Transit Administration hereby notifies all bidders that in regard to any contract entered into pursuant to this advertisement, monority business enterprises will be afforded full opportunity to submit bids in response to this notice and will not be subjected to discrimination on the basis of race, color, sex or national origin in consideration for an award.

Bidders on this Work will be required to comply with MTA Affirmative Action Requirements and all applicable Equal Employment Opportunity Laws and regulations and shall submit with their bid the Administration's minority Business Affirmative Action Certification.

All Bidders shall complete and submit the Contractor's Questionnaire-Pre-Award Evaluation Data Form.

Not less than 50 percent of the Work shall be performed by the Bidder with his own forces.

Any Bid received after the time and data specified shall not be considered.

Walter J. Addison
Mass Transit Administrator

NW-02-06
CONFORMED:
2/3/78

NTC 2 of 2

Exhibit A (*continued*).

BALTIMORE REGION RAPID TRANSIT SYSTEM
STATE OF MARYLAND DEPARTMENT OF TRANSPORTATION
MASS TRANSIT ADMINISTRATION

CONTRACTOR'S QUESTIONNAIRE
PRE-AWARD EVALUATION DATA

IMPORTANT

This questionnaire is intended as a basis for establishing the qualifications of Contractors for undertaking Construction, Maintenance and Repair Work under the jurisdiction of the Mass Transit Administration.

If a Contractor has submitted such a questionnaire to the Mass Transit Administration within six (6) months from the date the current Proposal is due, the Contractor may refer to that submittal, by date and proposal subject, in lieu of submitting a new questionnaire. The Mass Transit Administration requires that a current questionnaire be on file and Contractors shall submit a new questionnaire whenever major changes have occurred in their organization, financial position and experience.

This questionnaire forms a part of the Contractor's Bid and failure to submit it or lack of evidence of qualification may be a basis for rejection of a bid.

I. General

(a) Legal Title and Address of Organization

____

____

(b) Contractor's Local Representative's Name, Title and Address

____

____

(c) _______ Corporation _______ Partnership _______ Individual
(Check One)

(d) If a Corporation--State: ______________
Capital Paid in Cash $ ______________
Date of Incorporation ______________
State in which Incorporated ______________

Name and Title of Principal Officers	Date of Assuming Position

NW-02-06
4/29/77

CQ 1 of 6

Exhibit B.

CONTRACTOR'S QUESTIONNAIRE

(e) If Partnership--State:
Date of Organization ____________ Nature of Partnership
(General, Limited or Association) ________________________

Names and Addresses of Partners | Age

________________________________ ________
________________________________ ________
________________________________ ________

(f) If Individual--State:
Full Name and Address of Owner ______________________
____

(g) List major items of equipment fully owned by organization, giving approximate value and age. (If now fully owned, so state)

____
____
____
____
____
____
____
____
____

(h) Is any member of your organization employed by the State of Maryland, a member of any State Institution's Board or Managers or Trustees, or in any way officially connected with the State Government? ______________ If yes -- Explain ____________
____

(i) Give name and data about any construction projects you have failed to complete, including any terminations for default (use separate sheet if necessary)

____
____

(j) Has your organization or any of its Directors, Officers, Partners or Supervisory Personnel ever been party to any criminal action relating directly or indirectly to the general conduct of your business?
If yes -- Explain________________________________
____

(k) Has your organization ever been denied an award on which you were low bidder? _____ If yes -- Explain ________________
____
____

NW-02-06
4/29/77

CQ 2 of 6

Exhibit B (*continued*).

(1) Have you ever been assesed actual or liquidated damages for late completion?________If so, give full particulars ______

____

____

II. Financial

(a) Give value of all construction equipment fully owned by your organization ______________________________________

____

(b) Give value of total assets of organization (including equipment value in (IIa) above ______________________________

(c) Give value of total liabilities of organization ______________

(d) Give total contract value of work accomplished by your organization in each of the last three (3) years
____________ 19____ ____________ 19 ___ ______________ 19 ____

(e) Give contract value of work presently being accomplished by, or pending award to your organization ____________________

____
(Date)

(f) Give value of any judgments or liens outstanding against your organization ______________________________________

(g) Has any Bonding Company refused to write you a bond on any construction work? ____________ If yes--Explain ______________________

____

____

(h) Give maximum value of contract work for which you could obtain Bond

____

III. Experience

(a) Indicate type of contracting undertaken by your organization and years' experience:
General ____________ Sub _________ Type _____________________
Years Years Years

Type ____________________________Type _____________________
Years Years

(b) State construction experience of principal members of your organization:

NW-02-06
4/29/77

Exhibit B (*continued*).

Construction Experience

Name and Title (As Pres.,Mgr.,etc.)	Const. Experience Years	Type of Work (Such as Hospitals Apts., etc.)	In What Capacity (Foreman, Supt.,etc)
______	______	______	______
______	______	______	______
______	______	______	______
______	______	______	______
______	______	______	______
______	______	______	______
______	______	______	______
______	______	______	______
______	______	______	______
______	______	______	______
______	______	______	______

(c) Give any special qualifications of firm members (Registered Engineer, Surveyor, etc.) ______________________________
____
____
____

(Use Extension Sheet if Necessary)

(d) List some principal projects completed by your organization:

Name of Work	______	______	______	______
	______	______	______	______
General or Sub (If sub, what type of work)	______	______	______	______
	______	______	______	______
Your Contract Amount	______	______	______	______
	______	______	______	______

NW-02-06
4/29/77

CQ 4 of 6

Exhibit B (*continued*).

CONTRACTOR'S QUESTIONNAIRE

Year	______	______	______	______
Designing Architect or Engineer	______	______	______	______
	______	______	______	______
Owners' Name and Address	______	______	______	______
	______	______	______	______

(e) If General Contractor, list some sub-contractors in various fields who have worked under you: ______

(f) If Sub-contractor, list some General Constractors for whom you have worked: ______

(g) 1. What is the money value of the largest project accomplished by your Organization? ______

2. Maximum value in last three (3) years ______

3. Maximum value you prefer to undertake ______

4. Price range of work your organization is deemed best adapted to undertake ______

(h) Is your organization licensed in the State of Maryland for the current year? ______ Give date and number of license ______

The above statements are certified to be true and accurate.

Dated at ______ this ______ day of ______ 19 __

By ______

______
(Title of Person Signing)

______
(Name of Organization)

NW-02-06
4/29/77

CQ 5 of 6

Exhibit B (*continued*).

CONTRACTOR'S QUESTIONNAIRE

State of ______________________________

County of ______________ , S.S.______________________

______________________ Being duly sworn states that he is

______________________ of ______________________

and that the answers to the foregoing questions and all statements therein contained are true and correct.

Sworn to before me this ____________day of ______________ 19 __________

My Commission expires ______________ ______________________
(Notary Public)

NW-02-06
4/29/77

CQ 6 of 6

Exhibit B (*continued*).

143

SECTION H

BIDDER'S BOND

KNOW ALL MEN BY THESE PRESENTS, That we ____________________

____

as PRINCIPAL, and ______________________________________ as SURETY, are held and firmly bound unto the Port Authority of Allegheny County, hereinafter called the Port Authority in the penal sum of: ____________________________ DOLLARS, lawful money of the United States, for the payment of which sum well and truly to be made, we bind ourselves, our heirs, executors, administrators, and successors, jointly and severally, firmly by these presents.

THE CONDITION OF THIS OBLIGATION IS SUCH, that whereas the PRINCIPAL has submitted the accompanying Proposal dated ___________________, 19_____, for: ____.

NOW, THEREFORE, if the Principal shall not withdraw said proposal within either the appropriate sixty (60) day or one-hundred twenty (120) day period, as provided (in Section 2 of Act 317) after said opening, and shall within the period specified therefor, or if no period be specified, within ten (10) days after the prescribed forms are presented to him or it for signature, enter into a written Contract with the Port Authority in accordance with the proposal accepted, and execute and deliver to the Port Authority all bonds and other instruments required to be executed and delivered by the Principal in accordance with the proposal documents, or in the event of the withdrawal of said proposal within the period specified or the failure to enter in such Contract and to execute and deliver to the Port Authority all bonds and other instruments required to be executed and delivered by the Principal in accordance with the proposal documents within the time specified, or if no time is specified within ten (10) days after the prescribed forms are presented to him or it for signature, if the Principal shall pay the Port Authority the difference between the amount specified in said proposal and the amount for which the Port Authority may procure the required work, equipment or supplies, if the latter amount be in excess of the former, together with all other loss, damage or expense suffered by the Port Authority thereby, then the above obligation shall be void and of no effect, otherwise to remain in full force and virtue.

Said Surety, for value received, hereby stipulates and agrees that the obligations of said Surety under this Bond shall in no way be impaired or affected by an extension of the time within which said bid may be accepted and said Surety does hereby waive notice of any such extension.

Said Surety agrees that its liability hereunder shall be absolute regardless of any liability of the Principal hereunder whether by reason of any irregular or unauthorized execution of or failure to execute this Bond or otherwise.

Exhibit C.

143

IN WITNESS WHEREOF, the above-bounden parties have executed this instrument under their several seals this _________ day of _________________, 19______, the name and corporate seal of each corporate party being hereto affixed and these presents duly signed by its undersigned representative, pursuant to authority of its governing body.

IN PRESENCE OF

______________________________	______________________________ (SEAL) (Individual Principal)
______________________________ (Address)	______________________________ (Business Address)
______________________________	______________________________ (SEAL) (Individual Principal)
______________________________	______________________________ (Business Address)

ATTEST:

	______________________________ (Corporate Principal)
	______________________________ (Business Address)
______________________________	By: ___________________________ (AFFIX CORPORATE SEAL)
	______________________________

ATTEST:

	______________________________ (Corporate Surety)
	______________________________ (Business Address)
______________________________	By: ___________________________ (AFFIX CORPORATE SEAL)
	______________________________

CONSTRUCTION H-2

Exhibit C (*continued*).

143

CERTIFICATE AS TO CORPORATE PRINCIPAL

I, ______________________________, certify that I am the (Assistant (Secretary) of the corporation named as Principal in the within bond; that ______________________________, who signed the said bond on behalf of the Principal was then ______________________________ of said corporation, that I know his signature and his signature thereto is genuine and that said bond was duly signed, sealed and attested for and in behalf of said corporation by authority of its governing body.

____

(Corporate Seal)

Exhibit C (*continued*).

143 PERFORMANCE BOND

SECTION J

PERFORMANCE BOND

KNOW ALL MEN BY THESE PRESENTS, That we ____________________

____

as PRINCIPAL, and ____________________________________ as SURETY, are held firmly bound unto the Port Authority of Allegheny County hereinafter called the Owner, in the sum of: ______________________________ DOLLARS, 100% of contract price, for the payment of which sum well and truly be made, we bind ourselves, our heirs, executors, administrators, and successors, jointly and severally, firmly by these presents.

THE CONDITION OF THIS OBLIGATION IS SUCH, that whereas the PRINCIPAL entered into a certain Contract, hereto attached, with the Owner, dated ____________________, 19____, for: ____________________________ ____________________________.

NOW, THEREFORE, if the Principal shall well and truly perform and fulfill all the undertakings, covenants, terms, conditions, and agreements of said Contract during the original term of said Contract and any extensions thereof that may be granted by the Owner, with or without notice to the Surety, and during the life of any guaranty required under the Contract, and shall also well and truly perform and fulfill all the undertakings, covenants, terms, conditions, and agreements of any and all duly authorized modifications of said Contract that may hereafter be made, including changes which result in increases in the original Contract price without notice to the Surety, then, this obligation to be void, otherwise to remain in full force and virtue.

It is hereby further stipulated and agreed that if the Principal is a non-Pennsylvania corporation neither the Principal nor the Surety shall be discharged from liability on this bond, nor the bond surrendered, until such Principal files with the Port Authority of Allegheny County a certificate from the Pennsylvania Department of Revenue evidencing the payment in full of all bonus taxes, penalties and interest, and a certificate from the Bureau of Employment and Unemployment Compensation of the Pennsylvania Department of Labor and Industry, evidencing the payment of all unemployment compensation contributions, penalties, and interest due the Commonwealth of Pennsylvania from the said Principal, or any non-Pennsylvania corporation, sub-contractor thereunder, or for which liability has accrued but the time for payment has not arrived, as required by the Act of June 10, 1947, P.L. 493, 8 P.S. sec. 23, or as amended or superseded.

IN WITNESS WHEREOF, the above bounden parties have executed this instrument under their several seals this ______________ day of ______________, 19______, the name and corporate seal of each corporate party being hereto affixed and these presents duly signed by its undersigned representatives, pursuant to authority of its governing body.

J-1

Exhibit D.

143 PERFORMANCE BOND

IN PRESENCE OF

__________________ __________________ (SEAL)
(Individual Principal)

(Address) (Business Address)

__________________ __________________ (SEAL)
(Individual Principal)

(Address) (Business Address)

ATTEST:

(Corporate Principal)

(Business Address)

By:
(AFFIX CORPORATE SEAL)

ATTEST:

(Corporate Surety)

(Business Address)

By:
(AFFIX CORPORATE SEAL)

J-2

Exhibit D (*continued*).

SECTION K

LABOR AND MATERIALMAN'S BOND

KNOW ALL MEN BY THESE PRESENTS, That we, ____________________
____ as
PRINCIPAL, and ____
____________________________ as SURETY, are held firmly bound unto the Port Authority of Allegheny County hereinafter called the Owner, in the sum of ______________________________ DOLLARS, 100% of contract price, for the payment of which sum well and truly to be made, we bind ourselves, our heirs, executors, administrators, and successors, jointly and severally, firmly by these presents.

THE CONDITION OF THIS OBLIGATION IS SUCH, that whereas the Principal entered into a certain Contract, hereto attached, with the Owner, dated _______, 19____, for: ____
____

NOW, THEREFORE, if the Principal shall promptly make payment to all persons supplying labor and material in the prosecution of the work provided for in said Contract, and any duly authorized modifications of said Contract that may hereafter be made, except that no change will be made which increases the total Contract price more than twenty percent (20%) in excess of the original Contract price without notice to the Surety, then this obligation to be void, otherwise, to remain in full force and virtue.

IT IS FURTHER PROVIDED that every person, co-partnership, association or corporation who, whether as subcontractor or otherwise, has furnished material or supplied or performed labor in the prosecution of the work, and who has not been paid therefor, may sue in assumpsit on this bond, in his, their or its own name and prosecute the same to final judgment for such sum or sums as may be justly due him, them, or it, and have execution thereon, as provided by law, including the provisions of Act No. 385, 1967 of the Commonwealth of Pennsylvania, known as "The Public Works Contractors' Bond Law of 1967".

Exhibit E.

143 LABOR AND MATERIALMAN'S BOND

IN WITNESS WHEREOF, the above bounden parties have executed this instrument under their several seals this __________ day of _______, 19 _____, the name and corporate seal of each corporate party being hereto affixed and these presents duly signed by its undersigned representatives, pursuant to authority of its governing body.

IN PRESENCE OF

_______________ _______________ (SEAL)
(Individual Principal)

_______________ _______________
(Business Address)

_______________ _______________ (SEAL)
(Individual Principal)

_______________ _______________
(Business Address)

ATTEST:

____
(Corporate Principal)

____
(Business Address)

_______________ By: _______________
(AFFIX CORPORATE SEAL)

____

ATTEST:

____
(Corporate Surety)

____
(Business Address)

_______________ By: _______________
(AFFIX CORPORATE SEAL)

____

K-2

Exhibit E (*continued*).

Appendix B

Sources of Information for the Specification Writer

Requests to a source of information may concern a specific item or a list of available publications. The sources of information presented herein have been grouped into general classifications. Chapter 7, National Reference Standards, offers information on the more commonly used sources of reference.

1. Wood

American Institute of Timber Construction
333 West Hampden Avenue
Englewood, Colorado 80110

American Lumber Standards Committee
P.O. Box 210
Germantown, Maryland 20874

American Plywood Association
P.O. Box 11700
Tacoma, Washington 98411

American Wood Preservers Association
1625 I Street, N.W.
Washington, D.C. 20006

American Wood Preservers Bureau
2772 South Randolph Street
Arlington, Virginia 22206

American Wood Preservers Institute
1945 Gallows Road, Suite 405
Vienna, Virginia 22180

California Redwood Association
591 Redwood Highway, Suite 3100
Mill Valley, California 94941

Forest Products Research Society
2801 Marshall Court
Madison, Wisconsin 53705

Hardwood Plywood Manufacturers Association
Box 2789
Reston, Virginia 22090

National Coalition Against Misuse of Pesticides
530 Seventh Street, S.E.
Washington, D.C. 20003

National Forest Products Association
1619 Massachusetts Avenue, N.W.
Washington, D.C. 20036

Society of American Wood Preservers
7297 Lee Highway, Unit P
Falls Church, Virginia 22042

Southern Forest Products Association
P.O. Box 42468
New Orleans, Louisiana 70152

Southern Pine Inspection Bureau
4709 Scenic Highway
Pensacola, Florida 32504

Western Wood Products Association
1500 Yeon Building
Portland, Oregon 97204

2. Concrete and Reinforcement

Allentown Pneumatic Gun, Inc.
P. O. Box 185
Allentown, Pennsylvania 18105
(Shotcrete and equipment)

American Coal Ash Association
1819 H Street, N.W., No. 510
Washington, D.C. 20006

American Concrete Institute
Box 19150
Detroit, Michigan 48219-0150

American Concrete Pumping Association
1034 Tennessee Street
Vallejo, California 94590

American Fly Ash Company
606 Potter Road
Des Plaines, Illinois 60016

American Society for Concrete Construction
3330 Dundee Road
Northbrook, Illinois 60062

Cellular Concrete Association
715 Boylston Street
Boston, Massachusetts 02116

Concrete Construction Publications
426 South Westgate
Addison, Illinois 60101

Concrete Reinforcing Steel Institute
933 North Plum Grove Road
Schaumburg, Illinois 60195

Concrete Sawing and Drilling Association
3130 Maple Drive, Suite 7
Atlanta, Georgia 30305

Expanded Shale, Clay and Slate Institute
6218 Montrose Road
Rockville, Maryland 20852

Gunite Contractors Association
2837 Newell Street
Los Angeles, California 90039

International Grooving and Grinding Association
P.O. Box 1750
Briarcliff Manor, New York 10510

Lightweight Aggregate Producers Association
546 Hamilton Street
Allentown, Pennsylvania 18101

National Association of Reinforcing
Steel Contractors
10382 Main Street, P.O. Box 225
Fairfax, Virginia 22030

National Ash Association, Inc.
1819 H Street, N.W., Suite 350
Washington, D.C. 20006

National Crushed Stone Association
1415 Elliot Place, N.W.
Washington, D.C. 20007

National Precast Concrete Association
825 East 64th Street
Indianapolis, Indiana 46220

National Ready Mixed Concrete Association
900 Spring Street
Silver Spring, Maryland 20910

National Sand and Gravel Association
900 Spring Street
Silver Spring, Maryland 20910

National Slag Association
300 South Washington Street
Alexandria, Virginia 22314

Perlite Institute, Inc.
6268 Jericho Turnpike
Commack, New York 11725

Portland Cement Association
5420 Old Orchard Road
Skokie, Illinois 60077

Post-Tensioning Institute
301 West Osborn, Suite 3500
Phoenix, Arizona 85013

Prestressed Concrete Institute
Suite 1410, Wells Tower Building
201 North Wells Street
Chicago, Illinois 60606

Producers' Council, Inc.
1717 Massachusetts Avenue, N.W.
Washington, D.C. 20036

Reinforced Concrete Research Council
5420 Old Orchard Road
Skokie, Illinois 60077

Scaffold Industry Association
14039 Sherman Way
Van Nuys, California 91405

Scaffolding and Shoring and Forming Institute
1230 Keith Building
Cleveland, Ohio 44115

Tilt-Up Concrete Association
5420 Old Orchard Road
Skokie, Illinois 60077

Wire Reinforcement Institute, Inc.
8361A Greensboro Drive
McLean, Virginia 22120

3. Pipelines

American Concrete Pipe Association
8320 Old Courthouse Road
Vienna, Virginia 22180

American Concrete Pressure Pipe Association
8320 Old Courthouse Road
Vienna, Virginia 22180

American Water Works Association (AWWA)
6666 West Quincy Avenue
Denver, Colorado 80235

Asbestos Cement Pipe Producers Association
1600 Wilson Boulevard, No. 1008
Arlington, Virginia 22209

Cast Iron Soil Pipe Institute
1499 Chain Bridge Road, Suite 203
McLean, Virginia 22101

Ductile Iron Pipe Research Association
245 Riverchase Parkway East, Suite O
Birmingham, Alabama 35244

National Association of Sewer Service Companies, Inc.
110 East Morse Boulevard, Room 3
Winter Park, Florida 32789

National Clay Pipe Institute
350 West Terra Cotta Avenue
Crystal Lake, Illinois 60014

National Corrugated Steel Pipe Association
1750 Pennsylvania Avenue, N.W., Suite 1303
Washington, D.C. 20006

National Utility Contractors Association
Suite 606
1235 Jefferson-Davis Highway
Arlington, Virginia 22202

Pipe Line Contractors Association
4100 First City Center
1700 Pacific Avenue
Dallas, Texas 75201

Plastics Pipe Institute
355 Lexington Avenue
New York, New York 10017

Underground Contractors Association
8550 West Bryn Mawr Avenue, Suite 704
Chicago, Illinois 60631

Unibell Plastic Pipe Association
2655 Villa Creek Drive, Suite 164
Dallas, Texas 75234

4. Pavements

American Concrete Pavement Association
2625 Clearbrook Drive
Arlington Heights, Illinois 60005

Asphalt Institute
Asphalt Institute Building
College Park, Maryland 20740

Association of Asphalt Paving Technologies
155 Experimental Engineering Building
Minneapolis, Minnesota 55455

American Association of State Highway and Transportation Officials
444 North Capitol Street, N.W.
Suite 225
Washington, D.C. 20001

Highway Research Board
2101 Constitution Avenue, N.W.
Washington, D.C. 20418

International Grooving and Grinding Association
310 Madison Avenue, Suite 602
New York, New York 10017

National Asphalt Pavement Association
6811 Kenilworth Avenue
Riverdale, Maryland 20737

5. Metals

Aluminum Association
818 Connecticut Avenue, N.W.
Washington, D.C. 20006

American Hot Dip Galvanizers Association, Inc.
1101 Connecticut Avenue, N.W., Suite 700
Washington, D.C. 20036-4303

American Institute of Steel Construction
400 North Michigan Avenue
Chicago, Illinois 60611-4185

American Iron and Steel Institute
1000 16th Street, N.W.
Washington, D.C. 20036

American Society of Metals
Metals Park, Ohio 44073

American Welding Society (AWS)
2501 N.W. 7th Street
Miami, Florida 33125

Bethlehem Steel Corporation
Bethlehem, Pennsylvania 18016

Iron Castings Society
455 State Street
Des Plaines, Illinois 60016

Lead Industries Association
292 Madison Avenue
New York, New York 10017

Metal Building Manufacturers Association
1230 Keith Building
Cleveland, Ohio 44115

National Association of Architectural Metal Manufacturers
221 North LaSalle, Suite 2026
Chicago, Illinois 60601

National Association of Corrosion Engineers
Box 218340
Houston, Texas 77218

National Erectors Association
1501 Lee Highway, Suite 202
Arlington, Virginia 22209

Steel Deck Institute
P.O. Box 9506
Canton, Ohio 44711

Steel Joist Institute
1205 48th Avenue North, Suite A
Myrtle Beach, South Carolina 29577

Steel Structures Painting Council (SSPC)
4400 Fifth Avenue
Pittsburgh, Pennsylvania 15213-2683

Steel Tank Institute
728 Anthony Trail
Northbrook, Illinois 60062

Wire Reinforcement Institute
8361A Greensboro Drive
McLean, Virginia 22102

Woven Wire Products Association
2515 North Nordica Avenue
Chicago, Illinois 60635

6. Miscellaneous Work

American Colloid Company
5100 Suffield Court
Skokie, Illinois 60077

American Railway Engineering Association (AREA)
50 F Street, N.W., Room 7702E
Washington, D.C. 20001

Association of Soil and Foundation Engineers
8811 Colesville Road, Suite 225
Silver Spring, Maryland 20910

Brick Institute of America
1750 Old Meadow Road
McLean, Virginia 22102

International Association of Foundation Drilling Contractors
P. O. Box 280 379
Dallas, Texas, 75228

International Masonry Institute
823 15th Street, N.W., Suite 1001
Washington, D.C. 20005

Mason Contractors Association of America
17W601 14th Street
Oakbrook, Terrace, Illinois 60181

Masonry Advisory Council
1550 Northwest Highway 201
Park Ridge, Illinois 60068

National Association of Brick Distributors
11490 Commerce Park Drive, Suite 300
Reston, Virginia 22091

National Association of Dredging Contractors
16251 I Street, N.W., Suite 321
Washington, D.C. 20006

National Concrete Masonry Association
P.O. Box 781
2302 Horse Pen Road
Herndon, Virginia 22070

National Lime Association
3601 North Fairfax Drive
Arlington, Virginia 22201

National Roofing Contractors Association
8600 Bryn Mawr Avenue
Chicago, Illinois 60631

Sealant and Waterproofers Institute
3101 Broadway
Kansas City, Missouri 64111

Slurry Technology Association
1800 Connecticut Avenue, N.W.
Washington, D.C. 20009

7. General

American Arbitration Association
140 West 51st Street
New York, New York 10020

American Association for Laboratory Accreditation
2045 North 15th Street, Suite 1000
Arlington, Virginia 22201

American Association of State Highway and Transportation Officials (AASHTO)
444 North Capitol Street, N.W.
Suite 225
Washington, D.C. 20001

American Consulting Engineers Council
1015 15th Street, N.W., Suite 802
Washington, D.C. 20005

American Council of Independent Laboratories
1725 K Street, N.W., Suite 301
Washington, D.C. 20006

American Gas Association
1515 Wilson Boulevard
Arlington, Virginia 22209

American National Standards Institute (ANSI)
1430 Broadway
New York, New York 10018

American Public Works Association
1313 East 60th Street
Chicago, Illinois 60637

American Road and Transportation Builders Association
ARTBA Building
525 School Street, S.W.
Washington, D.C. 20024

American Society of Highway Engineers
151 Old Ford Drive
Camp Hill, Pennsylvania 17011

American Society of Safety Engineers
1800 East Oakton Street
Des Plaines, Illinois 60018

American Society for Testing and Materials (ASTM)
1916 Race Street
Philadelphia, Pennsylvania 19103

American Subcontractors Association
1004 Duke Street
Alexandria, Virginia 22314

Associated Builders and Contractors, Inc.
729 15th Street, N.W.
Washington, D.C. 20005

Associated Construction Publications
16231 West Ryerson Road
New Berlin, Wisconsin 53151

Associated General Contractors of America (AGC)
1957 E Street, N.W.
Washington, D.C. 20006

Building Officials and Code Administrators International
17926 South Halstead Street
Homewood, Illinois 60430

Building Research Institute
2101 Constitution Avenue, N.W.
Washington, D.C. 20418

Construction Industry Affairs Committee
228 North LaSalle
Chicago, Illinois 60601

Construction Information Source and Reference Guide
Jack W. Ward
Construction Publications
4552 East Palomino Road
Phoenix, Arizona 85018

Construction Specifications Institute, Inc. (CSI)
601 Madison Street
Alexandria, Virginia 22314

Deep Foundations Institute
66 Morris Avenue, Box 359
Springfield, New Jersey 07081

Directory of State Building Codes and Regulations
National Conference of States on Building Codes and Standards, Inc.
481 Carlisle Drive
Herndon, Virginia 22070

General Services Administration (GSA)
Specifications and Consumer Information Distribution Section
Washington Navy Yard, Building 197
Washington, D.C. 20407

International Conference of Building Officials
5360 South Workman Mill Road
Whittier, California 90601

National Association of Demolition Contractors
4415 West Harrison Street
Hillside, Illinois 60162

National Fire Protection Association
Batterymarch Park
Quincy, Massachusetts 02169

National Institute of Building Sciences
1015 15th Street, N.W., Suite 700
Washington, D.C. 20005

National Society of Professional Engineers (NSPE)
1420 King Street
Alexandria, Virginia 22314

Naval Publications and Forms Center
5801 Tabor Avenue
Philadelphia, Pennsylvania 19120
(Index of Specifications and Standards)

Nielsen/Dataquest, Inc.
2800 West Bayshore Road
P.O. Box 10113
Palo Alto, California 94303
(Rental Rate Blue Book of Construction Equipment)

Occupational Safety and Health Administration (OSHA)
Office of Information—Room N-3649
200 Constitution Avenue
Washington, D.C. 20210

Southern Building Code Congress International, Inc.
900 Montclair Road
Birmingham, Alabama 35213

Superintendent of Documents
U.S. Government Printing Office
Washington, D.C. 20402

Underwriters' Laboratories, Inc. (UL)
333 Pfingsten Road
Northbrook, Illinois 60062

U.S. Department of Commerce
National Bureau of Standards
Standards Development Services Section
Washington, D.C. 20234

Bibliography

Sources listed below have been consulted by the author during the preparation of this book. It is not a complete listing of all sources consulted; nor does it represent all the sources available on Engineering Construction Specifications.

Abdallah, Eli T., M.ASCE: "Guidelines for Producing Better Specifications", ASCE Journal of the Construction Division, Paper 17303, September 1982.

Abdun-Nur, Edward A., F.ASCE:

1. "Designing Specifications—A Challenge", ASCE Journal of the Construction Division, Paper 4315, May 1965.
2. "Control of Quality—A System", ASCE Journal of the Construction Division, Paper 7576, October 1970.

Abdun-Nur, Edward A. and Tuthill, Lewis H.: "Criteria for Modern Specifications and Control," Journal of the American Concrete Institute, Title No. 55–49, January 1959.

American Arbitration Association, New York, New York:

1. "Construction Industry Arbitration Rules", February 1, 1983.
2. "Construction Industry Mediation Rules", April 1, 1982.

American Association of State Highway and Transportation Officials, Washington, D.C.:

1. "Guide Specifications for Highway Construction," 1985.
2. "Standard Specifications for Highway Bridges," 1983.

American Concrete Institute, Detroit, Michigan; "ACI Manual of Concrete Practice," 1984.

Antonino, Ronald A., P.E.: "Construction Claims—An Avoidance Primer," The Construction Specifier (CSI), July 1980.

ASCE and AGC Joint Cooperative Committee: "A Recommended Guide to Bidding Procedure on Engineering Construction," March 1959.

Baltimore Region Rapid Transit System, State of Maryland Department of Transportation, Mass Transit Administration, Baltimore, Maryland:

1. "Standard Specifications", June 1976.
2. "General Provisions for Construction Contracts," 1976.
3. "Supplementary General Provisions for Construction Contracts," 1976.
4. "Contract No. NW-02-06, Lexington Market Station Structure, Contract Specifications Book," June 10, 1977.

Birdsall, Blair, Chairman, ASCE Committee on Contract Administration: "Who Pays for the Unexpected in Construction?," Journal of the Construction Division, Paper 3635, September 1963.

Byrne, Wayne S., FASCE, Chairman, Construction Committee of the United States Committee on Large Dams: "Responsibilities of the Engineer and the Contractor Under Fixed-Price Construction Contracts," ASCE Journal of the Construction Division, Paper 8781, March 1972.

City of Newark, New Jersey; Charlotteburg Reservoir Project: "Contract No. CRP-5, Specifications for Constructing Dam and Reservoir," October 1958.

Clark, John R. Esq.: "Commentary on Agreements for Engineering Services and Contract Documents," prepared for and published by Engineers' Joint Contract Documents Committee, 1981.

Commonwealth of Massachusetts, Department of Public Works, Boston, Massachusetts: "Standard Specifications for Highways and Bridges," 1973.

Commonwealth of Pennsylvania, Department of Transportation, Harrisburg, Pennsylvania: "Form 408 Standard Specifications," 1976.

Commonwealth of Virginia, Department of Highways and Transportation, Richmond, Virginia: "Road and Bridge Specifications," July 1, 1982.

Compton, G. R. Jr.: "Selecting Pile Installation Equipment," presented at Piletalk Seminar, San Francisco 1977; sponsored by Associated Pile and Fitting Corporation, Clifton, New Jersey.

Construction Industry Affairs Committee of Chicago; Chicago, Illinois:

1. "Recommendation #9—Substitutions," August 8, 1973.
2. "Recommendation #20—Unit Prices," December 5, 1973.

Construction Specifications Institute (CSI), Alexandria, Virginia: "Manual of Practice; Chapter 6, Section Format and Chapter 12, Specification Language."

Douglas, Walter S.: "Present and Future Risk Distribution in Construction Contracts"; delivered at the ASCE National Convention, November 3–7, 1975; Meeting Preprint 2555.

Dunham, Clarence W., CE and Young, Robert D., AB, LLB: "Contracts, Specifications and Law for Engineers,", second edition 1971; McGraw-Hill Book Company, New York, New York.

DuPont, E. I. DeNemours and Company, Wilmington, Delaware: "Methods of Controlled Blasting," 1964.

Federal Highway Administration, U.S. Department of Transportation: "Standard Specifications for Construction of Roads and Bridges on Federal Highway Projects," FP-79 (revised June 1981); Superintendent of Documents, U.S. Government Printing Office,Washington, D.C.

Fisk, Edward R., FASCE: "Evaluation of Owner Comments on Specifications"; ASCE Journal of the Construction Division, Paper 15875, December 1980.

Fox, George A., FASCE:

1. "Subsurface Construction Contracts—A Contractor's View"; ASCE Journal of the Construction Division, Paper 10608. June 1974.
2. Report of Conference on "Construction Risks and Liability Sharing," held in Scottsdale, Arizona, January 1979; sponsored by ASCE Committees on Contract Administration, and on Tunneling and Underground Construction.

Glidden, H. K.: "Reports, Technical Writing, and Specifications," 1964; McGraw-Hill Book Company, New York, New York.

Goldbloom, Joseph, FASCE, Chairman:

1. "Summary Report of Questionnaire on Construction Inspection," ASCE Task Committee on Inspection; Journal of the Construction Division, Paper 9192, September 1972.
2. "Summary Report of Questionnaire on Specifications (Contractor Returns)," ASCE Committee on Specifications; Journal of the Construction Division, Paper 14001, September 1978.
3. "Summary Report of Questionnaire on Specifications (Owner and Owner Representative Returns)," ASCE Committee on Specifications; Journal of the Construction Division, Paper 14799, September 1979.

Greenberg, Max E., Attorney and Fox, George A., FASCE, Contractor: "Are Construction Contracts Fair?"; Civil Engineering Magazine—ASCE, May 1975.

Greenberg, Max E., Attorney: "Role of Contract and Specifications in Foundations Construction"; ASCE Journal of the Construction Division, Paper 10607, June 1974.

Greenfield, Seymour S.: "Turnkey Construction in the United States"; ASCE Journal of the Construction Division, Paper 17121, June 1982.

Hammond, David G., Conference Chairman, Proceedings of Specialty Conference: "Reducing Risk and Liability Through Better Specifications and Inspection," 1982; sponsored by the Committees on Specifications and on Inspection of the ASCE Construction Division.

Hohns, Murray H.: "Preventing and Solving Construction Contract Disputes," 1979; Van Nostrand Reinhold Company, Inc., New York, New York.

Hunt, Hal W., PE: "Design and Installation of Driven Pile Foundations," second printing 1980; Associated Pile and Fitting Corporation, Clifton, New Jersey.

Interstate Division for Baltimore City, Baltimore, Maryland; Interstate Route 95, Contract No. 2547: "Specifications for Construction of Fort McHenry Tunnel, Trench Tunnel," January 25, 1980.

Jacobs, J. Donovan, FASCE: "Better Specifications for Underground Work," Civil Engineering Magazine—ASCE, June 1971.

Jellinger, Thomas C.: "Construction Contract Documents and Specifications," 1981; Addison-Wesley Publishing Company, Reading, Massachusetts.

Jessup, Edgar W. Jr., American Bar Association and Jessup, Walter E., FASCE: "Law and Specifications for Engineers and Scientists," 1963, Prentice-Hall Inc., Englewood Cliffs, New Jersey.

Kellogg Corporation: "Construction Trends and Problems Through 1990," 1981; Construction Sciences Research Foundation, Inc., Washington, D.C.

Kentucky Department of Highways, Frankfort, Kentucky: "Standard Specifications for Road and Bridge Construction," edition of 1983.

Kuesel, Thomas R.: "Pre-Contract Considerations Ease Construction of Underground Works"; Public Works Magazine, February 1975.

Metropolitan Atlanta Rapid Transit Authority, Atlanta, Georgia: "Standard Specifications," August 1974.

Meier, Hans W., FCSI: "Construction Specifications Handbook," second edition 1978; Prentice-Hall Inc., Englewood Cliffs, New Jersey.

Merritt, Frederick S., Editor: "Standard Handbook for Civil Engineers," third edition 1983, McGraw-Hill Book Company, New York, New York.

National Clay Pipe Institute, Washington, D.C.: "Building a Sewer—A Team Concept."

National Committee on Tunneling Technology: "Better Contracting for Underground Construction," November 1974; National Technical Information Service,Springfield, Virginia.

New Jersey Department of Transportation, Trenton, New Jersey: "Standard Specifications for Road and Bridge Construction," 1983.

Nichols, John W., Chairman: "Report on ImprovingConstruction Arbitrations," ASCE Committee on Contract Administration; Journal of the Construction Division, Paper 14414, March 1979.

Port Authority of Allegheny County, Pittsburgh,Pennsylvania, Light Rail Transit System:"Mt. Lebanon Tunnel, Contract No. CA 260,Contract Forms and Specifications," April 1983.

Rosen, Harold J., PE; FCSI: "Construction Specifications Writing," second edition 1981; John Wiley and Sons, New York, New York.

Rowland, Robert, Chairman ASCE Committee on Contract Administration:

1. "Report on Recommended Amendments to AIA General Conditions (A-201) Eleventh Edition, September 1967"; Journal of the Construction Division, Paper 7640, October 1970.
2. "Recommended Endorsement and Comments on NSPE/ACEC 1910-8 (1978) General Conditions"; Journal of the Construction Division, Paper 14641, June 1979.

Rubin, Robert A., MASCE: "Fifty Years of Construction Law"; ASCE Journal of the Construction Division, Paper 11758, December 1975.

Rubin, Robert A.; Guy, Sammie D.; Maevis, Alfred C.; and Fairweather, Virginia: "Construction Claims, Analysis, Presentation, Defense," 1983; Van Nostrand Reinhold Company, Inc., New York, New York.

Seelye, Elwin E.: "Specifications and Costs," Data Book for Civil Engineers, Volume II, third edition May 1957; John Wiley and Sons, New York, New York.

State of California Department of Transportation, Sacramento, California: "Standard Specifications," July 1984.

State of Maryland, State Roads Commission, Baltimore, Maryland:

1. "Specifications for Materials, Highways, Bridges and Incidental Structures," second edition March 1968.

2. "Baltimore Harbor Outer Crossing—Contract No. OT-2-2, Curtis Creek Bridge Substructure, Special Provisions," June 1971.
3. "Supplement to the Specifications, for Interstate Division for Baltimore City," July 1979.

Stukhart, George, MASCE: "Contractual Incentives," ASCE Journal of Construction Engineering and Management, Paper 18618, March 1984.

Technical Activities Committee, ASCE Metropolitan Section: "Contract Award Practices"; Journal of the Construction Division, Paper 4604, January 1966.

The Business Roundtable, New York, New York: "Administration and Enforcement of Building Codes and Regulations," Report E-1, October 1982.

Washington Metropolitan Area Transit Authority, Washington, D.C.: "Preparation Manual—Specifications and Special Provisions," 1973.

Washington State Department of Transportation, Olympia, Washington: "Standard Specifications for Road, Bridge and Municipal Construction," 1984.

Wilson, Roy L., MASCE: "Prevention and Resolution of Construction Claims"; ASCE Journal of the Construction Division, Paper 17309, September 1982.

Wilson, Woodrow W., Chairman ASCE Committee on Contract Administration:
1. "Model Form of Instructions to Bidders," Journal of the Construction Division, Paper 10407, March 1974.
2. "Model Form of Notice to Bidders," Journal of the Construction Division, Paper 10818, September 1974.

Index

FSC
www.fsc.org
MIX
Papier aus verantwortungsvollen Quellen
Paper from responsible sources
FSC® C105338

If you have any concerns about our products,
you can contact us on
ProductSafety@springernature.com

In case Publisher is established outside the EU,
the EU authorized representative is:
Springer Nature Customer Service Center GmbH
Europaplatz 3, 69115 Heidelberg, Germany

Printed by Libri Plureos GmbH
in Hamburg, Germany